Communicating
Pictures

Communicating Pictures

A Course in Image and Video Coding

David R. Bull

AMSTERDAM • BOSTON • HEIDELBERG • LONDON
NEW YORK • OXFORD • PARIS • SAN DIEGO
SAN FRANCISCO • SINGAPORE • SYDNEY • TOKYO
Academic Press is an imprint of Elsevier

Academic Press is an imprint of Elsevier
The Boulevard, Langford Lane, Kidlington, Oxford OX5 1GB, UK
225 Wyman Street, Waltham, MA 02451, USA

First edition 2014

Notice

No responsibility is assumed by the publisher for any injury and/or damage to persons or property as a matter of products liability, negligence or otherwise, or from any use or operation of any methods, products, instructions or ideas contained in the material herein. Because of rapid advances in the medical sciences, in particular, independent verification of diagnoses and drug dosages should be made.

British Library Cataloguing in Publication Data
A catalogue record for this book is available from the British Library

Library of Congress Cataloging-in-Publication Data
A catalog record for this book is available from the Library of Congress

ISBN–13: 978-0-12-405906-1

For information on all Academic Press publications
visit our web site at store.elsevier.com

Transferred to Digital Printing, 2016

Working together
to grow libraries in
developing countries

www.elsevier.com • www.bookaid.org

About the Author

David R. Bull currently holds the Chair in Signal Processing at the University of Bristol, where he is head of the Visual Information Laboratory and Director of Bristol Vision Institute (BVI).

David has worked widely in the fields of image and video communications. He has published some 450 academic papers, various articles, and 3 books and has given numerous invited/keynote lectures and tutorials. He has been the recipient of major awards for this work. He has also published several patents, many of which have been exploited commercially.

In 1996 David helped to establish the UK DTI Virtual Centre of Excellence in Digital Broadcasting and Multimedia Technology and was one of its Directors from 1997 to 2000. He has been a member of UK Government's Defence Scientific Advisory Council (DSAC) and has also advised Government through membership of the UK Foresight Panel and the HEFCE Research Evaluation Framework.

David has acted as a consultant to many companies throughout his career in the field of video technology and, in 2001, he co-founded ProVision Communication Technologies Ltd., a company specializing in wireless video communications.

In 2007 he co-founded Bristol Vision Institute (BVI). BVI functions as a virtual research institute and has been highly successful in stimulating research interaction and collaboration across the University of Bristol. It brings together engineers and scientists from a range of academic disciplines across vision science and the visual arts and currently represents a grouping of 120 researchers.

David is a Fellow of the IEEE and a Fellow of the IET.

Contents

Preface...xix
List of Figures...xxix
List of Tables...xli
List of Algorithms...xliii
Acknowledgments...xlv

CHAPTER 1 Introduction 1
    1.1 Communicating pictures: the need for compression 2
        1.1.1 What is compression? 2
        1.1.2 Why do we need compression?....................................3
    1.2 Applications and drivers 6
        1.2.1 Generic drivers 6
        1.2.2 Application drivers and markets 8
    1.3 Requirements and trade-offs in a compression system.................9
        1.3.1 The benefits of a digital solution 9
        1.3.2 Requirements 10
        1.3.3 Trade-offs..10
    1.4 The basics of compression..11
        1.4.1 Still image encoding 11
        1.4.2 Encoding video 12
        1.4.3 Measuring visual quality..12
    1.5 The need for standards..12
        1.5.1 Some basic facts about standards................................12
        1.5.2 A brief history of video encoding standards 13
    1.6 Summary...14
        References..15

CHAPTER 2 The Human Visual System..................................... 17
    2.1 Principles and theories of human vision......................................19
        2.1.1 Theories of vision 19
    2.2 Acquisition: the human eye 20
        2.2.1 Retinal tissue layers 21
        2.2.2 Optical processing..23
        2.2.3 Retinal photoreceptors and their distribution..................24
        2.2.4 Visual processing in the retina 26
    2.3 The visual cortex..28
        2.3.1 Opponent processes 29
        2.3.2 Biased competition..29

2.3.3 Adaptation processes 29
2.3.4 V1—the primary visual cortex 30
2.3.5 V2—the prestriate cortex 30
2.3.6 Dorsal and ventral streams 30
2.3.7 Extrastriate areas 31
2.4 Visual fields and acuity 31
2.4.1 Field of view 31
2.4.2 Acuity 32
2.4.3 Light, luminance, and brightness 32
2.4.4 Light level adaptation 35
2.5 Color processing 36
2.5.1 Opponent theories of color 37
2.5.2 CIE 1931 chromaticity chart 39
2.6 Spatial processing 39
2.6.1 Just noticeable difference, contrast, and Weber's law 40
2.6.2 Frequency-dependent contrast sensitivity 40
2.6.3 Multiscale edges 44
2.6.4 Perception of textures 44
2.6.5 Shape and object recognition 44
2.6.6 The importance of phase information 45
2.7 Perception of scale and depth 46
2.7.1 Size or scale 46
2.7.2 Depth cues 47
2.7.3 Depth cues and 3-D entertainment 48
2.8 Temporal and spatio-temporal response 49
2.8.1 Temporal CSF 50
2.8.2 Spatio-temporal CSF 50
2.8.3 Flicker and peripheral vision 50
2.9 Attention and eye movements 52
2.9.1 Saliency and attention 52
2.9.2 Eye movements 53
2.10 Visual masking 53
2.10.1 Texture masking 53
2.10.2 Edge masking 55
2.10.3 Temporal masking 55
2.11 Summary: a perceptual basis for image and
video compression 56
2.11.1 Influential factors 56
2.11.2 What have we learnt? 56
References 59

CHAPTER 3 Discrete-Time Analysis for Images and Video **63**

3.1 Signal and picture sampling...64
    3.1.1 The sampling theorem...65
    3.1.2 Multidimensional sampling lattices...............................69

3.2 Statistics of images 70
    3.2.1 Histograms and distributions 70
    3.2.2 Mean values 70
    3.2.3 Correlation in natural images....................................71

3.3 Filtering and transforms...76
    3.3.1 Discrete-time linear systems 76
    3.3.2 Convolution..76
    3.3.3 Linear filters...77
    3.3.4 Filter frequency response 79
    3.3.5 Examples of practical filters....................................80
    3.3.6 Non-linear filters 81
    3.3.7 Linear transforms and the DFT................................82

3.4 Quantization...84
    3.4.1 Basic theory of quantization 84
    3.4.2 Adaptation to signal statistics 86
    3.4.3 HVS weighting..87
    3.4.4 Vector quantization 87

3.5 Linear prediction..87
    3.5.1 Basic feedforward linear predictive coding 87
    3.5.2 Linear prediction with the predictor in the feedback loop 90
    3.5.3 Wiener Hopf equations and the Wiener filter.................92

3.6 Information and entropy 93
    3.6.1 Self-information...93
    3.6.2 Entropy..94
    3.6.3 Symbols and statistics 96

3.7 Summary...97
    References...97

CHAPTER 4 Digital Picture Formats and Representations **99**

4.1 Pixels, blocks, and pictures..100
    4.1.1 Pixels, samples, or pels 100
    4.1.2 Moving pictures 102
    4.1.3 Coding units and macroblocks....................................103
    4.1.4 Picture types and groups of pictures 104

4.2 Formats and aspect ratios...106

    4.2.1 Aspect ratios..106

    4.2.2 Displaying different formats 110

4.3 Picture scanning...111

    4.3.1 Interlaced vs progressive scanning 111

    4.3.2 Standards conversion 112

4.4 Gamma correction 114

4.5 Color spaces and color transformations...............................116

    4.5.1 Color descriptions and the HVS 116

    4.5.2 Sub-sampled color spaces.....................................121

    4.5.3 Bayer filtering 123

4.6 Measuring and comparing picture quality 124

    4.6.1 Compression ratio and bit rate............................124

    4.6.2 Objective distortion and quality metrics 125

    4.6.3 Subjective assessment 128

4.7 Rates and distortions..129

    4.7.1 Rate–distortion characteristics 129

    4.7.2 Rate–distortion optimization................................131

4.8 Summary..132

    References..132

CHAPTER 5   Transforms for Image and Video Coding 133

5.1 The principles of decorrelating transforms..............................134

    5.1.1 The basic building blocks 134

    5.1.2 Principal components and axis rotation 135

5.2 Unitary transforms...137

    5.2.1 Basis functions and linear combinations............137

    5.2.2 Orthogonality and normalization 138

    5.2.3 Extension to 2-D 139

5.3 Basic transforms 140

    5.3.1 The Haar transform 140

    5.3.2 The Walsh–Hadamard transform 141

    5.3.3 So why not use the discrete Fourier transform?............144

5.4 Optimum transforms...146

    5.4.1 Discarding coefficients..146

    5.4.2 The Karhunen–Loeve transform (KLT) 147

    5.4.3 The KLT in practice 148

5.5 Discrete cosine transform (DCT)...149

    5.5.1 Derivation of the DCT 149

    5.5.2 DCT basis functions..151

    5.5.3 Extension to 2-D: separability 153

5.6 Quantization of DCT coefficients 155
5.7 Performance comparisons 160
    5.7.1 DCT vs DFT revisited 160
    5.7.2 Comparison of transforms 160
    5.7.3 Rate–distortion performance of the DCT 161
5.8 DCT implementation 163
    5.8.1 Choice of transform block size 163
    5.8.2 DCT complexity reduction 164
    5.8.3 Field vs frame encoding for interlaced sequences 166
    5.8.4 Integer transforms 167
    5.8.5 DCT demo 167
5.9 JPEG 167
5.10 Summary 169
    References 169

CHAPTER 6 **Filter Banks and Wavelet Compression 171**
6.1 Introduction to multiscale processing 172
    6.1.1 The short-time Fourier transform
        and the Gabor transform 173
    6.1.2 So what is a wavelet? 174
    6.1.3 Wavelet and filter bank properties 177
6.2 Perfect reconstruction filter banks 178
    6.2.1 Filter and decomposition requirements 178
    6.2.2 The 1-D filter bank structure 178
6.3 Multirate filtering 182
    6.3.1 Upsampling 183
    6.3.2 Downsampling 183
    6.3.3 System transfer function 186
    6.3.4 Perfect reconstruction 189
    6.3.5 Spectral effects of the two-channel
        decomposition 191
6.4 Useful filters and filter banks 191
    6.4.1 Quadrature mirror filters 192
    6.4.2 Wavelet filters 194
    6.4.3 Multistage (multiscale) decompositions 197
    6.4.4 Separability and extension to 2-D 200
    6.4.5 Finite length sequences, edge artifacts,
        and boundary extension 202
    6.4.6 Wavelet compression performance 203
6.5 Coefficient quantization and bit allocation 204

6.5.1 Bit allocation and zonal coding.....................204

6.5.2 Hierarchical coding......................................207

6.6 JPEG2000..207

    6.6.1 Overview..207

    6.6.2 Architecture—bit planes and scalable coding...............207

    6.6.3 Coding performance..................................208

    6.6.4 Region of interest coding 208

    6.6.5 Benefits and status...................................210

6.7 Summary..210

    References...210

CHAPTER 7 Lossless Compression Methods 213

7.1 Motivation for lossless image compression.............214

    7.1.1 Applications 215

    7.1.2 Approaches 216

    7.1.3 Dictionary methods..................................216

7.2 Symbol encoding 217

    7.2.1 A generic model for lossless compression...............217

    7.2.2 Entropy, efficiency, and redundancy 217

    7.2.3 Prefix codes and unique decodability 218

7.3 Huffman coding 219

    7.3.1 The basic algorithm...................................219

    7.3.2 Minimum variance Huffman coding.............221

    7.3.3 Huffman decoding....................................223

    7.3.4 Modified Huffman coding..........................224

    7.3.5 Properties of Huffman codes.......................224

    7.3.6 Adaptive methods for Huffman encoding......................225

7.4 Symbol formation and encoding..........................229

    7.4.1 Dealing with sparse matrices 229

    7.4.2 Symbol encoding in JPEG 230

    7.4.3 JPEG performance examples 235

    7.4.4 Lossless JPEG mode..................................235

    7.4.5 Context-Adaptive Variable Length Coding (CAVLC) in H.264/AVC..............................236

7.5 Golomb coding 239

    7.5.1 Unary codes 239

    7.5.2 Golomb and Golomb–Rice codes 239

    7.5.3 Exponential Golomb codes 240

7.6 Arithmetic coding 241

    7.6.1 The basic arithmetic encoding algorithm.....................241

7.6.2 The basic arithmetic decoding algorithm......................244
7.6.3 Advantages of arithmetic coding 246
7.6.4 Binary arithmetic coding...247
7.6.5 Tag generation with scaling 249
7.6.6 Context-adaptive arithmetic coding 250
7.7 Performance comparisons—Huffman vs arithmetic coding.......251
7.8 Summary..252
References...252

CHAPTER 8 Coding Moving Pictures: Motion Prediction.......... 255
8.1 Temporal correlation and exploiting temporal
redundancy...256
8.1.1 Why motion estimation?...256
8.1.2 Projected motion and apparent motion 258
8.1.3 Understanding temporal correlation 259
8.1.4 How to form the prediction...260
8.1.5 Approaches to motion estimation 262
8.2 Motion models and motion estimation 263
8.2.1 Problem formulation 263
8.2.2 Affine and high order models 263
8.2.3 Translation-only models 265
8.2.4 Pixel-recursive methods 266
8.2.5 Frequency domain motion estimation using
phase correlation (PhC)...267
8.3 Block matching motion estimation (BMME) 269
8.3.1 Translational block matching......................................269
8.3.2 Matching criteria...270
8.3.3 Full search algorithm 272
8.3.4 Properties of block motion fields and error surfaces 272
8.3.5 Motion failure 276
8.3.6 Restricted and unrestricted vectors 277
8.4 Reduced complexity motion estimation 277
8.4.1 Pixel grids and search grids 277
8.4.2 Complexity of full search...278
8.4.3 Reducing search complexity 279
8.4.4 2-D logarithmic search (TDL) 280
8.4.5 *N*-Step Search (NSS)...281
8.4.6 Diamond search 282
8.4.7 Hexagonal search 284
8.4.8 Initialization and termination criteria............................285

8.4.9 Reduced complexity block matching
algorithm performance comparisons.............................286
8.5 Motion vector coding...287
8.5.1 Motion vector prediction...287
8.5.2 Entropy coding of motion vectors..............................287
8.6 Summary...288
References...288

CHAPTER 9 The Block-Based Hybrid Video Codec 291
9.1 The block-based hybrid model for video compression..............292
9.1.1 Picture types and prediction modes 292
9.1.2 Properties of the DFD signal..293
9.1.3 Operation of the video encoding loop..........................294
9.1.4 Operation of the video decoder....................................295
9.2 Intra-frame prediction..298
9.2.1 Intra-prediction for small luminance blocks.................298
9.2.2 Intra-prediction for larger blocks 300
9.3 Sub-pixel motion estimation..301
9.3.1 Sub-pixel matching 301
9.3.2 Interpolation methods 303
9.3.3 Performance 305
9.3.4 Interpolation-free methods...306
9.4 Multiple reference frame motion estimation 307
9.4.1 Justification 307
9.4.2 Properties, complexity, and performance
of MRF-ME 308
9.4.3 Reduced complexity MRF-ME......................................308
9.5 Variable block sizes for motion estimation...............................309
9.5.1 Influence of block size 309
9.5.2 Variable block sizes in practice....................................310
9.6 Variable sized transforms...310
9.6.1 Integer transforms 310
9.6.2 DC coefficient transforms..313
9.7 In-loop deblocking operations 313
9.8 Summary...315
References...315

CHAPTER 10 Measuring and Managing Picture Quality............. 317
10.1 General considerations and influences....................................318
10.1.1 What do we want to assess?.......................................319
10.1.2 Influences on perceived quality..................................319

10.2 Subjective testing 321
    10.2.1 Justification 321
    10.2.2 Test sequences and conditions 321
    10.2.3 Choosing subjects 323
    10.2.4 Testing environment..324
    10.2.5 Testing methodology and recording of results..............324
    10.2.6 Statistical analysis and significance testing 328
    10.2.7 The complete evaluation process 329
10.3 Test data sets and how to use them 330
    10.3.1 Databases 330
    10.3.2 The relationship between Mean Opinion Score
           and an objective metric 331
    10.3.3 Evaluating metrics using public (or private) databases..332
10.4 Objective quality metrics 333
    10.4.1 Why do we need quality metrics?................................333
    10.4.2 A characterization of PSNR...335
    10.4.3 A perceptual basis for metric development...................337
    10.4.4 Perception-based image and video quality metrics........338
    10.4.5 The future of metrics...341
10.5 Rate–distortion optimization 343
    10.5.1 Classical rate–distortion theory....................................343
    10.5.2 Practical rate–distortion optimization 345
    10.5.3 The influence of additional coding modes
           and parameters 348
    10.5.4 From Rate–Distortion to Rate–Quality Optimization....350
10.6 Rate control..350
    10.6.1 Buffering and HRD..352
    10.6.2 Rate control in practice 354
    10.6.3 Regions of interest and rate control 357
10.7 Summary...357
    References...358

**CHAPTER 11 Communicating Pictures: Delivery
               Across Networks 361**
11.1 The operating environment..362
    11.1.1 Characteristics of modern networks.............................362
    11.1.2 Transmission types..364
    11.1.3 Operating constraints 365
    11.1.4 Error characteristics 365
    11.1.5 The challenges and a solution framework.....................367
11.2 The effects of loss...368

11.2.1 Synchronization failure 368

11.2.2 Header loss 370

11.2.3 Spatial error propagation 370

11.2.4 Temporal error propagation 372

11.3 Mitigating the effect of bitstream errors 373

11.3.1 Video is not the same as data! 373

11.3.2 Error-resilient solutions 374

11.4 Transport layer solutions 375

11.4.1 Automatic Repeat request (ARQ) 375

11.4.2 FEC channel coding 376

11.4.3 Hybrid ARQ (HARQ) 379

11.4.4 Packetization strategies 379

11.5 Application layer solutions 379

11.5.1 Network abstraction 379

11.5.2 The influence of frame type 380

11.5.3 Synchronization codewords 382

11.5.4 Reversible VLC 382

11.5.5 Slice structuring 384

11.5.6 Error tracking 385

11.6 Cross layer solutions 386

11.6.1 Link adaptation 386

11.7 Inherently robust coding strategies 389

11.7.1 Error-Resilient Entropy Coding (EREC) 389

11.7.2 Pyramid Vector Quantization (PVQ) 394

11.8 Error concealment 397

11.8.1 Detecting missing information 397

11.8.2 Spatial error concealment (SEC) 398

11.8.3 Temporal error concealment 399

11.8.4 Hybrid methods with mode selection 403

11.9 Congestion management and scalable video coding 405

11.10 Summary 408

References 408

CHAPTER 12 Video Coding Standards **411**

12.1 The need for and role of standards 412

12.1.1 The focus of video standardization 412

12.1.2 The standardization process 413

12.1.3 Intellectual property and licensing 414

12.2 H.120 414

12.2.1 Brief history 414

12.2.2 Primary features 415

12.3 H.261 415
    12.3.1 Brief history 415
    12.3.2 Picture types and primary features................................415
12.4 MPEG-2/DVB...417
    12.4.1 Brief history 417
    12.4.2 Picture types and primary features................................417
    12.4.3 MPEG-2 profiles and levels 418
12.5 H.263 419
    12.5.1 Brief history 419
    12.5.2 Picture types and primary features................................419
12.6 MPEG-4 423
    12.6.1 Brief history 423
    12.6.2 Picture types and primary features................................424
12.7 H.264/AVC 426
    12.7.1 Brief history 426
    12.7.2 Primary features 426
    12.7.3 Network abstraction and bitstream syntax 427
    12.7.4 Pictures and partitions..429
    12.7.5 The video coding layer...430
    12.7.6 Profiles and levels 434
    12.7.7 Performance 435
    12.7.8 Scalable extensions 435
    12.7.9 Multiview extensions 435
12.8 H.265/HEVC 436
    12.8.1 Brief background 436
    12.8.2 Primary features 436
    12.8.3 Network abstraction and high level syntax 437
    12.8.4 Pictures and partitions..438
    12.8.5 The video coding layer (VCL)....................................440
    12.8.6 Profiles and levels 446
    12.8.7 Extensions 446
    12.8.8 Performance gains for HEVC over recent standards 446
12.9 Other de-facto standards and proprietary codecs........................447
    12.9.1 VP9 447
    12.9.2 VC-1..447
    12.9.3 RealVideo..447
    12.9.4 Dirac..447
12.10 Summary...447
    References..448

CHAPTER 13 Communicating Pictures—The Future **451**
  13.1 The motivation: more immersive experiences 452
  13.2 Emerging formats and extended video parameter spaces 452
      13.2.1 Influences 453
      13.2.2 Spatial resolution 453
      13.2.3 Temporal resolution 455
      13.2.4 Dynamic range 459
      13.2.5 Parameter interactions and the creative continuum 461
  13.3 Challenges for compression 462
  13.4 Parametric video compression 464
  13.5 Context-based video compression 466
  13.6 Summary 467
          References 468

Appendix A: Glossary of Terms **471**

Appendix B: Tutorial Problems **479**

INDEX **499**

Preface

Throughout history, pictures have played a major role in communicating information, emotion, and entertainment. Images are now everywhere and the way we exchange and interpret them is central to everyday life.

A very large proportion, approximately 50%, of the human brain's volume is dedicated to vision and related tasks. Comparing this with around 20% for motor functions and only 10% for problem solving, helps to explain why we can build computers that beat us at chess but cannot yet design machines that come close to replicating the human vision system. It also reflects the importance of vision for function and survival, and explains its capacity to entertain, challenge, and inform.

The eye is a relatively simple organ which captures two inverted and distorted impressions of the outside world, created by light reflected from external objects. It is interesting to note that optical images were unknown before the 10th century and it was not until the 17th century that it was realized that "images" existed in the eyes—via light projected onto our retinas.

So why can such a relatively simple receptor be responsible for creating such a rich interface to the outside world? The brain processes the crude images on the retina to detect, assimilate, and predict threats, to produce perceptions of beauty, and to perceive other means for survival and reproduction. These perceptions are not just derived from instantaneous images, but by the temporal and spatial relationships which they represent, placed in the context of a wealth of knowledge obtained from prior experiences. This combination of sensing, stored data, and dynamic neural processing uniquely enables animals, in varying degrees, to perform recognition, prediction, and planning tasks.

Although many contemporary applications use computer vision to aid image understanding (for example in manufacturing, medical diagnosis, robotics, and surveillance), in most cases, the final consumer and interpreter is a human being. In many ways, this captures the essence of communicating pictures—the convergence of information engineering and mathematics with psychology, biology, and the creative arts.

Early visual communications

Visual communication has a very long history, from early cave paintings to Egyptian hieroglyphs, through the inventions of paper, moveable type, photography, and film, to modern art, television, and "new digital media." To give some feel for its beginnings, Figure 1 shows a simple piece of ochre which was discovered in a cave in South Africa. While there is no evidence of drawings in the cave, the markings on the ochre strongly indicate that it was used for drawing—some 200,000 years ago. This is the earliest evidence of humans using tools for visual communication.

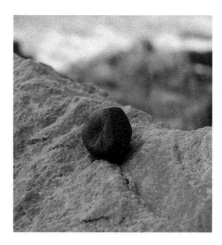

FIGURE 1

Drawing Ochre: c. 200,000 BC, South Africa. (Courtesy of Professor Alice Roberts.)

Exploration and exploitation of different media has grown ever more sophisticated over the ensuing 200,000 years. The first evidence for humans engraving patterns dates from the mid stone age (South Africa c. 60,000 BC) and exists in the form of hatched bands on ostrich eggshells. Another equally famous example of prehistoric art, known as Venus of Willendorf, is a stone-carved sculpture representing the female form and dates from the Paleolithic Period (c. 24,000 BC). Later examples of paintings on cave walls, presumed to be a means of recording significant events (but surely also admired for their beauty), can be traced back as far as c. 15,000 BC. Particularly fine examples can be found in Lascaux, France.

Mobility and print

Mobility in image communications started with the need to exchange information. Early Mesopotamian writing systems (c. 4000 BC) used a stylus and clay for recording numbers whereas Chinese tortoise-shell carvings date back even further to c. 6000 BC. Papyrus was first manufactured in Egypt as far back as the third millennium BC and this was rivaled by parchment from c. 600 BC, which was prepared from animal skins. Paper, made from wood-pulp, is attributed to the Chinese c. 200 BC, and provided a cheaper and more flexible alternative to silk. The availability of this new medium and its impact on cheap mass communication was unimaginable at the time.

Printing (leading to the mass reproduction of text and images, typically with ink on paper) was clearly one of the most significant developments in image and textual communications, supporting widespread and consistent dissemination of art and information. Over time, print provided knowledge to the masses and enabled an increased accessibility to what was previously only in the domain of the privileged.

The origins of printing can be found in the cylinder seals used in Mesopotamia around 3500 BC. The invention of woodblock printing is attributed to the Chinese c. 200 AD, as is the introduction of moveable type in 1040.

Perhaps the most significant step forward was taken by Johannes Gutenberg in 1454, with the development of the moveable-type printing press. This high quality process led to the production of the Gutenberg Bible (1455) which rapidly spread across Europe, triggering the Renaissance. Lithography—the use of a flat stone or metal plate to transfer ink to paper—was introduced in 1796 by Alois Senefelder, followed by chromolithography in 1837. It was not until 1907 that screen printing was invented, with phototypesetting in 1960. The printing process has developed dramatically over the past 50 years with the introduction of the laser printer in 1969 and the digital press in 1993.

Artistic impression, painting, and perspective

Byzantine and Gothic art from the Early Modern Era and Middle Ages primarily reflects the dominance of the church through the expression of religious stories and icons—depiction of the material world or everyday life was deemed unnecessary at the time. There was extensive use of gold in paintings for religious emphasis and this has been shown through contemporary studies to influence the way eyes are drawn toward or away from key areas (especially when viewed under the prevailing candlelit conditions).

Figures were normally idealized and drawn without perspective. This, however, all changed with the Renaissance (14th–17th centuries) which began in Firenze and then spread to the rest of Europe. During this period, art returned to the material world, with painters such as Holbein depicting real people with emotions, and Filippo Brunelleschi and Leon Battista Alberti formalizing linear perspective to create depth. Leonardo Da Vinci (scientist, engineer, musician, anatomist, cartographer, and painter) was perhaps the most famous contributor to Renaissance art. His painting, Mona Lisa (completed 1519), is worthy of special mention—a half-length portrait of a woman whose facial expression possessed an enigmatic quality that still fascinates today.

The separation of the Church of England from Rome was used, first by Henry VIII in his Great Bible (1539—based on Tyndale's earlier incomplete version) and then by James I in the King James Bible of 1611, to bring an English version of the bible to the population, reinforcing the message of an independent church. Two thousand and five hundred copies of the Great Bible were produced and the King James Bible is widely acknowledged as the best selling book in history.

In the 18th century (the so-called Enlightenment Period), inspiration in art was derived more from our expanding knowledge of the universe and an associated deterministic interpretation of the world. This was then progressively transformed and challenged in the 19th and 20th centuries leading to more abstract forms of art such as Impressionism, Expressionism, Cubism, and Surrealism, all influenced by experiments with perception, the distortion of reality and uncertainty.

Photographic film and motion pictures

The process of creating art on canvas was transformed in 1840 when Louis Daguerre (France) and William Fox Talbot (UK) independently invented the photographic process. Like the printing press 400 years before, this revolutionized the way we see and communicate pictures. At once, it opened up new opportunities for artistic creation as well as, for the first time, enabling communication of the "true" representation of people and events—"the photograph never lies."

A number of other key inventions emerged in the late 19th century that facilitated the creation and public display of stored moving images. Firstly, US inventor Thomas Edison is credited with inventing the electric light bulb (although this was the culmination of many years of work by him and others) in New Jersey in 1879. It was probably Edison who was the first to realize that light was the key to display moving images. Around the same time, George Eastman in Rochester, USA, founded the Eastman Kodak Company and invented roll film, bringing photography into mainstream use. This was used in 1888 by the world's first filmmaker Louis Le Prince and followed in 1895 by the first public motion picture presentation created by the Lumiere brothers in Lyon, France. Louis Lumiere was the first to realize that the grab-advance mechanism of a sewing machine could be used to advance, pause, and expose Eastman's film, illuminating it with one of Edison's light bulbs and projecting the light via a lens (i.e. a back to front camera) onto a screen in front of an audience.

Leon Bouly and Thomas Edison himself were also among the world's first film makers. One of the earliest films by the Lumiere brothers (1896) was called *Arrival of a Train at Ciotat*. This is said to have unnerved the audience due to the realism of the train coming toward them "through" the screen! An impact perhaps more profound or "immersive" at the time than some of today's 3-D movies! We think of widescreen formats as a recent innovation, in cinema since 1953, but only recently in TV. However, the first use of widescreen was in 1897 when Enoch Rector replaced conventional 35 mm film with a 63 mm alternative to form a broader image of a boxing match, in order to show more of the action.

Movies in the early 20th century moved from using technology for capturing reality to story telling and innovations such as special effects, which both mirrored and distorted reality. The development of cinematography followed, exploring close-ups, camera angles, and narrative flow. Sound arrived in the 1920s and color movies in the 1930s.

As Howard Hughes found when making the film *Hells Angels* (1930), communicating pictures can become very expensive. With production costs of about $3 million, the film (which told the story of WWI fighter pilots) and its director were ridiculed by other Hollywood producers, such as Louis Mayer, for its lavish use of real aircraft, props, and cameras. He was vindicated, however, by taking nearly $8 million at the box office, a record at the time.

By way of comparison, Martin Scorsese's film *The Aviator* (2004), about the life of Hughes, was made with a budget of some $116 million. The film is of particular interest as, for the first 50 min, scenes are rendered only in shades of red and cyan-blue—green objects are rendered as blue. This emulated the early bipack color movies, in particular the Multicolor process, which Hughes himself owned. Scenes in the film depicting events after 1935 emulate the saturated appearance of three-strip Technicolor. It is also important to note that, unlike *Hells Angels*, almost all the flying scenes in *The Aviator* were created using scale models and computer graphics. The film won five academy awards.

As a young man, Hughes set himself three major ambitions: to be the world's best pilot, to be the world's best filmmaker, and to be the richest man in the world. While many might argue with his success in the first two, there is certainly no doubt that he achieved the last one!

Television

Television was perhaps the technology that brought the most dramatic change to the way we acquired information and how we used our leisure time in the 20th century. Television has its roots in the discovery of selenium's photoconductivity (Willoughby Smith, 1873). Early mechanical versions started with German student, Paul Nipkow, who invented the scanning disk–image rasterizer in 1884 (although he never made a working model). This scanning method solved the key problem of how to map a 2-D image onto a 1-D radio signal. Constantin Perskyi is accredited with being the first to use the word "television" in his presentation to the International Electricity Congress in Paris in 1900.

After contributions by DeForest, Korn, and others in the US and Russia, a crude still image was transmitted by Rignoux and Fournier in Paris in 1909. Their scanner employed a rotating mirror with a receiver based on 64 selenium cells. This technology was developed through the early 20th century, culminating in the first public demonstration of TV in 1925 by John Logie Baird in London. AT&T's Bell Telephone Laboratories transmitted halftone still images later the same year. None of these, however, really achieved psychovisually acceptable moving images since they were only transmitted at around five frames per second.

The first demonstration of motion television is attributed to Baird in January 1926 in London, when he presented his 30-line electromechanical system to members of the Royal Institution. Baird's work continued apace, with transmission over fixed and wireless channels. The first transatlantic transmission, between London and New York, was achieved in 1928. It is interesting to note that Baird also worked on low light imaging and 3-D TV at that time!

In the USA, Philo Farnsworth transmitted the first fully electronic television picture in 1927—an image of a straight line—later that year followed by his *Puff of Smoke* sequence. The first experimental TV service was established in Germany in 1929 and later that year a similar operation was established in France. Although Baird's system provided only 240 scanning lines, it did form the basis of the world's first public service, launched by the BBC from Alexandra Palace in 1936. Baird's method was, however, rapidly replaced by Marconi-EMI's "high definition" 405-line all-electronic system.

625-line monochrome broadcasts were introduced in 1964 and the BBC launched its BBC2 service on UHF using a 625-line system with a PAL color system in 1967. 405-line transmissions continued until 1985 when the frequencies were reused for Digital Audio Broadcasting (DAB) and Private Mobile Radio (PMR) services. Color TV was actually introduced much earlier in the US (1953) but was slow to take off due to a lack of content and high prices. The US networks converted to color in 1965, coincident with the introduction of GE's Porta-Color TV set, and it gained popularity rapidly. From the earliest days, operators realized that television broadcasting was expensive in terms of radio bandwidth, especially as frame rates and resolution (or the numbers of scanning lines) increased. This was addressed in a number of ingenious ways—through the use of interlaced fields and through the use of color sub-carriers.

The digital video revolution really took off when television (both satellite and terrestrial) went digital (in the UK) in 1998. This was only made possible by extensive worldwide effort in digital video broadcasting, in particular by the Motion Picture Experts Group (MPEG), chaired by Leonardo Chiariglione. This led to the ubiquitous MPEG-2 Audiovisual Coding framework which still forms the basis for most digital TV in the world. Analog transmission has now been discontinued in much of the world (change-over was completed in the UK in 2012).

The first public high definition television (HDTV) broadcast in the United States occurred in July 1996. Although HDTV broadcasts had been demonstrated in Europe since the early 1990s, the first regular broadcasts were started in 2004 by the Belgian company, Euro1080.

Pervasive media and the internet

Alongside the development of visual media through advances in film and television, the invention of advanced telescopes, microscopes, and computer-based imaging technology allows us to see what used to be too far away, too small, or simply invisible. Now, we can "see" almost everything. But… the challenge is to store and transmit the captured information in an efficient manner without compromising quality. We therefore need to exploit knowledge of the human visual system, to identify redundancy in the pictures and, hence, to compress them efficiently.

Video is now one of the main drivers for advances in communication technology and video content accounts for the majority of internet traffic. Real Networks were one of the pioneers of streaming media, introducing RealPlayer in 1997. Streaming was first incorporated into Windows media Player 6.4 in 1999 and Apple introduced streaming in Quicktime 4 in 1999. The demand for a common streaming format was settled through widespread adoption of Adobe Flash which is used by sites such as YouTube.

Internet video has been enabled by advances in video compression, most notably H.264/AVC, which was developed by the ITU-T Video Coding Experts Group (VCEG) together with the International Organization for Standardization (ISO)/International

Electrotechnical Commission (IEC) Moving Picture Experts Group (MPEG). H.264/ AVC was co-chaired by Gary Sullivan and Thomas Wiegand and ended up doubling the performance of its predecessor, MPEG-2.

Sites like YouTube have transformed how we access and share video. Created in 2005 by three former PayPal employees, it enabled users to upload, view, and share videos. It uses Adobe Flash Video and HTML5 technology with advanced codecs such as H.264/AVC to display a wide variety of content such as movie clips, TV clips, music videos, amateur footage, and blogs. Google bought YouTube in 2006 for $1.65 billion. At the time of writing, it is reported that over 6 billion hours of video are watched each month on YouTube, with over 25% of this on mobile devices.[1] In 2013 YouTube and Netflix together accounted for approximately half of North America's peak download traffic.[2]

It is currently a very exciting and challenging time for video compression. The predicted growth in demand for bandwidth, driven largely by video applications, is probably greater now than it has ever been. There are four primary drivers for this:

1. Recently introduced formats such as 3-D and multiview, coupled with pressures for increased dynamic range, spatial resolution and frame rate, all require increased bit rate to deliver improved levels of immersion or interactivity.
2. Video-based web traffic continues to grow and dominate the internet through social networking and catch-up TV. In recent years YouTube has accounted for 27% of all video traffic and, by 2015, it is predicted that there will be 700 billion minutes of video downloaded.
3. User expectations continue to drive flexibility and quality, with a move from linear to non-linear delivery. Users are demanding "my-time" rather than "prime time" viewing.
4. New services, in particular mobile delivery through 4G/LTE to smart phones, has led some mobile network operators to predict that the demand for bandwidth will double every year for the next 10 years! While this can, to some extent, be addressed through efficiency improvements in network and physical layer technology, because video is the prime driver, the role of video compression is also enormously important.

The recent and ongoing standards activity, referred to as *High Efficiency Video Coding* (HEVC) chaired by Jens-Rainer Ohm, has driven the video community to challenge existing methods, delivering still further improvements. HEVC clearly represents the immediate future for compression and has produced impressive rate-quality gains. In the longer term, alternative approaches—perhaps merging conventional compression techniques with computer graphics—have the potential to create a new content-driven rate–quality optimization framework for video compression.

---

[1] YouTube Statistics: http://youtube.com/yt/press/en-GB/statistics.html.

[2] Sandvine Global Internet Phenomenon Report 1H 2013: https://www.sandvine.com/trends/global-internet-phenomena/.

This book

The importance of video compression and communications has increased enormously in recent years as more and more products and services are being designed around video content. And, with the emergence of new immersive services, coupled with increased user numbers and expectations, the tension between available bandwidth and picture quality has never been greater. Graduates with experience in video compression are, therefore, highly sought-after in industry across the world and most leading EEE, ECE, and CS departments now include coverage of this topic in their programmes.

Several good books have been written on image and video compression but few have taken the holistic approach presented here. *Communicating Pictures* is targeted primarily at 4th (Senior) year and Masters courses on image and/or video compression or coding. I have tried to balance theory with practice, so I hope it will also be of significant value to practitioners and those wishing to update their skills in industry.

In *Communicating Pictures* I have tried to place the topic in an application framework while also providing some historical context which I hope the reader will find interesting. The book is intended to provide sufficient mathematical background to enable a proper understanding of the material, but without detracting from the design and implementation challenges that are faced. The linkage between coding and assessment methods and our understanding of visual perception is also a key and is emphasized throughout. An algorithmic approach is adopted where appropriate, to capture the essence of the methods presented. While I have, on the whole, kept algorithmic developments independent of specific standards, I have used current standards (H.264/AVC and HEVC) as a reference where appropriate, especially in the context of hybrid motion compensated, block-based compression.

After an introduction to the motivation for compression and the description of a generic coding framework in Chapter 1, the book describes some basic concepts that underpin the topic in Chapters 2–4. Chapter 2 looks at the perceptual aspects of visual information, characterizing the limits and attributes of our visual system. This chapter concludes with a comprehensive listing of these attributes and how we can exploit redundancy to reduce bit rate and improve rate–quality performance. Chapter 3 provides a grounding in some of the more fundamental mathematical concepts that underpin discrete time signal and image processing, including: signal statistics, sampling, filters, transforms, information theory, quantization, and linear prediction. Chapter 4 looks, in more detail, at the practical basics, covering acquisition formats, coding structures, and color spaces, as well as providing an introduction to quality assessment and rate–distortion theory.

The most common basis for decorrelating still images is the block transform. This approach is described in Chapter 5, where decorrelating transforms are derived and their properties investigated. The primary focus of this chapter is on the workhorse of compression, the Discrete Cosine Transform (DCT), including associated algorithmic and implementation issues. Coverage of lossless coding is given in Chapter 6—both in its own right and in association with the entropy coding stages of lossy encoders.

This culminates in an operational description of a full image encoder–decoder (or codec) using JPEG as an example. An important alternative to block-based transform coding, based on wavelet filter banks, is addressed in Chapter 7. This explains the benefits and practical issues of the approach such as boundary extension, filter selection, and quantization.

Chapter 8 begins the journey into coding moving pictures, with a detailed description of motion estimation. The focus here is on the trade-offs between accuracy and complexity and a range of fast algorithms are described algorithmically and compared in terms of their efficiencies. This leads on to Chapter 9 which integrates the concepts and methods from previous chapters into a description of the ubiquitous hybrid motion compensated block-based codec. This chapter also introduces many of the codec features and refinements, such as sub-pixel and multiple reference frame estimation, variable block sizes, loop filtering, and intra-prediction, that have facilitated significant coding gains in recent years.

Chapter 10 addresses the important and challenging topic of measuring and managing the quality of digital pictures. This is related, throughout, to human perception, starting with an analysis of the advantages and disadvantages of mean squared error, then continuing to described various perceptually inspired objective metrics. This chapter also includes a detailed description of how to conduct subjective assessment trials, including aspects related to subject screening, test environment, test conditions, and post test analysis. Rate–Quality Optimization (RQO) is also covered here. It is only through effective RQO that modern codecs can make informed local mode decisions during coding.

Now that all of the components of the video codec have been described in Chapters 1–10. Chapter 11 offers insight into the problems associated with the transmission of coded video in real time over lossy and bandwidth-limited networks (wireless and fixed). It explains the conflicts between conventional video compression, based on variable length coding and spatio-temporal prediction, and the requirements for error-resilient transmission. It describes how the presence of packet and bit errors will influence a codec's performance and how we can mitigate their impact through packetization strategies, cross layer optimization, resilient entropy coding, pyramid vector quantization, and through the addition of source and channel coding redundancy.

The final two Chapters 12 and 13, provide a view of the current state of the art in standards and coding efficiency. Chapter 12 explains how the methods described throughout the book have been integrated into today's standards, specifically H.264/AVC and the most recent H.265/HEVC standard. The book concludes in Chapter 13 with a glimpse of how we might code and communicate pictures in the future, including examples of synthesis-based parametric and region-based compression methods. These are the topic of ongoing research and are candidates for future standardization.

I hope you find this book clear and concise. I have tried to make it informative and relevant, and to present concepts in a style that is rigorous yet interesting and accessible. I have included numerous worked examples together with many additional tutorial questions—hopefully these will help to reinforce the important messages and place them in a practical context.

I have always found picture processing to be a popular topic that students enjoy, perhaps because it so intimately relates technology to our everyday personal experiences. This connection between the visual senses and technology, where psychology, biology, and aesthetics meet mathematics, engineering, and computer science, is what makes the subject so fascinating. Also, perhaps, because we can create the illusion of high quality while throwing away most of the original data.

I hope you will benefit from and have pleasure reading this book. Enjoy the magic!

Additional Resources

For the solutions manual, a list of errata and selected high quality colour images, please visit the accompanying resources website at http://booksite.elsevier.com/9780124059061 or see the electronic version of this text.

A software (Matlab) based teaching aid has also been developed at the University of Bristol by Steve Ierodiaconou, Aaron Zhang, Alex Mackin, Paul Hill and myself. This provides a range of demonstrations that will enable students to better understand some of the issues associated with image and video compression. It can be accessed via the above website or directly from www.bristol.ac.uk/vi-lab.

List of figures

Fig. 1.1   A geometric interpretation of compression.   3

Fig. 1.2   The multimedia communications jigsaw puzzle.   7

Fig. 1.3   Simplified high level video compression architecture.   11

Fig. 1.4   The scope of standardization.   13

Fig. 1.5   A chronology of video coding standards from 1990 to the present date.   14

Fig. 2.1   The visible spectrum.   20

Fig. 2.2   Cross-section of the human eye. (Public domain: http://commons.wikimedia.org/wiki/File:Schematic_diagram_of_the_human_eye_en.svg.)   21

Fig. 2.3   Fundus image of a healthy retina. (Public domain from: http://commons.wikimedia.org/wiki/File:Fundus_photograph_of_normal_right_eye.jpg.)   22

Fig. 2.4   The focal length of the lens.   23

Fig. 2.5   Photoreceptor distribution in the retina. (Reproduced with permission from: Mustafia et al. [12].)   24

Fig. 2.6   Normalized rod and cone responses for the human visual system. (Reproduced with permission from: Bowmaker and Dartnall [33]. //www.ncbi.nlm.nih.gov/pmc/articles/PMC1279132/. Avail Wikimedia Commons.)   25

Fig. 2.7   Retinal cell architecture (Public domain image adapted from http://commons.wikimedia.org/wiki/File:Retina-diagram.svg).   27

Fig. 2.8   Spatial opponency, showing center-surround on cell and its firing pattern due to excitation.   28

Fig. 2.9   The visual cortex.   28

Fig. 2.10   Mach band effect.   33

Fig. 2.11   Adelson's grid. (Reproduced with permission from: http://web.mit.edu/persci/people/adelson/checkershadow_illusion.html.)   33

Fig. 2.12   CIE luminous efficiency curve. (Public domain image: http://en.wikipedia.org/wiki/File:CIE_1931_Luminosity.png.)   34

Fig. 2.13   Dark adaptation of rods and cones.   35

Fig. 2.14   Increased immersion from color images.   36

Fig. 2.15   Opponent processing of color.   38

Fig. 2.16 Color dependence on context. The bottom picture is just a squashed version of the top one yet the green stripes become blue. 38

Fig. 2.17 The CIE 1931 chromaticity chart. (Reproduced with permission from Ref. [21].) 39

Fig. 2.18 Just-noticeable differences at different contrast increments. 40

Fig. 2.19 JND curve for human vision. 41

Fig. 2.20 Contrast sensitivity chart. 41

Fig. 2.21 Luminance and chrominance CSF responses. 42

Fig. 2.22 Luminance contrast sensitivity function. 43

Fig. 2.23 Texture change blindness. (Images courtesy of Tom Troscianko.) 45

Fig. 2.24 The importance of phase information in visual perception. Left: original. Right: phase distorted version using the complex wavelet transform (Reproduced with permission from Vilankar et al. [14]). 46

Fig. 2.25 Perspective-based depth cues can be very compelling and misleading. 47

Fig. 2.26 Pits and bumps—deceptive depth from lighting. 48

Fig. 2.27 The hollow mask illusion. 49

Fig. 2.28 Spatio-temporal CSF. (Adapted from Kelly [18].) 51

Fig. 2.29 Variation of critical flicker frequency (Reproduced with permission from Tyler [23]). 52

Fig. 2.30 Eye movements in response to task (Public domain image from: http://commons.wikimedia.org/wiki/File:Yarbus_The_Visitor.jpg). 54

Fig. 2.31 Example of texture masking. 54

Fig. 2.32 Edge masking for high and low dynamic range content. 55

Fig. 2.33 Temporal masking effects for various edge step sizes. (Reproduced with permission from Girod [30].) 55

Fig. 3.1 Spectral characteristics of sampling and aliasing. 66

Fig. 3.2 Demonstration of aliasing for a 1-D signal. Top: sinusoid sampled below Nyquist frequency. Bottom: Fourier plot showing spectral aliasing. 67

Fig. 3.3 2-D spectral characteristics of sampling and aliasing. Left: Top—original signal spectrum; Bottom—sampled signal spectrum with no aliasing. Right: Top—original signal spectrum; Bottom—sampled signal spectrum with aliasing due to sub-Nyquist sampling. 68

Fig. 3.4 Hexagonal sampling lattice and its reciprocal as defined by equation (3.8). 69

Fig. 3.5 Example image histogram for 256×256 image Stampe_SV4. 70

Fig. 3.6   Autocorrelation plots for Acer image (512 × 512). Top left to bottom right: original image; autocorrelation function for row 100; autocorrelation function for whole image; 2-D autocorrelation surface.   72

Fig. 3.7   Autocorrelation plots for Stampe_SV4 image (512 × 512). Top left to bottom right: original image; autocorrelation function for row 100; autocorrelation function for whole image; 2-D autocorrelation surface.   73

Fig. 3.8   Autocorrelation plots for Stampe_SV4 image (256 × 256). Top left to bottom right: original image; autocorrelation function for row 100; autocorrelation function for whole image; 2-D autocorrelation surface.   74

Fig. 3.9   Temporal autocorrelation plots for Foreman (30 fps). Top to bottom right: sample frame showing selected 16 × 16 block used; temporal correlation for a single pixel; temporal correlation for a 16 × 16 block.   75

Fig. 3.10   Filterbank responses for the LeGall low-pass and high-pass analysis filters.   80

Fig. 3.11   Filter response for H.264 half-pixel interpolation filter.   81

Fig. 3.12   Common uniform quantizer characteristics.   85

Fig. 3.13   Common non-uniform quantizers. Left: center deadzone. Right: Lloyd Max quantizer.   86

Fig. 3.14   Feedforward linear prediction. Top: encoder. Bottom: decoder.   88

Fig. 3.15   Prediction signal dynamic range. Top: input signal. Bottom left: distribution of 1000 samples of input signal. Bottom right: distribution of 1000 samples of prediction residual.   89

Fig. 3.16   Feedback-based linear prediction.   90

Fig. 3.17   Feedback-based linear predictor with quantization noise modeling.   91

Fig. 3.18   Self-information and probability. Left: plot of self-information vs probability for a single event. Right: plot of the self-information of an event weighted by its probability.   95

Fig. 4.1   Image sample array.   100

Fig. 4.2   Image samples.   101

Fig. 4.3   Pixelation at varying resolutions. Top left to bottom right: 256 × 256; 64 × 64; 32 × 32; 16 × 16.   102

Fig. 4.4   Typical macroblock structure.   103

Fig. 4.5   Typical group of pictures structure and prediction modes.   105

Fig. 4.6   Aspect ratios of common formats, normalized according to resolution.   107

Fig. 4.7   Widescreen formats.   108

Fig. 4.8   Variation of field of view with viewing distance (aspect ratio = 16:9 here).   109

Fig. 4.9   Interlaced vs progressive frame scanning.   111

Fig. 4.10   Example of effects of interlaced scanning with poor deinterlacing.   112

Fig. 4.11   3:2 pull-down example.   113

Fig. 4.12   Gamma curves for $\gamma = \{0.45, 1, 2.2\}$.   114

Fig. 4.13   Examples of gamma correction.   115

Fig. 4.14   CIE 1931 chromaticity diagram. (Public domain image from: http://commons.wikimedia.org/wiki/File:CIExy1931_Rec_2020_and_Rec_709.svg.)   118

Fig. 4.15   Color difference images using the YC_bC_r decomposition. Top left: original. Right: luma channel. Bottom left: C_b channel. Right: C_r channel.   120

Fig. 4.16   Common chroma sub-sampling formats.   121

Fig. 4.17   Bayer filter mosaic.   124

Fig. 4.18   Quality comparisons for the same PSNR (16.5 dB). Top left to bottom right: original; AWGN (variance 0.24); grid lines; salt and pepper noise; spatial shift by 5 pixels vertically and horizontally.   127

Fig. 4.19   Rate–distortion plot for various coding parameter choices.   130

Fig. 4.20   Rate–distortion plot showing quantizer controlled operating points.   130

Fig. 4.21   Example rate–quality comparison curves for typical standard definition entertainment content.   131

Fig. 5.1   Typical transform-based image compression architecture.   135

Fig. 5.2   Plot of correlated adjacent pixel data (left) and decorrelation through rotation of principal axes (right).   136

Fig. 5.3   Relationship between original and transformed data.   136

Fig. 5.4   DWHT basis functions for $N = 4$.   142

Fig. 5.5   Basis functions for the 2-D DWHT.   143

Fig. 5.6   Spectral leakage due to windowing and periodic extension with the DFT: the rectangular window function and its magnitude spectrum (top); the input signal showing periodic extension and its spectrum (bottom).   145

Fig. 5.7   Symmetrical signal extension for the DCT.   150

Fig. 5.8   DCT basis functions for $N = 8$.   153

Fig. 5.9   2-D DCT basis functions for $N = 8$.   154

Fig. 5.10   Separable computation of the forward DCT.   154

Fig. 5.11   Effects of coefficient quantization on various types of data block.   159

Fig. 5.12   Comparison of DFT and DCT for compression. Top: input signal $x[n]$ and the symmetrically extended sequence $x_1[n]$. Middle: eight-point DFT$\{x[n]\}$ (magnitude only) and the eight-point DCT$\{x[n]\}$. Bottom: IDFT$\{X(k)\}$ (magnitude only) and IDCT$\{C(k)\}$.   160

Fig. 5.13   Comparison of KLT, DFT, DWHT, DST, and DCT on a natural image.   161

Fig. 5.14   Comparison of KLT and DCT performance.   162

Fig. 5.15   DCT performance. Left: Lena original 256 × 256. Right:
             compressed at 0.3 bpp.   162
Fig. 5.16   DCT performance. Left: Lena 256 × 256 0.3 bpp. Right: Lena
             512 × 512 0.3 bpp.   163
Fig. 5.17   DCT rate–distortion performance for 256 × 256 Lena image using
             Huffman encoding.   163
Fig. 5.18   Efficient DCT implementation: McGovern's algorithm.   166
Fig. 5.19   DCT demo GUI.   168
Fig. 5.20   JPEG baseline system diagram.   168
Fig. 6.1   Equivalences between subband filtering and wavelet nomenclature.   173
Fig. 6.2   Comparison of time domain (top right), frequency domain (bottom
             left), and time–frequency analysis (bottom right) applied to the
             combination of a sinusoidal signal and an impulse function (top
             left). (Example adapted from Vetterli and Kovacevic [12].)   174
Fig. 6.3   Wavelet translation and dilation.   175
Fig. 6.4   Two-channel perfect reconstruction filter bank.   176
Fig. 6.5   Complementary half-band filters from equations (6.4) (solid) and
             (6.5) (dotted).   180
Fig. 6.6   Basic system diagram for the two-channel filter bank.   181
Fig. 6.7   The two-channel filter bank.   181
Fig. 6.8   Upsampling by a factor L.   183
Fig. 6.9   Spectral effect of upsampling: (a) input; (b) output.   184
Fig. 6.10   Two-channel synthesis filter bank.   184
Fig. 6.11   Spectral effects of upsampling where $L = 2$.   185
Fig. 6.12   Downsampling by a factor of L.   185
Fig. 6.13   Spectral effects of downsampling: (a) input; (b) output.   187
Fig. 6.14   Two-channel analysis filter bank.   187
Fig. 6.15   Spectral effects of downsampling with $M = 2$.   188
Fig. 6.16   Spectral relationships within the two-channel perfect
             reconstruction filter bank. The spectral characteristics
             at signal points A–H are shown.   190
Fig. 6.17   QMF filter pair frequency response.   194
Fig. 6.18   LeGall filter frequency responses.   196
Fig. 6.19   1-D three stage wavelet decomposition. Top: block diagram.
             Bottom: associated time–frequency tiling.   197
Fig. 6.20   Time–frequency tiling for progressive stages of decomposition in
             a 1-D wavelet filter bank: (a) input of eight samples; (b) after the
             first stage; (c) after the second stage; (d) after the third stage. 198
Fig. 6.21   1-D three-stage wavelet decomposition—modified downsampling
             structure.   199
Fig. 6.22   2-D two-channel, two-stage wavelet decomposition. Data transpo-
             sition (not shown) is required between row and column processing.   200

Fig. 6.23	2-D wavelet—frequency plane tiling.	200
Fig. 6.24	Frame from Akiyo sequence (left) wavelet transformed (right) using a three-stage decomposition with LeGall 5/3 filters.	201
Fig. 6.25	Illustration of boundary extension.	201
Fig. 6.26	Compression performance, DCT vs wavelet, for the 256×256 luma Lena image.	204
Fig. 6.27	Comparisons of wavelet-coded images, 256×256 Lena. Left: LeGall 5/3 wavelets at 0.3 bpp. Right: LeGall 5/3 wavelets at 0.5 bpp.	204
Fig. 6.28	Subband weighting.	205
Fig. 6.29	Scanning of wavelet subbands.	205
Fig. 6.30	Tree structures for embedded wavelet coding.	206
Fig. 6.31	JPEG2000 encoding of *Puppet* test image (512×512). Top left to bottom right: JPEG at 64:1; JPEG2000 at 64:1; JPEG at 128:1; JPEG2000 at 128:1.	209
Fig. 6.32	JPEG2000 ROI coefficient coding using Maxshift.	209
Fig. 7.1	Symbol-code mappings for the 26 letters from the English alphabet in Morse code.	215
Fig. 7.2	Generic model of lossless coding.	217
Fig. 7.3	Adaptive entropy coding.	226
Fig. 7.4	Example of adaptive Huffman encoding tree evolution.	227
Fig. 7.5	Example of adaptive Huffman encoding showing node swapping (bottom) after weight updating symbol *a* by five counts.	228
Fig. 7.6	Zig-zag scanning prior to variable length coding.	229
Fig. 7.7	JPEG baseline image encoder architecture.	230
Fig. 7.8	Distribution of *run/size* values in a typical JPEG encoding.	231
Fig. 7.9	JPEG results for various compression ratios for 512×512 *maple* image. Top left: original. Top right: 0.16 bpp. Middle left: 0.22 bpp. Middle right: 0.5 bpp. Bottom left: 1 bpp. Bottom right: 2 bpp.	236
Fig. 7.10	JPEG results for various compression ratios for 512×512 *puppet* image. Top left: original. Top right: 0.16 bpp. Middle left: 0.22 bpp. Middle right: 0.5 bpp. Bottom left: 1 bpp. Bottom right: 2 bpp.	237
Fig. 7.11	Prediction samples for lossless JPEG.	238
Fig. 7.12	A graphical illustration of the arithmetic encoding and decoding processes. Top: probability intervals for three iterations. Bottom: codeword intervals. Shows $P(a) = 0.25$ and $P(b) = 0.75$ with example sequences *abb* and *bbb*. Adapted from Ref. [20].	247
Fig. 7.13	Context-Adaptive Binary Arithmetic Coding. Adapted from Ref. [17].	251
Fig. 7.14	Huffman vs arithmetic coding.	252
Fig. 8.1	Variation in the number of bits per frame and frame coding mode.	257

Fig. 8.2   Temporal correlation between two adjacent video frames.   259

Fig. 8.3   Frame differencing (left) vs motion-compensated prediction (right).   260

Fig. 8.4   Probability distributions for an original frame and FD and DFD frames from the *Football* sequence.   261

Fig. 8.5   Motion-based image deformations.   264

Fig. 8.6   Mesh-based warping.   265

Fig. 8.7   Phase correlation-based motion estimation.   267

Fig. 8.8   Example of phase correlation-based motion estimation. Top: input frames of a tarmac road surface taken with an approximately translating camera. Bottom: phase correlation surface.   269

Fig. 8.9   Translational block matching motion estimation (BMME).   270

Fig. 8.10   Effect of matching function: SAD vs SSD.   271

Fig. 8.11   Autocorrelation values for typical block motion fields (*Foreman* at 25 fps). (Adapted from Ref. [5].)   272

Fig. 8.12   The effect of block size on motion estimation performance.   273

Fig. 8.13   The influence of search range on prediction accuracy and coding bit rate.   274

Fig. 8.14   Typical motion residual error surface for a 16×16 block from the *Foreman* sequence (CIF at 25 fps, $\pm 16 \times \pm 16$ search window). Top left: search window. Top right: current block.   275

Fig. 8.15   Motion vector pdfs for *Akiyo*, *Coastguard* and *Football* sequences.   276

Fig. 8.16   Example of motion model failure for *Football* sequence. Top left: current frame, right: reference frame. Bottom left: full-pixel estimation, right: half-pixel estimation.   277

Fig. 8.17   The pixel grid and the search grid.   278

Fig. 8.18   2-D logarithmic search patterns.   280

Fig. 8.19   *N*-Step Search patterns.   281

Fig. 8.20   Diamond search patterns.   283

Fig. 8.21   Hexagonal search patterns.   284

Fig. 8.22   Predicted motion vectors for initialization.   285

Fig. 8.23   Fast motion estimation performance comparison for the CIF Bream sequence at 25 fps: 4PMV+0 vs ADZS-ER vs FS.   286

Fig. 9.1   Typical group of pictures structure.   293

Fig. 9.2   Video encoder structure (intra-frame mode). Dotted lines depict control signals relating to quantizer step size, etc.   294

Fig. 9.3   Video encoder structure (inter-frame mode).   294

Fig. 9.4   Video decoder structure.   296

Fig. 9.5   Intra-prediction modes in H.264/AVC.   299

Fig. 9.6 4×4 intra-prediction modes.   299

Fig. 9.7   Sub-pixel search with local refinement.   302

Fig. 9.8   Motion estimation search grid with half-pixel refinement.   302

Fig. 9.9   H.264 sub-pixel interpolation. (a) Half-pixel interpolation.
             (b) Quarter-pixel interpolation.   303

Fig. 9.10   Comparison of full, half-, and quarter-pixel motion estimation. 306

Fig. 9.11   Influence of sub-pixel motion estimation on residual error.
             Left: integer-pixel ME. Right: half-pixel ME.   306

Fig. 9.12   Multiple reference frame motion estimation.   308

Fig. 9.13   MRF-ME performance. Left: PSNR. Right: complexity.   309

Fig. 9.14   Variable block sizes supported by H.264/AVC. Top: 16×16 modes
             1–4. Bottom: 8×8 modes 1–4.   310

Fig. 9.15   Block size distribution (Foreman sequence coded using
             H.264/AVC). Produced using Ref. [14].   311

Fig. 9.16   One-dimensional edge example.   314

Fig. 9.17   Illustration of the effect of a deblocking filter. Left: without
             deblocking. Right: with deblocking filter applied.   314

Fig. 10.1   Spatial and temporal information coverage for a range of test
             sequences.   323

Fig. 10.2   Generic arrangement for subjective video evaluation.   325

Fig. 10.3   A section of the DSCQS assessment form.   326

Fig. 10.4   Interface for SAMVIQ testing. Courtesy: ProVision
             Communications.   326

Fig. 10.5   Typical timing of a DSCQS subjective test.   327

Fig. 10.6   Reduced reference video quality assessment.   334

Fig. 10.7   Scatter plots of DMOS vs PSNR predicted DMOS for various
             classes of sequence from the VQEG database [42].   336

Fig. 10.8   Comparison of MOVIE, ST-MAD, and PVM on LIVE and VQEG
             FRTV database [42].   342

Fig. 10.9   Rate–distortion plot for various coding parameter choices. The
             associated convex hull provides the operational R-D characteristic
             for the given codec–source combination.   344

Fig. 10.10   Lagrangian optimization for a coding unit, i.   347

Fig. 10.11   Managing bit rate in the context of changing source complexity. 351

Fig. 10.12   Transmission buffering at encoder and decoder.   351

Fig. 10.13   Streaming playout schedule on a picture by picture basis,
             corresponding to the points labeled in Figure 10.12.   352

Fig. 10.14   Example decoder buffer timing, indicating HRD parameters.   353

Fig. 10.15   Generic rate controller.   355

Fig. 11.1   Logged radio link parameter variations with channel conditions
             for a wireless video trial.   363

Fig. 11.2   Radiowave propagation modeling using ray tracing. (Courtesy of
             A. Nix.)   366

Fig. 11.3   Generic error-resilient video transmission architecture.   367

Fig. 11.4   The effect of a single bit error on a DCT/Huffman encoded Lena
             image.   370

Fig. 11.5   Impact of increased error rates on a DCT image codec. Left: 0.01% BER. Middle: 0.1% BER. Right: 1% BER.   371

Fig. 11.6   Temporal error propagation from a single corrupted block in frame $k - 1$ to multiple corrupted blocks in frame k.   372

Fig. 11.7   Example of temporal propagation; Top: across four frames of the table tennis sequence. Bottom: across 190 frames of Foreman.   373

Fig. 11.8   Subjective appearance of loss. PVQ encoded Riverbed sequence at 30 Mb/s. Left: encoded at 30 Mb/s (intra). Right: reconstructed after 25% PER. Bottom: magnified artifact from box shown in top right image.   374

Fig. 11.9   Idealized comparison of performance.   375

Fig. 11.10   Cross packet FEC.   377

Fig. 11.11   Performance of cross packet FEC for a coding depth of 8 (left) and 32 (right) for various coding rates.   378

Fig. 11.12   Error resilience based on reference picture selection.   381

Fig. 11.13   Periodic reference frames.   382

Fig. 11.14   Example slice structures.   384

Fig. 11.15   Use of dispersed mode slice groups as a basis for error concealment.   385

Fig. 11.16   802.11g MCS switching characteristics based on throughput. (Reproduced with permission from Ref. [20]).   387

Fig. 11.17   802.11g MCS switching characteristics based on video distortion. (Reproduced with permission from Ref. [20]).   387

Fig. 11.18   MCS switching characteristic comparison: video quality based (left) vs throughput based (right). (Reproduced with permission from Ref. [20]).   388

Fig. 11.19   MCS link adaptation results. (Reproduced with permission from Ref. [20]).   388

Fig. 11.20   EREC operation.   390

Fig. 11.21   EREC performance examples of reconstructed DCT/Huffman encoded images after erroneous transmission. Left: 0.1% BER. Right: 1% BER. (Courtesy of D. Redmill.)   393

Fig. 11.22   Graph of EREC performance for increasing BER compared to conventional DCT/VLC encoding. (Courtesy of D. Redmill.)   393

Fig. 11.23   RD optimized PVQ performance for Tennis. Left: RD performance in a clean channel. Right: performance in a lossy 802.11n channel with correlated errors (Reproduced with permission from Ref. [26]).   395

Fig. 11.24   Frame shots for Tennis at 25 Mb/s and 25% BER. Top: H.264/AVC (HPS). Bottom: RD optimized PVQ (Reproduced with permission from Ref. [26]).   395

Fig. 11.25   Variability in frame distortions for PVQ and H.264/AVC (Reproduced with permission from Ref. [26]).   396

Fig. 11.26 Packet error rate vs residual bit error rate for an 802.11n channel (Reproduced with permission from Ref. [26]). 396

Fig. 11.27 Spatial error concealment using nearest neighboring pixels. 398

Fig. 11.28 Spatial error concealment applied to a lost row of macroblocks in the Foreman sequence. Left: original with error shown. Middle: result of spatial error concealment. Right: amplified difference signal. MSE = 46.6. (Courtesy of D. Agrafiotis.) 399

Fig. 11.29 Temporal error concealment based on temporal copying (top) and motion compensated prediction (bottom). 400

Fig. 11.30 Temporal error concealment results for Foreman based on temporal copying. MSE = 19.86. (Courtesy of D. Agrafiotis). 401

Fig. 11.31 Boundary Matching Error measures. Left: Boundary Matching Error (BME). Right: External Boundary Matching Error (EBME). 401

Fig. 11.32 Temporal error concealment results for *Foreman* based on motion compensated replacement. MSE = 14.61. (Courtesy of D. Agrafiotis.) 402

Fig. 11.33 Enhanced error concealment with mode selection [27]. 404

Fig. 11.34 Scalable (spatial or temporal) video codec architecture. 406

Fig. 11.35 Frame types for scalable encoding. 406

Fig. 11.36 Multiple description codec architecture. 407

Fig. 12.1 A chronology of video coding standards from 1990 to the present date. 413

Fig. 12.2 The scope of standardization. 413

Fig. 12.3 H.261 macroblock, GOB, and CIF frame format. 416

Fig. 12.4 MPEG-2 encoder performance improvements between 1994 and 2004. (Adapted from an original presented by Tandberg.) 417

Fig. 12.5 MPEG-2 GOP structure. 418

Fig. 12.6 H.263 picture format (QCIF). 420

Fig. 12.7 Motion vector coding in H.263. 421

Fig. 12.8 An example of an MPEG-4 audiovisual scene. 424

Fig. 12.9 Generic representation of MPEG-4 Video Object Plane (VOP) coding. 425

Fig. 12.10 H.264/AVC layer structure. 427

Fig. 12.11 H.264 syntax. 428

Fig. 12.12 Example slice structures. 430

Fig. 12.13 Hadamard transformation of prediction residuals and coded block ordering. 431

Fig. 12.14 Variable block sizes supported by H.264/AVC. Top: 16×16 modes 1–4. Bottom: 8×8 modes 1–4. 432

Fig. 12.15 Example H.264/AVC macroblock partition for inter-coding. 432

Fig. 12.16 Hierarchical B-pictures in H.264.   433

Fig. 12.17 Selected H.264 profiles.   434

Fig. 12.18 Typical H.264/AVC (MP) performance relative to MPEG-2 for stan-
            dard definition entertainment content.   435

Fig. 12.19 HEVC video encoder architecture.   437

Fig. 12.20 CTU partitioning in HEVC—an example.   439

Fig. 12.21 HEVC intra-prediction modes (mode 0 = planar and mode 1 = DC).   440

Fig. 12.22 CB partitions for inter-prediction PBs.   442

Fig. 12.23 Coefficient scanning in HEVC.   443

Fig. 12.24 Significance flag contexts for 4×4 and 8×8 HEVC transform
            blocks.   444

Fig. 13.1   Does increased resolution mean increased quality?   454

Fig. 13.2   Frame rate and shutter angle. The figure illustrates the effect of
            a ball thrown from left to right across a static scene, captured at
            various shutter angles.   456

Fig. 13.3   The influence of frame rate (*Outdoor* sequence). Left: 25 fps.
            Right: 600 fps.   456

Fig. 13.4   Static vs dynamic resolution—the relationship between frame
            rate and spatial resolution (adapted from an original by Richard
            Salmon, BBC).   457

Fig. 13.5   Motion vector PDF comparisons for 25 fps and 600 fps.   458

Fig. 13.6   Motion vectors and residual content for high frame rate content
            (*Mouse* sequence). Top: 400 fps. Bottom: 25 fps. Left: DFD signal.
            Right: spatial distribution of motion vectors.   458

Fig. 13.7   Intensity-dependent quantization for HDR extensions to HEVC.   460

Fig. 13.8   The creative continuum.   462

Fig. 13.9   Parametric video compression architecture (Reproduced with
            permission from Zhang and Bull [15]).   465

Fig. 13.10 Texture classification results for parametric video coding. Left:
            original frame. Right: classified regions for coding: Red—static,
            blue—dynamic, black—structural, or non-textured (Reproduced
            with permission from Ref. [15]). (For interpretation of the refer-
            ences to color in this figure legend, the reader is referred to the
            web version of this book.)   465

Fig. 13.11 Coding results from parametric video coding. Left: H.264 frame.
            Right: proposed method. Center: coding mode employed
            (from Ref. [15]).   466

Fig. 13.12 SURF-based feature matching for context-based coding of
            planar regions. Left: reference frame. Right: current frame.
            Circles represent RANSAC inliers used for planar motion
            modeling.   467

Fig. 13.13 Context-based video compression results. Left: H.264 coded.
            Right: context-based coding.   467
            For colour versions of figures please refer to the electronic
            version or the website
            http://booksite.elsevier.com/9780124059061/

List of tables

Table 1.1 Typical parameters for common digital video formats and their (uncompressed) bit rate requirements. 4

Table 1.2 Theoretical bandwidth characteristics for common communication systems. 5

Table 4.1 Aspect ratios of film and TV. 107

Table 4.2 ITU-R Rec.BT.500 subjective evaluation impairment scale. 129

Table 6.1 Relationship between subband frequency range and number of samples for each subband in a five-stage decomposition. 199

Table 6.2 Technical overview of JPEG2000. 208

Table 7.1 Alternative codes for a given probability distribution. 218

Table 7.2 Table for luminance DC coefficient differences. 232

Table 7.3 Table for luminance AC coefficients, showing *run/size* (*R/S*) values, Huffman codes, and their length (*L*). 233

Table 7.4 JPEG coefficient size categories. 234

Table 7.5 Prediction modes for lossless JPEG. 238

Table 7.6 Golomb codes for index values $i = 0 \cdots 9$. 239

Table 8.1 Comparison of warping and translational models for three test sequences [5]. 266

Table 8.2 Reduced complexity BMME comparisons for the Bream CIF sequence at 25 fps. Top: checking points per macroblock. Bottom: SAD error relative to full search. 286

Table 9.1 Performance of simplex-based (SMS) multiple reference frame motion estimation. Results from Ref. [13]. 309

Table 10.1 Selected subjective testing viewing environment parameters. From Refs. [4,5]. 324

Table 11.1 FLC vs VLC encoding for various channel conditions. 256×256 Lena image at 2 bpp. 374

Table 11.2 Selected performance results for EECMS (from Ref. [27]). 404

Table 12.1 Primary features of H.261. 416

Table 12.2 MPEG-2 main profile@main level (MP@ML) bounds. 419

Table 12.3 Primary features of H.263. 420

Table 12.4 Primary features of H.264/AVC. 427

Table 12.5 Example HEVC levels. 438

Table 12.6 SAO edge classification. 445

Table 12.7 Comparison of video coding standards for entertainment
           applications. Average bit rate savings are shown for equal
           objective quality values measured using PSNR.   446
Table 13.1 Parameter set for UHDTV/ITU-R Rec.2020.   454

List of algorithms

Algorithm 4.1	3:2 pull-down.	113
Algorithm 7.1	Huffman tree formation.	220
Algorithm 7.2	Minimum variance Huffman tree formation.	222
Algorithm 7.3	Huffman decoding process.	223
Algorithm 7.4	Adaptive Huffman coding: tree updating.	226
Algorithm 7.5	JPEG DC coefficient symbol encoding.	231
Algorithm 7.6	JPEG AC coefficient symbol encoding.	231
Algorithm 7.7	Arithmetic encoding.	242
Algorithm 7.8	Arithmetic decoding.	245
Algorithm 8.1	TDL search.	280
Algorithm 8.2	NSS search.	281
Algorithm 8.3	Diamond search.	283
Algorithm 9.1	Intra-mode encoder operation.	295
Algorithm 9.2	Inter-mode encoder operation.	295
Algorithm 9.3	Intra-mode decoder operation.	296
Algorithm 9.4	Inter-mode decoder operation.	296
Algorithm 9.5	Deblocking operation.	314
Algorithm 10.1	Screening of observers in subjective trials using the β_2 test [4].	329
Algorithm 10.2	Subjective testing methodology.	330
Algorithm 11.1	EREC encoding.	390
Algorithm 11.2	EECMS.	404

Acknowledgments

This book would not have been possible without the support of many people over many years. As any teacher will know, a key contribution to our own education comes from our students, who take relish and pride, in equal proportion, from pointing out errors, inconsistencies, and ambiguities. I have been privileged to have taught (and be taught by!) many excellent students over the past 25 years.

I would like to thank all my outstanding colleagues in the Visual Information Laboratory at Bristol who have been supportive and professional. I would particularly like to highlight those who I have worked with on compression, who, in different ways have helped to shape this book: Nishan Canagarajah, Dimitris Agrafiotis, and Paul Hill. Much of my work on image communications has also benefited from rich interactions with colleagues including Andrew Nix, Angela Doufexi, and Joe McGeehan in the Center for Communications Research. Alongside these academic colleagues there is a long list of PhD students and research assistants without whom much of my work would not have happened and I am grateful to all of them. Those that have directly contributed results include Mohammed Al-Mualla, Mohsin Bokhari, Przemek Czerepinski, TK Chiew, James Chung-How, Pierre Ferre, Paul Hill, Francois Martin, Robert O'Callaghan, David Redmill, Oliver Sohm, Richard Vigars, Dong Wang, Fan Zhang, and Yang Zhang. Finally, I would also like to acknowledge colleagues and collaborators who I have worked with in other organizations across the world, too numerous to list, but who have, in different ways, been an inspiration.

In particular I would like to acknowledge the great help I have had in preparing some of the figures and results in this book from my colleague Fan (Aaron) Zhang. Several people have also helped to proofread this book. Alongside Aaron, I would like to highlight my PhD student, Alex Mackin, who's attention to detail has uncovered and fixed many errors and inconsistencies. I would also like to thank Stephen Kiely for his tips on grammar and hyphens.

If there is a unique feel from this book, it almost certainly comes from the influences of the wide range of people I work with in the Visual Information Laboratory and in Bristol Vision Institute (BVI). It has been an honor to be the first Director of BVI which has provided such rich interactions with colleagues in psychology, biology, engineering, computer science, mathematics, biochemistry, neuroscience, ophthalmology, medicine, and the arts. In particular I must highlight psychologists Tom Troscianko and Iain Gilchrist, biologist Innes Cuthill, and anatomist/mechanical engineer (with a degree in electronics!) Jeremy Burn. BVI, at least in part, owes its existence to the inspiration of Professor Richard Gregory FRS and Professor Tom Troscianko. Richard sadly died in 2010 and Tom passed away too early, in 2011.

Finally I would like to thank my family and in particular my partner, Janice, for her love, support and patience throughout—especially in the final months of writing when she didn't get a summer holiday (fortunately she's an academic so she understands!). I dedicate this book to her.

Introduction

CHAPTER OUTLINE

1.1 **Communicating pictures: the need for compression** **2**
    1.1.1 What is compression? 2
    1.1.2 Why do we need compression? 3
1.2 **Applications and drivers** **6**
    1.2.1 Generic drivers 6
    1.2.2 Application drivers and markets 8
1.3 **Requirements and trade-offs in a compression system** **9**
    1.3.1 The benefits of a digital solution 9
    1.3.2 Requirements 10
    1.3.3 Trade-offs 10
1.4 **The basics of compression** **11**
    1.4.1 Still image encoding 11
    1.4.2 Encoding video 12
    1.4.3 Measuring visual quality 12
1.5 **The need for standards** **12**
    1.5.1 Some basic facts about standards 12
    1.5.2 A brief history of video encoding standards 13
1.6 **Summary** **14**
References **15**

Visual communications have a rich history, from the earliest cave painting, through Egyptian hieroglyphs, the invention of paper, moveable type, photography and film, to modern art, television, and "new media." We now capture, store, access, transmit, and receive visual information as part of our everyday lives and we can visualize almost anything, from distant stars to the smallest organism.

Visual information is the primary consumer of communications bandwidth across all broadcast, internet, and surveillance networks. The demand for increased qualities and quantities of visual content is increasing daily and this is creating a major tension between available network capacity and required video bit rate. Network operators,

Communicating Pictures. http://dx.doi.org/10.1016/B978-0-12-405906-1.00001-5

content creators, and service providers all need to transmit the highest quality video at the lowest bit rate and this can only be achieved through the exploitation of perceptual redundancy to enable video compression.

This chapter provides a descriptive introduction to image and video compression. We first look at what compression means and why we need it. We then examine the primary drivers for video compression in today's world. Finally we consider the requirements of a video compression system and explain why standards are so important in supporting interoperability.

1.1 Communicating pictures: the need for compression

1.1.1 What is compression?

Before we look at why we need compression, let us consider the fundamental issue of what compression actually does. First let's define compression: there are a number of ways that we can express this, and these are covered in more detail later, but for now let us simply define it as the ratio of the size of (or number of bits in) the original image or video file B_i to that of the file used for storage or transmission B_o

$$\text{CR} = \frac{B_o}{B_i} \quad \text{(unitless)} \tag{1.1}$$

This simple definition ignores a lot about real-time aspects or dynamic channel conditions, but will suffice for now.

Jumping ahead a little to Section 1.1.2, we will see that typical video compression ratio requirements are currently between 100:1 and 200:1, with this increasing to many hundreds or even thousands to one as new, more demanding formats emerge. If we consider Figure 1.1, we can see what that means in terms of a geometric analogy—in order to achieve a compression ratio of 256:1 we must replace the large square on the left by the small one at the bottom right of the figure. There is a common expression "You can't squeeze a quart into a pint pot,"[1] but we can see that the compression process must do exactly that. In fact, for the example of the bottom square in the figure, it must squeeze 128 quarts into a pint pot!

So what does this mean for pictures? Consider a small image, of dimension 10×10 samples (known as picture elements, *pels* or *pixels*), where each pixel is stored as a single byte to produce a 10×10 matrix of byte values. After compression (for a compression ratio of 100:1) we expect that these 100 values will be replaced by just one single byte AND that this is achieved without introducing any noticeable distortion in the image! At first this appears ridiculous, but that is what we need to achieve! Actually it's only what we need to achieve on average—as the only way we could reduce 100 numbers to a single value was if they were all the same (or fitted exactly some other known model)—e.g. the image was of constant luminance.

---

[1] A quart is a unit of volume (for either the imperial or United States customary units) equal to 2 pints or one quarter of a gallon.

FIGURE 1.1

A geometric interpretation of compression.

In reality most images are not constant valued, but they are highly correlated spatially. Similarly, most video sequences are correlated temporally. As we will see, it is the modeling and exploitation of this spatio-temporal correlation, primarily through transformation and prediction, that enables us to compress images and video so effectively. Consider your Blu-ray disc—most people think of these as extremely high quality. However, they are encoded at around 20 Mbps—a compression ratio of around 60:1 compared to the original movie.[2]

So now let's look in a little more detail at why we need to compress pictures.

1.1.2 Why do we need compression?

Picture formats and bit rate requirements

We will consider picture formats in detail in Chapter 4, but introduce them briefly here in order to obtain an understanding of the requirements of a picture compression system. Pictures are normally acquired as an array of color samples, usually based on combinations of the red, green, and blue primaries. They are then often converted to some other more convenient color space that encodes luminance separately to two color difference signals (see Chapter 4).

Table 1.1 shows typical sampling parameters for a range of common video formats. Without any compression, it can be seen, even for the lower resolution formats, that the bit rate requirements are high—much higher than what is normally provided by today's communication channels. Note that the chrominance signals are encoded at a reduced resolution as indicated by the 4:2:2 and 4:2:0 labels.[3] Also note that two formats are included for the HDTV case (the same could be done for the other formats); for broadcast quality systems, the 4:2:2 format is actually more representative of the

---

[2]Assumes a 1080p24 4:2:2 format original at 12 bits per pixel. More about formats and color subsampling in Chapter 4.

[3]See Chapter 4 for more details on color spaces and sub-sampling.

Table 1.1 Typical parameters for common digital video formats and their (uncompressed) bit rate requirements.

Format	Spatial sampling (V × H)	Temporal sampling (fps)	Raw bit rate (30 fps, 8/10 bits) (Mbps)
UHDTV (4:2:0) (ITU-R 2020)	Lum: 7680 × 4320 Chrom: 3840 × 2160	24, 25, 30, 50, 60,120	14,930*
HDTV (4:2:0) (ITU-R 709)	Lum: 1920 × 1080 Chrom: 960 × 540	24, 25, 30, 50, 60	933.1*
HDTV (4:2:2) (ITU-R 709)	Lum: 1920 × 1080 Chrom: 960 × 1080	24, 25, 30, 50, 60	1244.2*
SDTV (ITU-R 601)	Lum: 720 × 576 Chrom: 360 × 288	25, 30	149.3
CIF	Lum: 352 × 288 Chrom: 176 × 144	10–30	36.5
QCIF	Lum: 176 × 144 Chrom: 88 × 72	5–30	9.1

*Assumes encoding at 10 bits.
UHDTV = Ultra High Definition Television; HDTV = High Definition Television; SDTV = Standard Definition Television; CIF = Common Intermediate Format; QCIF = Quarter CIF.

original bit rate as this is what is produced by most high quality cameras. The 4:2:0 format, on the other hand, is that normally employed for transmission after compression.

Finally, it is worth highlighting that the situation is actually worse than that shown in Table 1.1 especially for the new Ultra High Definition Television (UHDTV) standard where higher frame rates and longer wordlengths will normally be used. For example at 120 frames per second (fps) with a 10 bit wordlength for each sample, the raw bit rate increases to 60 Gbps for a single video stream! This will increase even further if 3-D or multiview formats are employed.

Available bandwidth

To see the other side of the inequality, let us now examine the bandwidths available in typical communication channels. Some common communication systems for broadcast and mobile applications are characterized in Table 1.2. This table should however be read with significant caution as it provides the theoretical maximum bit rates under optimum operating conditions. These are rarely, if ever, achieved in practice. Bandwidth limitations are particularly stringent in wireless environments because the usable radio spectrum is limited and the transmission conditions are variable and data loss is commonplace.

The bit rates available to an individual user at the application layer (which is after all what we are interested in) will normally be grossly reduced from the figures

Table 1.2 Theoretical bandwidth characteristics for common communication systems.

Communication system	Maximum bandwidth
3G mobile (UMTS)	384 kbps
4G mobile (4 × 4 LTE)	326 Mbps
Broadband (ADSL2)	24 Mbps
Broadband (VSDL2)	100 Mbps
WiFi (IEEE 802.11n)	600 Mbps
Terrestrial TV (DVB-T2 (8 MHz))	50 Mbps

quoted in Table 1.2. The effective throughput (sometimes referred to as *goodput*) is influenced by a large range of internal and external factors. These include:

- Overheads due to link layer and application layer protocols.
- Network contention and flow control.
- Network congestion and numbers of users.
- Asymmetry between download and upload rates.
- Network channel conditions.
- Hardware and software implementations that do not support all functions needed to achieve optimum throughput.

In particular, as channel conditions deteriorate, modulation and coding schemes will need to be increasingly robust. This will create lower spectral efficiency with increased coding overhead needed in order to maintain a given quality. The number of retransmissions will also inevitably increase as the channel worsens. As an example, DVB-T2 will reduce from 50 Mbps (256QAM @ 5/6 code-rate) to around 7.5 Mbps when channel conditions dictate a change in modulation and coding mode down to 1/2 rate QPSK. Similarly for 802.11n, realistic bandwidths per user can easily reduce well below 10 Mbps. Typical broadband download speeds where the author lives (in a semi-rural area of the UK) are, at the time of writing, 3 Mbps on a good day. 3G download speeds never offer 384 kbps—more frequently they will be less than 100 kbps. In summary, useful bandwidths per user at the application layer will rarely exceed 50% of the theoretical maximum and are more likely to be in the range 10–20%.

On that basis, let us consider a simple example which relates the raw bit rates in Table 1.1 to the realistic bandwidth available. Consider the example of a digital HDTV transmission at 30 fps using DVB-T2, where the average bit rate allowed in the multiplex (per channel) is 15 Mbps. The raw bit rate, assuming a 4:2:2 original at 10 bits, is approximately 1.244 Gbps, while the actual bandwidth available dictates a bit rate of 15 Mbps. This represents a compression ratio of approximately 83:1.

Download sites such as YouTube typically support up to 6 Mbps for HD 1080p[4] format, but more often video downloads will use 360p or 480p (640×480 pixels) formats at 30 fps, with a bit rate between 0.5 and 1 Mbps encoded using the H.264/AVC

---

[4]p stands for progressive transmission—see Chapter 4 for further details.

standard. In this case the raw bit rate, assuming color sub-sampling in 4:2:0 format, will be 110.6 Mbps. As we can see, this is between 100 and 200 times the bit rate supported for transmission.

Example 1.1 (Compression ratio for UHDTV)

Consider the case of 8K UHDTV with the original video in 4:2:2 format (a luminance signal of 7680 × 4320 and two chrominance signals of 3840 × 4320) at 10 bits per sample and a frame rate of 60 fps. Calculate the compression ratio if this video is to be transmitted over a DVB-T2 link with an average bandwidth of 15 Mbps.

Solution. The 4:2:2 color sub-sampling method has, on average, the equivalent of two 10 bit samples for each pixel (see Chapter 4). Thus, in its uncompressed form the bit rate is calculated as follows:

$$R = 7680\ (H) \times 4320\ (V) \times 2\ \text{(samples/pixel)} \times 10\ \text{(bits)} \times 60\ \text{(fps)}$$
$$= 39{,}813{,}120{,}000\ \text{b/s}$$

i.e. a raw bit rate approaching 40 Gbps. Assuming this needs to be transmitted in the same bandwidth as for conventional HDTV (i.e. 15 Mbps) then a compression ratio of 2700:1 would be required!

$$CR = \frac{39{,}813{,}120{,}000}{15{,}000{,}000} \approx 2700$$

Hopefully this section has been convincing in terms of the need for compression. A tension between user expectations, in terms of quality and ease of access on one hand, and available bandwidth on the other, has existed since the first video transmissions and this has promoted vigorous research in the fields of both coding and networks. Fortunately the advances in communications technology have mirrored those in video compression, enabling the transmission of good (and in most cases very good) quality video that meets user expectations.

In the next section we examine the applications that are currently driving video compression performance as well as those that are likely do so in the future.

1.2 Applications and drivers

1.2.1 Generic drivers

By 2020 it is predicted that the number of network-connected devices will reach 1000 times the world's population; there will be 7 trillion connected devices for 7 billion people [1]. Cisco predict [2] that this will result in 1.3 ZB of global internet traffic by 2016, with over 80% of this being video traffic. The generic drivers for this explosion in video technology are as follows:

- Increased numbers of users with increased expectations of quality and accessibility.
- The ubiquity of user-generated image, video, and multimedia available through social networking and download sites.
- The emergence of new ways of working using distributed applications and environments such as the cloud.
- The demand for rapid and remote access to information, driven by the need for improved productivity, security, and responsiveness.
- Emerging immersive and interactive entertainment formats for film, television, and streaming.

In order to satisfy the emerging demands for accessibility and quality, video communication solutions require a high performance core network, an adaptive wireless infrastructure, robust data transport protocols, and efficient content representations. The interactions between these components in the delivery chain can be thought of as a jigsaw puzzle (Figure 1.2) where complex and dynamically changing pieces need to fit together. Only in combination can these provide a truly end-to-end flexible and robust solution to the delivery challenges of the future. Video compression is a key component in the solution.

FIGURE 1.2

The multimedia communications jigsaw puzzle.

1.2.2 Application drivers and markets

The range of applications which rely wholly or partially on video technology is enormous; as a consequence, the market for video compression technology is growing rapidly. Some examples are provided below.

Consumer video

Entertainment, personal communications, and social interaction provide the primary applications in consumer video, and these will dominate the video landscape of the future. As mentioned previously, Cisco predict [2] that by 2016, 80% of all internet traffic (1.3 ZB) will be video. According to a recent Frost and Sullivan report, more than 1 billion consumer video devices were sold in 2013 and of these, 59% were smart phones [3]. There has been a massive increase in the consumption and sharing of content on mobile devices and this is likely to be the major driver over the coming years. The Quarterly Global TV Shipment and Forecast Report by NPD DisplaySearch [4] predicts global TV sales to reach 270 million in 2016.

The cinema projector market has been predicted [5] to be $0.8 billion in 2018, with market growth being driven by the shift toward digital cinema technology and the availability of 3-D content. The key drivers in this sector are:

- Broadcast television, digital cinema, and the demand for more immersive content (3-D, multiview, higher resolution, frame rate, and dynamic range).
- Internet streaming, peer to peer distribution, and personal mobile communication systems.
- Social networking, user-generated content, and content-based search and retrieval.
- In-home wireless content distribution systems and gaming.

Business, manufacturing, and automation

Computer vision and visual communications are playing an increasingly important role in business. Particularly, since the perceived security threats associated with air travel have emerged, the demand for higher quality video conferencing and the sharing of visual content have increased. Similarly in the field of automation, vision-based systems are playing a key role in transportation systems and are now underpinning many manufacturing processes, often demanding the storage or distribution of compressed video content. The drivers in this case can be summarized as:

- Video conferencing, tele-working, and other interactive services.
- Publicity, advertising, news, and journalism.
- Design, modeling, and simulation.
- Transport systems, including vehicle guidance, assistance, and protection.
- Automated manufacturing and robotic systems.

Security and surveillance

In an uncertain world, we have become increasingly aware of our safety and security, and video monitoring is playing an increasingly important role in this respect. It is estimated that the market for networked cameras (non-consumer) [6] will grow to around $4.5 billion in 2017 (19% annual growth from 2012 to 2017), with 28 million units shipped in 2017. Aligned with this, the same source predicts an even larger growth in video analytics to $1.2 billion in 2017. The key drivers in this sector are:

- Surveillance of public spaces and high profile events.
- National security.
- Battlefield situational awareness, threat detection, classification, and tracking.
- Emergency services, including police, ambulance, and fire.

Healthcare

The healthcare market is becoming increasingly reliant on imaging methods to aid diagnoses and to monitor the progression of disease states. 3-D imaging modalities such as CT and MRI produce enormous amounts of data for each scan and these need to be stored as efficiently as possible while retaining the highest quality. Video is also becoming increasingly important as a point-of-care technology for monitoring patients in their own homes to reduce costs and limit hospital admissions. The primary healthcare drivers for compression are:

- Point-of-care monitoring.
- Emergency services and remote diagnoses.
- Tele-surgery.
- Medical imaging.

It is clear that all of the above application areas require considerable trade-offs to be made between cost, complexity, robustness, and performance. These issues are addressed further in the following section.

1.3 Requirements and trade-offs in a compression system

1.3.1 The benefits of a digital solution

We live in a world where digital technology dominates. In the context of video compression, digital solutions offer many significant advantages over their analog counterparts. These include:

- Ease of editing, enhancement, special effects, transcoding, and multigeneration processing.
- Ease of indexing, metadata extraction, annotation search, and retrieval.
- Ease of content protection, conditional access, rights management, and encryption.
- Support for scalability and interactivity.
- Source and channel coding can be separated to provide better management of rate, distortion, and robustness to errors.
- Flexible support for new formats such as 3DTV and multiview.

1.3.2 Requirements

The basic requirement of a video compression system can be stated quite simply: we want the highest quality at the lowest bit rate. However, in addition to this, a number of other desirable features can be listed. These include:

- **Robustness to loss:** We want to maintain high quality when signals are transmitted over error-prone channels.
- **Reconfigurability and flexibility:** To support delivery over time-varying channels or heterogeneous networks.
- **Low complexity:** Particularly for low power portable implementations.
- **Low delay:** To support interactivity.
- **Authentication and rights management:** To support conditional access, content ownership verification or to detect tampering.
- **Standardization:** To support interoperability.

1.3.3 Trade-offs

In practice we cannot satisfy all of the above requirements, either due to cost or complexity constraints or because of limited bandwidth or lossy channels. We therefore must compromise and make design trade-offs that provide the best solution given the prevailing constraints. Areas of possible compromise include:

- **Lossy vs lossless compression:** Ideally, we do not want the video to be distorted at all compared to the original version. In practice this is not usually possible as perfectly lossless (reversible) compression places severe limitations on the compression ratio achievable. We must therefore exploit any redundancy in the image or video signal in such a way that it delivers the desired compression with the minimum perceived distortion.
- **Rate vs quality:** In order to compromise between bit rate and quality, we can adjust a number of parameters. These include: frame rate, spatial resolution (luma and chroma), dynamic range, prediction mode, and latency. The influence of these parameters on the final video quality is however highly content dependent. For example, with a fixed camera and a completely static scene, frame rate will have little or no influence on the perceived quality. In contrast, for a high motion scene, a low frame rate will result in unpleasant jerkiness.
- **Complexity and power consumption vs cost and performance:** In general, a more complex video encoder will support more advanced features and thus produce a higher quality result than a lower complexity version. However, more complex architectures invariably are more expensive and will often introduce more delay. In addition, they may not be realizable using a software implementation, possibly demanding custom hardware solutions.
- **Delay vs performance:** Low latency is key in interactive and conversational applications. However, if an application is tolerant to latency in encoding or decoding, then performance gains can be made elsewhere. For example, increased error

resilience features could be incorporated to retransmit data corrupted during transmission or, alternatively, more reference frames could be used to improve motion prediction performance (see Chapters 8 and 9).

1.4 The basics of compression

A simplified block diagram of a video compression system is shown in Figure 1.3. This shows an input digital signal (using for example one of the formats from Table 1.1) being encoded, transmitted, and decoded. Not shown are possible pre-processing stages such as filtering, interlacing, segmenting, color sub-sampling, gamma correction; these are covered in Chapter 4. A full description of the operation of this so-called block-based hybrid motion compensated video encoder is provided in Chapter 9.

1.4.1 Still image encoding

If we ignore the blocks labeled as motion compensation, the diagram in Figure 1.3 describes a still image encoding system, such as that used in JPEG. The intra-frame encoder performs (as its name suggests) coding of the picture without reference to any other frames. This is normally achieved by exploiting spatial redundancy through transform-based decorrelation (Chapters 5, 6, and 9) followed by variable length symbol encoding (VLC) (Chapter 7). The image is then conditioned for transmission using some means of error resilient coding that makes the encoded bitstream more robust to channel errors. Methods for achieving error resilience are described in Chapter 11.

At the decoder, the inverse operations are performed and the original image is reconstructed at the output.

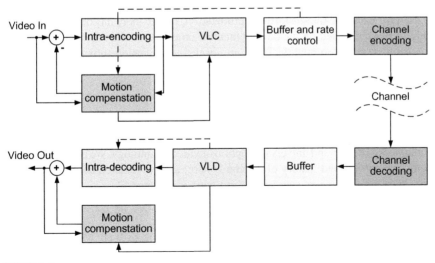

FIGURE 1.3

Simplified high level video compression architecture.

1.4.2 Encoding video

A video signal is simply a sequence of still images, acquired typically at a rate of 24, 25, 30, 50, or 60 fps. Video can of course be encoded as a series of still images using intra-frame methods as described above. However, we can achieve significant gains in coding efficiency if we also exploit the temporal redundancy that exists in most natural video sequences. This is achieved using inter-frame motion prediction as represented by the motion compensation block in Figure 1.3. This block predicts the structure of the incoming video frame based on the contents of previously encoded frames (Chapter 8). The encoding continues as for the intra-frame case, except this time the intra-frame encoder block processes the low energy residual signal remaining after prediction, rather than the original frame.

After variable length encoding, the encoded signal will be buffered prior to transmission. The buffer serves to match the output bit rate to the instantaneous capacity of the channel and this is normally achieved using a rate control algorithm which adjusts coding parameters as described in Chapter 10.

Because of the reliance on both spatial and temporal prediction, compressed video bitstreams are more prone to channel errors than still images, suffering from temporal as well as spatial error propagation. Methods of mitigating this, making the bitstream more robust and correcting or concealing the resulting artifacts are described in Chapter 11.

1.4.3 Measuring visual quality

In the final analysis, we need to assess how good our encoding system is, and the absolute arbiter of this is the human visual system. Because of this, subjective testing methodologies have become an important component in the design and optimization of new compression systems. However, such tests are very time consuming and cannot be used for real-time rate-distortion optimization. Hence there has been a significant body of work reported on the development of objective metrics that provide a realistic estimate of video quality. These are discussed alongside subjective evaluation methods in Chapter 10.

1.5 The need for standards

Standardization of image and video formats and compression methods has been instrumental in the success and universal adoption of video technology. We briefly introduce the motivation for and history of standards here; a more detailed description of the primary features of recent standards is provided in Chapter 12.

1.5.1 Some basic facts about standards

A few basic facts about image and video compression standards are worth mentioning first:

1. Standards are essential for interoperability, enabling material from different sources to be processed and transmitted over a wide range of networks or stored

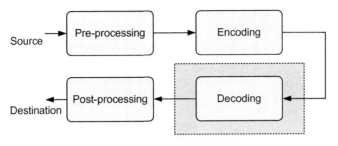

FIGURE 1.4

The scope of standardization.

on a wide range of devices. This interoperability opens up an enormous market for video equipment, which can exploit the advantages of volume manufacturing, while also providing the widest possible range of services for users.

2. Standards define the bitstream format and decoding process, not (for the most part) the encoding process. This is illustrated in Figure 1.4. A standard-compliant encoder is thus one that produces a compliant bitstream and a standard-compliant decoder is one that can decode a standard-compliant bitstream, i.e. a standard specifies the syntax that a bitstream must follow and the corresponding decoding process, not explicitly the encoding process.

3. The real challenge lies in the bitstream generation, i.e. the encoding. This is where manufacturers can differentiate their products in terms of coding efficiency, complexity, or other attributes.

4. Many encoders are standard-compliant, but some are better than others! Compliance is no guarantee of absolute quality.

1.5.2 A brief history of video encoding standards

A chronology of video coding standards is represented in Figure 1.5. This shows how the International Standards Organization (ISO) and the International Telecommunications Union (ITU-T) have worked both independently and in collaboration on various standards. In recent years, most ventures have benefited from close collaborative working.

Study Group SG.XV of the CCITT (now ITU-T) produced the first international video coding standard, H.120, in 1984. H.120 addressed videoconferencing applications at 2.048 Mbps and 1.544 Mbps for 625/50 and 525/60 TV systems respectively. This standard was never a commercial success. H.261 [7] followed this in 1989 with a codec based on a $p \times 64$ kbps ($p = 1 \cdots 30$) targeted at ISDN conferencing applications. This was the first block-based hybrid compression algorithm using a combination of transformation (the Discrete Cosine Transform (DCT)), temporal Differential Pulse Code Modulation (DPCM) and motion compensation. This architecture has stood the test of time as all major video coding standards since have been based on it.

In 1988 the Moving Picture Experts Group (MPEG) was founded, delivering a video coding algorithm targeted at digital storage media at 1.5 Mbs/s in 1992.

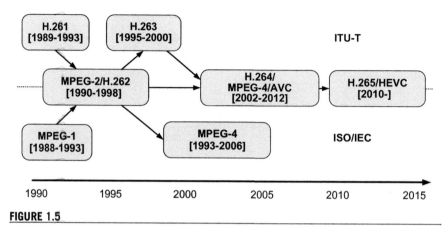

FIGURE 1.5

A chronology of video coding standards from 1990 to the present date.

This was followed in 1994 by MPEG-2 [8], specifically targeted at the emerging digital video broadcasting market. MPEG-2 was instrumental, through its inclusion in all set-top boxes for more than a decade, in truly underpinning the digital broadcasting revolution. A little later in the 1990s ITU-T produced the H.263 standard [9]. This addressed the emerging mobile telephony, internet, and conferencing markets at the time. Although mobile applications were slower than expected to take off, H.263 had a significant impact in conferencing, surveillance, and applications based on the then-new Internet Protocol.

MPEG-4 [10] was a hugely ambitious project that sought to introduce new approaches based on object-based as well as, or instead of, waveform-based methods. It was found to be too complex and only its Advanced Simple Profile (ASP) was used in practice. This formed the basis for the emerging digital camera technology of the time.

Around the same time ITU-T started its work on H.264 and this delivered its standard, in partnership with ISO/IEC, in 2004 [11]. In the same way that MPEG-2 transformed the digital broadcasting landscape, so has H.264/AVC transformed the mobile communications and internet video domains. H.264/AVC is by far the most ubiquitous video coding standard to date. Most recently in 2013, the joint activities of ISO and ITU-T delivered the HEVC standard [12–14], offering bit rate reductions of up to 50% compared with H.264/AVC.

Further details on coding standards are provided in Chapter 12.

1.6 Summary

This chapter has introduced the context and the primary drivers for image and video compression technology in today's world. We have seen that compression exists because of the tensions between, on one hand, the increasing user demands and expectations for increased quality, mobility and immersion, and on the other hand, the limited bandwidth and high traffic volumes that characterize most communications networks. Communicating pictures is a key part of modern life and compression is an essential element in ensuring that they are delivered faithfully, reliably, and in a timely manner.

We have examined the requirements of a compression system and have seen that these represent a compromise between coding rate, picture quality, and implementation complexity. Finally the justification for universal compression standards has been presented, as these are essential in order to enable interoperability across networks, terminals, and storage devices. The standardization process is also a primary driver for continued research in the field and creates a productive interaction between industry and academia.

References

[1] <http://www.wireless-world-research.org/fileadmin/sites/default/files/publications/-Out look/Outlook4.pdf>.

[2] Cisco Visual Networking Index: Forecast and Methodology, 2011–16 (update 2012–17). <http://www.cisco.com/en/US/netsol/ns827/networking_solutions_sub_solution.html>.

[3] Analysis of the Global Consumer Video Devices Market: Apps, QoE and Transformative User-Experiences Key to Growth in Price-Challenged Industry, Frost & Sullivan Report NC68-01, 2013.

[4] Global TV Shipment and Forecast Report, NPD DisplaySearch, 2013.

[5] Digital Cinema Projectors: A Global Strategic Business Report, Global Industry Analysts Inc., 2012.

[6] Network Camera and Video Analytics Market—Global Forecast, Trend & Analysis—Segmentation by Technology, Function, Resolution, Product & Service Type, System Architecture, Verticals, Application and Geography (2012–2017) Report by marketsandmarkets.com, Report Code: SE 1238, 2012.

[7] International Telecommunication Union-Telecommunication (ITU-T), Recommendation H.261, Video codec for audiovisual services at p × 64 kbit/s, version 1, 1990, version 2, 1993.

[8] ITU-T and ISO/IEC JTC 1, Generic Coding of Moving Pictures and Associated Audio Information—Part 2: Video, ITU-T Rec. H.262 and ISO/IEC 13818-2 (MPEG-2 Video), version 1, 1994.

[9] ITU-T, Video Coding for Low Bitrate Communication, ITU-T Rec. H.263, version 1, 1995, version 2, 1998, version 3, 2000.

[10] ISO/IEC JTC 1, Coding of Audio-Visual Objects—Part 2: Visual, ISO/IEC 14496-2 (MPEG-4 Visual), version 1, 1999, version 2, 2000, version 3, 2004.

[11] ITU-T and ISO/IEC JTC 1, Advanced Video Coding for Generic Audiovisual Services, ITU-T Rec. H.264 and ISO/IEC 14496-10 (AVC), version 1, 2003, version 2, 2004, versions 3, 4, 2005, versions 5, 6, 2006, versions 7, 8, 2007, versions 9, 10, 11, 2009, versions 12, 13, 2010, versions 14, 15, 2011, version 16, 2012.

[12] J.-R. Ohm, G. Sullivan, H. Schwartz, T. Tan, T. Wiegand, Comparison of the coding efficiency of video coding standards—including high efficiency video coding (HEVC), IEEE Transactions on Circuits and Systems for Video Technology 22 (12) (2012) 1669–1684.

[13] G. Sullivan, J.-R. Ohm, W. Han, T. Wiegand, Overview of the high efficiency video coding (HEVC) standard, IEEE Transactions on Circuits and Systems for Video Technology 22 (12) (2012) 1648–1667.

[14] Joint Collaborative Team on Video Coding (JCT-VC) of ITU-T SG 16 WP 3 and ISO/IEC JTC 1/SC 29/WG 11 ISO/IEC 23008-2 and ITU-T Recommendation H.265, High Efficiency Video Coding (HEVC), January 2013.

The Human Visual System

2

CHAPTER OUTLINE

2.1   **Principles and theories of human vision** **19**
    2.1.1   Theories of vision 19
2.2   **Acquisition: the human eye** **20**
    2.2.1   Retinal tissue layers 21
    2.2.2   Optical processing 23
    2.2.3   Retinal photoreceptors and their distribution 24
    2.2.4   Visual processing in the retina 26
2.3   **The visual cortex** **28**
    2.3.1   Opponent processes 29
    2.3.2   Biased competition 29
    2.3.3   Adaptation processes 29
    2.3.4   V1—the primary visual cortex 30
    2.3.5   V2—the prestriate cortex 30
    2.3.6   Dorsal and ventral streams 30
    2.3.7   Extrastriate areas 31
2.4   **Visual fields and acuity** **31**
    2.4.1   Field of view 31
    2.4.2   Acuity 32
    2.4.3   Light, luminance, and brightness 32
    2.4.4   Light level adaptation 35
2.5   **Color processing** **36**
    2.5.1   Opponent theories of color 37
    2.5.2   CIE 1931 chromaticity chart 39
2.6   **Spatial processing** **39**
    2.6.1   Just noticeable difference, contrast, and Weber's law 40
    2.6.2   Frequency-dependent contrast sensitivity 40
    2.6.3   Multiscale edges 44
    2.6.4   Perception of textures 44
    2.6.5   Shape and object recognition 44
    2.6.6   The importance of phase information 45

Communicating Pictures. http://dx.doi.org/10.1016/B978-0-12-405906-1.00002-7

2.7	**Perception of scale and depth**	**46**
	2.7.1 Size or scale	46
	2.7.2 Depth cues	47
	2.7.3 Depth cues and 3-D entertainment	48
2.8	**Temporal and spatio-temporal response**	**49**
	2.8.1 Temporal CSF	50
	2.8.2 Spatio-temporal CSF	50
	2.8.3 Flicker and peripheral vision	50
2.9	**Attention and eye movements**	**52**
	2.9.1 Saliency and attention	52
	2.9.2 Eye movements	53
2.10	**Visual masking**	**53**
	2.10.1 Texture masking	53
	2.10.2 Edge masking	55
	2.10.3 Temporal masking	55
2.11	**Summary: a perceptual basis for image and video compression**	**56**
	2.11.1 Influential factors	56
	2.11.2 What have we learnt?	56
References		**59**

A study of the structure, function, and perceptual limitations of the human visual system (HVS) provides us with clues as to how we can exploit redundancy in visual information in order to compress it with minimum degradation in perceived quality. The way humans view and perceive digital images and video relates strongly to how we perceive the world in our everyday lives. While a full understanding of the HVS is still a long way off, we do know enough to be able to create the illusion of high quality in an image sequence where, perhaps, 199 out of every 200 bits from the original are discarded.

This chapter is intended to provide the reader with a basic understanding of the human visual system and, where possible, to relate its characteristics to visual redundancy and ultimately to a means of compressing image and video signals. We first consider the physical architecture of the HVS and the constraints it imposes on the way we see the world. We then review perceptual aspects of vision related to brightness, contrast, texture, color, and motion, indicating the physical limits of key visual parameters. Visual masking is a key component in perception-based compression, so we end by looking at how certain spatio-temporal visual features can be masked by certain types of content.

This chapter is primarily a review of the work of others who are much more expert in the biological, neurological, and psychological aspects of the subject than the author. For further information, the reader is directed to many excellent texts on this topic including those by Snowden et al. [1], Mather [2], Wandell [3], and Marr [4].

For colour versions of Figures 2.1, 2.11, 2.14, 2.16 and 2.17 please refer to the electronic version or the website.

2.1 **Principles and theories of human vision**

The vision system is the most powerful and complex of our senses and we have only began to understand it since the pioneering work of Helmholtz in the 1890s [5].

Between 40% and 50% of the neurons in the human brain are estimated to be associated with visual information processing. This compares to approximately 10% for hearing, 10% for the other senses, and 20% for motor functions, leaving 10–20% for everything else—things like playing chess [1]. This is not entirely surprising because visual activity is linked to almost everything we do. In basic terms we interpret colors, classify shapes, detect and assess motion, estimate distances, and do a pretty good job at creating a three-dimensional world out of the 2-D images that fall on our retinas.

So what is the difference between what we see and what we perceive? Quite a lot actually. For example, we fill in the blind spots where the optic nerve passes through the back of the eye, we correct the distorted imperfect images that fall on the retina, and ignore things that get in the way like our nose and the blood vessels in the eye. At a higher level, it is clear that the visual system does not exist to faithfully record images, but instead to interpret the world and survive in it—to recognize friends, to assess threats, to navigate, to forage, and to find mates. Humans have also used visual stimuli to entertain, communicate, and inform, probably over most of our 200,000 year history—from cave paintings to modern media.

A typical retina contains some 130 million light-sensitive rod and cone cells. Its neural processors are an extension of the visual cortex and it is the only part of the brain visible from outside the body. Each of the optic nerves, which carries signals from the eye to the brain, consists of around one million fibers and it is these that transmit the signals from the retina to the rest of the visual cortex. This contrasts starkly with other senses—for example, each auditory nerve comprises about 30,000 fibers.

We will examine the structure of the retina and the visual cortex in Sections 2.2 and 2.3, but first let us take a brief look at the theories that exist about how our visual system operates.

2.1.1 **Theories of vision**

Visual perception is highly specialized and complex. Much of the research historically into the biological and anatomical aspects of the vision system has been based on primate brains, which appear to be very similar to the HVS. More recently, however, technologies such as fMRI have revolutionized neuroscience and neuropsychology, providing ever increasing insight into the workings of the human system.

Until the late 1970s there were two main competing theories of vision—from the *Empiricists* and the *Rationalists*. Their views on the workings of the visual system differed significantly. Empiricists believed in innate, low level, bottom-up processing, allocating less importance to higher level functions. Julesz [6] for example showed that depth perception can arise from random dot stereograms with no high level meaning, thus inferring that this could not be driven by a top-down mechanism.

Empiricist models were largely based on the extraction of features which facilitated the generation of connected patterns. The pioneering work by Hubel and Wiesel [7] showed the orientation and scale selectivity of the low level visual system and Uhr and Vossler in 1961 demonstrated character recognition from such low level features. Empiricists however did not consider context.

In contrast, Rationalists focused on high level control of vision and the construction of percepts by the brain. They hypothesized mental models that are used to influence our search for and verification of lower level visual features [8]. Early work in this area developed *picture grammars* that described 2-D properties using 2-D structures [9]. These, however, described scene syntax rather than scene semantics and again did not show how 3-D visual inputs could be represented. This in turn led to a *scene-grammar* approach which attempted to map 3-D scenes to 2-D descriptors. This was achieved using either high level shapes, such as cubes, or by using low level structural elements, such as vertices and edges. Such models still failed to describe how humans can generate perceptions of unfamiliar objects since the models rely on prior knowledge of "grammars" to describe the visual input in a high level fashion.

From the Rationalist approaches emerged a hybrid *connectionist* approach. It was not until Marr's contribution in 1982 [4] that this approach was formalized. Connectionist theories emphasized interconnections between processing units embodying both the bottom-up (data driven) extraction of features combined with the top-down (interpretative) influence that generates perceptions based on this low level information. This can be strongly modulated by processes such as attention. This connectionist distributed processing type of approach is still influential and draws heavily on biological evidence providing close correlations with experimental data.

2.2 Acquisition: the human eye

Photons from the sun or other sources of light reflect off physical objects, and some of these photons enter our eyes. Our visual system senses energy radiated in the part of the electromagnetic spectrum from about 380 to 720 nm (420–790 THz) as shown in Figure 2.1. There is nothing particularly special about this range except that our physiology has adapted to it, and it has adapted to it because it is useful for things

400   450   500   550   600   650   700
Wavelength (nm)

FIGURE 2.1

The visible spectrum.

like finding food and detecting threats and potential mates. Some animals see other ranges, for example many birds and insects see in the UV range as this enables them to see certain features in, for example, flowers. The human eye has a peak sensitivity around 555 nm in the green region of the spectrum—perhaps not a coincidence since the Earth's surface mostly reflects light in the green band from 550 to 560 nm due to foliage cover [1].

The human eye senses brightness approximately logarithmically over a broad range of luminance values and can also see colors that are not present in the spectrum. Unsaturated colors, such as pink, are not present and neither are colors like magenta which is formed from a combination of multiple wavelengths. A diagram of the human eye in cross-section is shown in Figure 2.2. Let us now look in a little more detail at its main components.

2.2.1 Retinal tissue layers

The eye is an approximately spherical volume filled with fluid, comprising four layers of tissue—the sclera, the retina, the choroid, and the ciliary body.

The sclera

The sclera is the layer of white tissue that provides the main strength of the eye's structure. It becomes transparent at the front of the eye where light enters through the cornea.

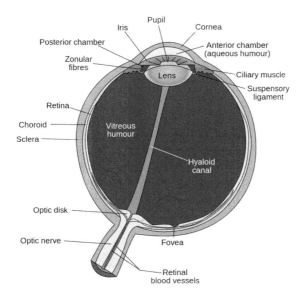

FIGURE 2.2

Cross-section of the human eye. (Public domain: http://commons.wikimedia.org/wiki/File: Schematic_diagram_of_the_human_eye_en.svg.)

The ciliary body

The ciliary body is a ring of tissue that surrounds the eye. It contains the ciliary muscles that adjust the refractive power of the lens by changing its shape.

The retina

The retina contains the photoreceptors that are sensitive to light as well as several other types of neuron that process and combine signals from the photoreceptors and transmit these to the main brain areas via the optic nerve. A fundus photograph of a healthy human retina is shown in Figure 2.3. The macula is the darker area in the center and the optic disk, where the nerve fibers pass through the back of the eye, can be seen as the brighter area on the left. Interestingly, and despite its size, our vision system does a pretty good job of creating an impression of vision at this blind spot even though there are no photoreceptors present! Major nerve pathways are seen as white striped patterns radiating from the optic disk and blood vessels can also be clearly seen.

The choroid

The choroid is the capillary bed that provides nourishment to the photoreceptors. It also contains the light sensitive, UV absorbing pigment—melanin.

Let us now look at the front of the eye, where the light enters, in a little more detail.

FIGURE 2.3

Fundus image of a healthy retina. (Public domain from: http://commons.wikimedia.org/wiki/File:Fundus_photograph_of_normal_right_eye.jpg.)

2.2.2 **Optical processing**

The cornea

The cornea is the sensitive transparent circular region at the front of the eye where light enters. It refracts the light onto the lens which then focuses it onto the retina.

The lens

The lens is the transparent structure located behind the pupil that, in combination with the cornea, refracts the incident light to focus it on the retina. For distant objects, the ciliary muscles cause it to become thinner and for close-up objects it becomes fatter. Interestingly the lens is only responsible for around 25% of the refractive power of the eye's optics, the rest is provided by the cornea. Although the lens has much less refractive power than the cornea, it enables changes in focal length that enable us to accommodate objects of interest at various distances.

The relationship between object distance d_o, retinal distance d_r and focal length f is illustrated in Figure 2.4 and equation (2.1). The focal length of the human eye is about 17 mm and the optical power is about 60 diopters.

$$\frac{1}{f} = \frac{1}{d_o} + \frac{1}{d_r} \tag{2.1}$$

The iris

The iris is the individually colored circular region at the front of the eye containing the ciliary muscles that control the aperture of the pupil.

The pupil

The pupil is the aperture that controls how much light enters the eye and becomes smaller in brighter environments. The pupil however only adjusts by a factor of around 16:1 (dilation from 2 mm to 8 mm). Most of the compensation for varying light levels

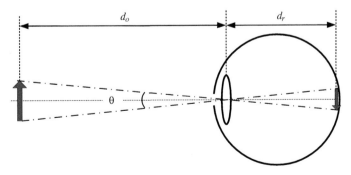

FIGURE 2.4

The focal length of the lens.

that allows us to adapt to some 8–9 orders of luminance magnitude is done in the retina by the photoreceptors and in other areas of the brain.

Now let us move to the transducer at the back of the eye—the retina.

2.2.3 Retinal photoreceptors and their distribution

When light hits the retina it is turned into electrical pulses by the responses of the photoreceptors. There are typically around 130 million sensor cells in the retina—125 million rods and 6–7 million cones. The distribution of these is as shown in Figure 2.5 [12]. Although the packing of cones is at its densest in the fovea, there are still several thousand per square millimeter in the periphery.

Rod cells

There are approximately 125 million rods in the retina. These are responsible for vision at low light levels (scotopic vision) and are so sensitive that they become saturated or bleached at mid light levels (mesopic vision). Rods do not mediate color vision and provide lower spatial acuity than foveal cone cells. The photopigment, rhodopsin, is most sensitive to green light of wavelength around 498 nm. The central fovea—the region of highest visual acuity—has a very low density of rods which is why, at night, we can see objects better by looking slightly away from them.

Cone cells

Cones operate at higher light levels than rods and provide what is known as photopic vision. In contrast to rods, they offer a range of different spectral sensitivity characteristics and they mediate color vision. Three types of cone exist and these

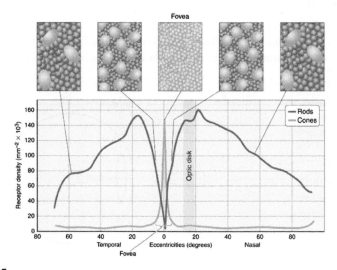

FIGURE 2.5

Photoreceptor distribution in the retina. (Reproduced with permission from: Mustafia et al. [12].)

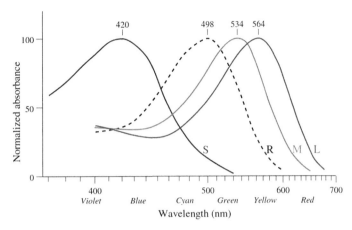

FIGURE 2.6

Normalized rod and cone responses for the human visual system. (Reproduced with permission from: Bowmaker and Dartnall [33].   //www.ncbi.nlm.nih.gov/pmc/articles/PMC1279132/. Avail Wikimedia Commons.)

are generally referred to as short wavelength (S) or blue; medium wavelength (M) or green; and long wavelength (L) or red. These have broad overlapping spectral responses with peaks at around 420, 534 and 564 nm respectively. Light at blue wavelengths tends to be out of focus since it is refracted more than red and green light. The normalized frequency response characteristics of the rods and cones is shown in Figure 2.6.

As shown in Figure 2.5, the central fovea has the highest cone density and thus provides the region of highest visual acuity.

Macula

The macula is a depressed region of diameter 5.5 mm near the center of the retina which surrounds the fovea. It is in this region that the number of cone cells starts to increase dramatically. It has two or more layers of ganglion cells, and at its center lies the fovea. Because of its yellow color the macula absorbs excess blue and ultraviolet light and acts as a natural sunblock.

Fovea

The fovea is the depression at the center of the macula about 1.5 mm in diameter that is responsible for central, high resolution vision. As shown in Figure 2.5, it has a very high spatial density of cones and very few rods. It also coincides with a region of the retina that is void of blood vessels and this further improves acuity as dispersion and loss are minimized. The center of the fovea, the foveola, is about 0.2 mm in diameter and is entirely formed of very compact, hexagonally packed, thin cones that are rod-like in structure. The fovea contains mostly M and L cones with around 5% of S cones.

The sampling resolution of the retinal image is highly non-uniform and falls off rapidly with increasing eccentricity. This creates a more blurred image in the peripheral vision which we do not perceive due to rapid eye movements and higher level visual processing. The effective angle subtended at the fovea is around 2°, hence we really do only "see" a small proportion of the visual scene sharply, yet we still perceive it all to be sharp. The average spacing of cone receptors in the fovea is 2–3 μm (in the periphery this increases to around 10 μm), and it is this that effectively limits our spatial resolution to approximately 1 arcmin. This is important when we assess the relative merits of high resolution TV formats in terms of viewing distance. The spacing also gives us the maximum spatial frequency that can be perceived. In practice the eye cannot resolve spatial frequencies of more than about 60 cycles per degree (cpd).

Optic disk and nerve

The optic disk is the portion of the optic nerve that can be seen on the retina, where the signals from the ganglion cells in the retina leave the retina along axons on their way to the visual cortex. About half of the nerve fibers in the optic nerve originate from the fovea with the remainder carrying information from the rest of the retina.

The retina contains a number of other cell types, which provide early vision processing and reduce the amount of information (HVS compression!) that is transmitted to the main visual cortex. These cells are briefly explained in Section 2.2.4.

2.2.4 Visual processing in the retina

The responses from the retina's photoreceptors are processed by a complex (and far from perfectly understood) network of neurons. This network comprises: bipolar, horizontal, amacrine, and retinal ganglion cells. Together, these provide early visual pre-processing, feature extraction and detection, resulting in significantly reduced information flow through the optic nerve. This structure is shown in Figure 2.7.

As depicted in Figure 2.7 bipolar cells connect to either rods or cones and are activated by an increase in the photons incident on the associated photoreceptors. The horizontal cells connect laterally across the retina to a greater spatial extent and, because of their stimulation from a neighborhood of photoreceptors, enable lateral inhibition. Whereas the response of bipolar cells in isolation is rather crude, the influence of the horizontal cells is to add an opponent signal where the response of one or more photoreceptors can influence the response of surrounding receptors, thus shaping the receptive field. The horizontal cells are also known to modulate the photoreceptor signal under different illumination conditions [10].

Relatively little is known about the roles of amacrine cells, except that they can have extensive dendritic trees and, in a similar way to horizontal cells, contribute feedback to the receptive fields of both bipolar and ganglion cells.

There are approximately 1.6 million retinal ganglion cells and these create the earliest receptive fields in the vision system, providing a basis for low level feature

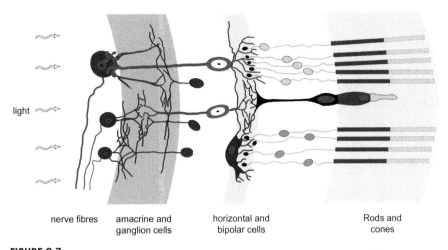

FIGURE 2.7

Retinal cell architecture (Public domain image adapted from http://commons.wikimedia.org/wiki/File:Retina-diagram.svg).

detection and the opponent coding of color (see Section 2.5). The axons from these form the optic nerve and transmit the electrochemical signals to the higher functions of the visual cortex. There are approximately 100 times more photoreceptors than ganglion cells in the eye—providing evidence of the compression performed by the retina.

The receptive field of a cell is a region within which a stimulus must occur if it is to influence the response of that cell. Receptive fields are formed using spatial opponency, i.e. responses from one part of the receptive field are opposed or inhibited by other parts of the receptive field. The interactions between the bipolar and the horizontal cells provide an orientation insensitive center-surround organization of the ganglion cell receptive field, where the center and surrounding areas act in opposition. This could comprise an excitatory center and an inhibitory surround as shown in Figure 2.8 or vice versa. Ganglion cells act as change detectors, responding only to differences in light intensities (or contrast) across their receptive field and not to absolute light levels. They are therefore sensitive to edges and can be modeled as a difference of Gaussian functions (DoG).

A scene is thus effectively encoded as intensity changes in the different color channels. However, it is clear that we see more than just edges, so this provides evidence that higher levels of the vision system perform in-filling between receptive field responses with averages in intensity levels. This is interesting as it can be viewed as a simple form of image synthesis, providing a biological analog to perceptual redundancy removal. A particularly striking example of this is the visual in-filling of our blind spot which contains no photoreceptors.

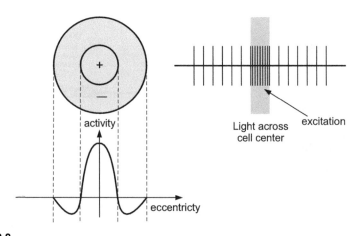

FIGURE 2.8
Spatial opponency, showing center-surround on cell and its firing pattern due to excitation.

2.3 The visual cortex

A simple diagram showing the major visual pathways from the retinas to the primary visual cortex is shown in Figure 2.9. The optic nerve axons from the retinal ganglion cells in each eye meet at the optic chiasm where all the fibers from the nasal side of each eye cross to join those from the temporal side of the other eye and form two optic tracts. The left visual field is thus processed by the right hemisphere and vice versa. The fibers in the optic tract then synapse at the lateral geniculate nucleus (LGN)

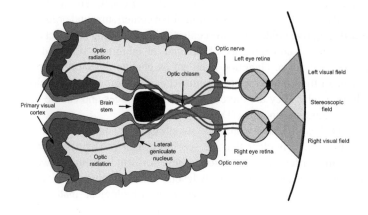

FIGURE 2.9
The visual cortex.

where they are distributed to other areas of the visual cortex. After the LGN, signals pass to the primary visual cortex, V1, located at the rear of the skull. They then pass to V2, and branch out to at least 20 other centers, each providing specialized functions such as detecting motion, recognising faces, or interpreting color and shape.

The HVS provides its functionality largely through the combined processes of inhibition, excitation, biased competition, input races, opponent coding, and adaptation. These are considered in more detail below.

2.3.1 Opponent processes

As discussed above and depicted in Figure 2.8, many neural functions in the vision system appear to encode data using opponent coding. Opponent coding is a process in which neurons encode opposite or opponent features in their single output. For example, a motion sensitive cell has a base firing rate which reduces when motion in one direction is detected and which increases with motion in the opposite direction. The opponent processes of inhibition and excitation appear crucial to function, with inhibitory responses often being used to control a normally excitatory state.

We also interpret color by processing signals from cones and rods in opponency. The L, M, and S cone spectral responses overlap in terms of their wavelengths so it is more efficient for the visual system to use differences between their responses rather than individual responses. This suggests that there are three opponent channels: red versus green, blue versus yellow, and black versus white. We will consider this further in Section 2.5.

2.3.2 Biased competition

Biased competition refers to the way in which information has to compete for limited resources in the visual pathways and cortices. Inputs from feature detectors in the visual system compete to be encoded into short term visual memory. The competition between such inputs is weighted according to higher level feedback, called attention, to ensure that those features relevant to the current task are prioritized. For example, during the Stroop task (subjects are asked to state the color of a series of words rather than read the words themselves) it has been observed that task-based attention causes enhancement of the relevant color feature sensitive areas of the brain while suppressing responses from those areas that perform word processing.

2.3.3 Adaptation processes

Adaptation is the process by which a cell's response reduces over time when its input is constant. Adaptation to varying degrees is common in most cortical cells and, when coupled with opponent coding processes, is responsible for the after-effects we experience related to brightness, color, or motion. After an opponent-coding feature-tuned cell adapts to a particular stimulus and that stimulus is removed, the cell reacts—briefly signaling the opposite feature to which it has adapted. Examples of such after-effects include the sensation of slowness after traveling at high speeds, and the perception that static objects move in the opposite direction to that of a previously viewed moving object. Dramatic brightness and color after-effects are also frequently used as the basis for visual illusions. Even observing smiles for a prolonged period can make expressionless faces seem sadder!

2.3.4 V1—the primary visual cortex

The primary visual or striate (layered) cortex (otherwise known as V1) is probably the best understood area of the HVS. The role of V1 is to extract basic visual features from a scene and it has been shown that V1 contains receptive fields that are sensitive to line orientations, color, and spatial frequency. This effects visual data compaction through abstraction of visual stimuli into higher level constructs. The majority of V1 connections are from the retina but it is also known to receive feedback connectivity from higher functional areas, which influence its cell's receptive fields.

Hubel and Wiesel [7] identified three primary detectors in V1:

- **Simple cells:** Tuned to specific orientations of edges.
- **Complex cells:** Phase insensitive versions of simple cells, i.e. the response to an appropriately oriented stimulus is the same no matter where it falls within the receptive field.
- **Hypercomplex cells:** That show stimulus length sensitivity, i.e. the cell response increases as the stimulus length increases.

Neurons in V1 are generally grouped in columns that respond to a particular low level feature such as line orientation, spatial frequency, or color. V1 is known to preserve the retinotopic mapping of the ganglion cells on the retina in its organization. This is however not a linear or uniform mapping since V1 exhibits cortical magnification, dedicating a disproportionate percentage of its cells to processing foveal information. This emphasizes the importance of the higher resolution central part of the visual field.

2.3.5 V2—the prestriate cortex

V2 is strongly linked to V1 with both feedforward and feedback connections and it also connects strongly to V3, V4, and V5. Although less well understood than V1, V2 cells appear to operate in a similar manner, combining V1 features tuned to orientation, spatial frequency, and color as well as responding to more complex patterns. Damage in V2 can lead to poor texture and shape discrimination. There is also evidence of V2 cell modulation by binocular disparity.

2.3.6 Dorsal and ventral streams

Connections from V1 to the V2 area are generally considered to diverge into two streams—dorsal and ventral (otherwise referred to as "where" and "what" or "action" and "perception"). The dorsal stream relates to motion and depth perception, originating in V2 and passing dorsally through the brain to the motion sensitive parts of the visual cortex. The ventral stream on the other hand is associated with the perception of shapes and object recognition, again originating in V2, but in this case passing ventrally through V4 to the inferior temporal cortex (IT).

2.3.7 Extrastriate areas

The areas of the visual cortex beyond V1 are generally referred to as the *extrastriate* areas. These possess larger receptive fields and generally compute more complex and abstract features. They also generally exhibit strong attentional modulation. Although the function of these areas is not well understood, their connectivity is informative.

V3 is strongly connected to V2 with additional inputs from V1. It processes both dorsal and ventral streams and has been shown to respond to global motions.

V4 lies on the ventral stream and has been shown to respond to shapes of relatively high complexity. It also shows strong attentional modulation and a significant amount of color sensitivity. Damage to V4 can cause impairments to shape and texture discrimination.

V5 contains cells that respond to complex motion patterns and is believed to be the main area dedicated to the perception of motion.

The *medial superior temporal* (MST) area is fed primarily from the medial temporal (MT) area, and is also sensitive to motion. There is evidence that the MST computes optic flow, especially in the context of global expansions and rotations.

The *visual short term memory* (VSTM) can be considered to be a short term buffer where prioritized data is stored for further processing. VSTM has been shown to persist across visual interruptions such as saccades and blinks and thus provides a link between the previously viewed scene and the new one after interruption.

The *inferior temporal cortex* (IT) provides complex shape recognition by parsing simpler shapes from lower levels in the HVS. It contains cells that trigger in response to specific shapes and patterns and hosts cells that are known to respond to faces. It fuses information from both halves of the visual field (that up to this point was separated across the hemispheres of the brain). At higher stages of visual processing, the receptive fields become larger and more complex in their connectivity. For example, in the IT some receptive fields are thought to cover the entire visual field.

Many other important areas of the visual cortex have been identified and investigated, including the *fusiform face area* (FFA), largely responsible for recognition of faces and other complex entities. The *frontal eye field* (FEF) is a further important area that controls voluntary eye movements. It has been proposed that this is based on some form of visual saliency map, encoded with behaviorally significant features.

2.4 Visual fields and acuity

2.4.1 Field of view

The field of view of the human eye is approximately 95° in the temporal direction (away from the nose), 60° in the nasal direction (toward the nose), 75° downward, and 60° upward, enabling a horizontal field of view in excess of 180°. Obviously, when rotation of the eye is included, the horizontal field of view increases still further, even without head movement.

2.4.2 Acuity

It is not too difficult to estimate the spatial acuity of the retina based simply on the distribution of cones in the fovea. We illustrate this in Example 2.1. The measured Nyquist frequency for the HVS is normally between 40 and 60 cpd. This upper value has been attributed to the photoreceptor sampling grid pattern which is hexagonal rather than rectangular and to other local enhancement techniques. No aliasing effects are normally experienced by the HVS and this is due to the fact that the optics of the eye perform low pass filtering, acting as an anti-aliasing filter.

Example 2.1 (Visual acuity of the retina).
Assuming the photoreceptor densities indicated in Figure 2.5, and that the distance from the lens to the retina is 16.5 mm, estimate the acuity of the human visual system in the fovea.

Solution. We can see from Figure 2.5 that the cone density in the central fovea peaks at around 150,000 cones/mm^2. If we assume a very simple packing arrangement this gives around 400 cones per linear mm. To be conservative let us take the distance between cone centers as 0.003 mm. The optics of the eye dictate [2] that, for a lens to retina distance of 16.5 mm, there are 0.29 mm/deg subtended. The angular resolution is thus 0.003/0.29 = 0.0103° between cone centers. Now, since we need at least two samples per cycle to satisfy the Nqyuist criterion, this means that there are 0.0206° per cycle. Hence the estimate for the maximum detectable spatial frequency is 49 cpd subtended.

2.4.3 Light, luminance, and brightness

Sensations of brightness are not direct analogs of photometric intensity (luminance). The number of photons that enter the eye and the perception of brightness are only related indirectly. For example, the apparent brightness of an achromatic target, under equi-luminant conditions, will depend on the frequency content of the electromagnetic radiation reflected from its surroundings. Thus, a gray target on a relatively dark background looks brighter than the same target on a lighter background. As a result of this context-based perception, it is generally thought that the visual system computes brightness using luminance ratios across the contrast boundaries in a scene. The influence of color on brightness also provides evidence that perceptions of luminance are related to electromagnetic frequency [11].

The influence of local contrast can be seen, in combination with the effects of lateral inhibition, in the Mach bands shown in Figure 2.10. Here the intensity is uniform over the width of each bar, yet the visual appearance is that each strip is

FIGURE 2.10

Mach band effect.

FIGURE 2.11

Adelson's grid. (Reproduced with permission from: http://web.mit.edu/persci/people/
adelson/checkershadow_illusion.html.)

darker at its right side than its left. Another striking example is provided by the
Adelson grid as shown in Figure 2.11, where squares A and B both appear very
different due to shadowing, yet are actually identical.

There is significant confusion about the terms used in connection with light radiation, so let us define some terms related to light and how we perceive it [13].

Radiant intensity and radiance

The radiant energy (joules) is the energy propagating from a source and the radiant
flux Φ is the radiant energy per unit time (joules/sec or watts). Radiant intensity I
is defined as the radiated light power in a given direction and is measured in terms
of the amount of light passing through a solid angle (watts/steridian). Radiance L is
defined as the light intensity that falls on a unit projected area (watts/steridian/m^2).

Luminance

None of the above measures take account of the composition of the light in terms
of its wavelength or wavelengths. Luminance does this by weighting the radiance
value according to the human visual response. The luminance, Y, is thus normally
represented by equation (2.2), where $L(\lambda)$ is the incident light intensity as a function of
wavelength at a given position and time and $\bar{y}(\lambda)$ is a wavelength-dependent weighting
(or relative luminous efficiency) that reflects the perceived luminance of the incident

FIGURE 2.12

CIE luminous efficiency curve. (Public domain image: http://en.wikipedia.org/wiki/File: CIE_1931_Luminosity.png.)

flux. Luminance is measured in candelas/m^2 (cd/m^2):

$$Y = K \int_{320}^{720} L(\lambda)\bar{y}(\lambda)\mathrm{d}\lambda \tag{2.2}$$

The constant $K = 685$ lumens/W. The function $\bar{y}(\lambda)$ will vary from person to person, but was measured by the CIE[1] in 1929 for a standard observer as shown in Figure 2.12. If in the integral of equation (2.2), the term $L(\lambda)$ is replaced by radiant flux $\Phi(\lambda)$ then the result is termed luminous flux and has the units of lumens.

Brightness

Brightness is the perception elicited by the luminance of a visual target and, as such, represents a non-quantitative indication of the physiological sensation of light. It is normally assessed subjectively by an observer with reference to a white area adjusted to have the same brightness as the target viewed.

Luma

An image sensor normally produces output values proportional to radiance, with the incident light typically filtered into R, G, and B bands. However, before these emerge from the camera, they are usually non-linearly mapped according to the response of the HVS, using a technique known as gamma correction. These outputs are normally labeled as R', G', and B'. Luma is the term used to denote the combination of these corrected components that produces the gamma-corrected luminance signal, according to the specification of the color space used. Gamma correction is considered further in Chapter 4.

---

[1] The International Commission on Illumination.

2.4.4 **Light level adaptation**

Our ability to adapt to a wide range of light levels is critical for function and survival. The photoreceptors, indeed the whole of the visual system, must remain sensitive as the ambient light intensity varies. This variation is quite dramatic as we are able to see in conditions from starlight (10^{-3} cd/m^2), through indoor lighting (10^2 cd/m^2) to bright sunlight (10^5 cd/m^2). The eye can thus function (at least to some degree) across some eight orders of magnitude in luminance level—a ratio of 100,000,000:1.

The eye, of course, cannot cope with this dynamic range instantaneously. If we consider our eye's instantaneous dynamic range (where the pupil opening is fixed), then this is around 10 f-stops. If we include rapid pupil adaptation this increases to about 14 f-stops. The pupil diameter typically ranges from about 2 mm in bright conditions to around 8 mm in dark conditions, giving a 16:1 dynamic range or 4 f-stops.

Adaptation also takes place in both rods and cones. The cones adapt quicker than the rods and this explains why it takes much longer to adapt to darkness (dark adaptation) than to brightness.

A typical dark adaptation characteristic is shown in Figure 2.13. This illustrates the variation over time of an individual's threshold after adaptation to a bright light. This figure shows that the dark adaptation process is quite slow, taking several minutes for complete adaptation, but also that the adaptation occurs in two stages. The first of these is due to the cones adapting and the second is due to the rods. The retinal photoreceptors adapt through pigment bleaching—but also by feedback from horizontal cells.

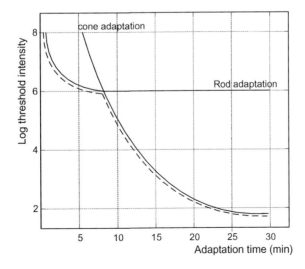

FIGURE 2.13

Dark adaptation of rods and cones.

2.5 Color processing

If you ask someone if they prefer to watch a color rather than a monochrome TV programme or movie, invariably they would say yes on the basis that it is closer to reality. This is obvious—consider Figure 2.14 which, on the right, shows a color image of Yellow Water in Kakadu, Australia and, on the left, a monochrome version of the same image. Most would agree that the color image is easier to interpret, providing improved differentiation between natural objects and features in the scene and a better sense of depth and engagement.

Yet, despite few people seeing the world in black and white, we readily accept monochrome images and are perfectly happy to suspend disbelief, for the sake of entertainment or information. In some cases (for example in Steven Spielberg's *Schindler's List*) directors actually choose to use monochrome imagery to effect mood and increase impact in films. This acceptance of, and occasional preference for, luminance-only signaling may be associated with the way in which we process our trichromatic sensors.

A final point worth highlighting here is that the choice of primary Red, Green, and Blue colors in displays is not based on the sensitivities of the S, M, and L cones. Instead they are selected to be well spaced spectrally, thus allowing a large gamut of colors to be produced.

Following from Section 2.4 and equation (2.2) we can generate average cone responses R, G, B to incident radiation as follows, where $\bar{y}_r(\lambda)$, $\bar{y}_g(\lambda)$, and $\bar{y}_b(\lambda)$ are the luminous efficiencies for the red, green, and blue color channels respectively:

$$R = K_r \int_{380}^{720} L(\lambda)\bar{y}_r(\lambda)\mathrm{d}\lambda$$

FIGURE 2.14

Increased immersion from color images.

$$G = K_g \int_{380}^{720} L(\lambda) \bar{y}_g(\lambda) d\lambda$$

$$B = K_b \int_{380}^{720} L(\lambda) \bar{y}_b(\lambda) d\lambda \tag{2.3}$$

2.5.1 Opponent theories of color

We have three types of cone. However, most mammals only have two types—evolved from a single yellow-green cone centered around 550 nm (corresponding to chlorophyll reflectivity maximum). A blue cone was then added centered around 430 nm—which assists with plant discrimination. Humans and other primates added a third cone type (actually the yellow-green cone probably split into two types)—at the red end of the spectrum. This enabled its owners to forage more effectively (red fruit can be more easily differentiated) and signal more dramatically. As a point of interest, some birds possess four channels and the mantis shrimp has 12!

As described earlier, color perception in the HVS employs an opponent-based approach. The reason for this becomes clear if we consider color constancy requirements. Clearly if the light reflected from a certain object varies, we would ideally like to perceive it as a brighter or darker version of the same color. This can be achieved if we consider the ratio of the outputs from the photoreceptors rather than their absolute outputs. This was first identified by Herring in 1878, who realized that certain hues could not co-exist in a single color sensation. For example, we can experience a red-yellow = orange and blue-green = cyan but cannot readily experience red-green or blue-yellow. Hence the pairs *red-green* and *blue-yellow* are termed opponent pairs. Herring argued that we could not experience both colors in a pair simultaneously as they are encoded in the same pathway. This observation has been validated from psychophysics experiments and it is easy to convince yourself of it. For example, if you stare at green long enough you get a red after-effect (and vice versa) and if you stare at blue long enough you get a yellow after-effect (and vice versa).

A diagram showing a possible opponent model is given in Figure 2.15. This shows a luminance signal generated from the sum of the signals from green and red cones and two chrominance channels generated from ratios of red to green and yellow to blue. Three things should be noted here. Firstly expert opinion varies on the nature of the combining processes and there is some evidence for the chrominance channels being based on color differences (i.e. *R-G* and *Y-B*). However, because the sensitivities of the cone spectral responses overlap, most wavelengths will stimulate at least two types of cone. Some tristimulus values therefore cannot physically exist as an additive color space and imply negative values for at least one of the three primaries. Secondly it is well known that color sensations vary significantly with context and spatial frequency (see Figure 2.16 and try this for yourself—if you move the page closer, the blue bars will turn green). Theories to explain aspects of this have been proposed by authors such as Shapley and Hawken [20]. Finally, the observant reader will notice that there is no contribution from blue cones in the luminance channel. The main reason for this is that there are relatively few blue cones in the fovea and almost

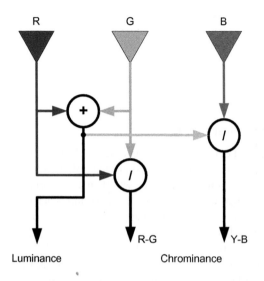

FIGURE 2.15

Opponent processing of color.

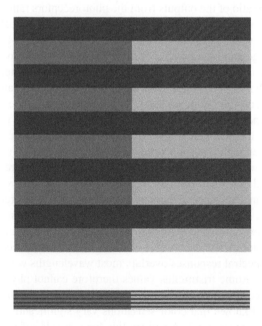

FIGURE 2.16

Color dependence on context. The bottom picture is just a squashed version of the top one yet the green stripes become blue. (For interpretation of the references to color in this figure legend, the reader is referred to the web version of this book.)

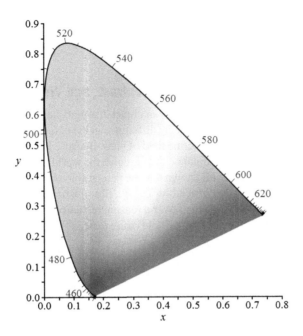

FIGURE 2.17
The CIE 1931 chromaticity chart. (Reproduced with permission from Ref. [21].)

none at its center. This is probably because of the effects of chromatic aberration which would significantly reduce acuity.

2.5.2 CIE 1931 chromaticity chart

CIE, the International Commission on Illumination, in 1931 defined a color-mapping function based on a standard observer, representing an average human's chromatic response within a $2°$ arc, to primaries at $R = 435.8$ nm, $G = 546.1$ nm, and $B = 700$ nm. Figure 2.17 shows the CIE 1931 chromaticity chart. The boundary represents maximum saturation for the spectral colors, and the diagram forms the boundary of all perceivable hues. The colors that can be created through combinations of any three primary colors (such as RGB) can be represented on the chromaticity diagram by a triangle joining the coordinates for the three colors. Color spaces are discussed in more detail in Chapter 4.

2.6 Spatial processing

Spatial resolution is important as it influences how sharply we see objects. As discussed in Section 2.4, the key parameter is not the number of pixels in each row or column of the display, but the angle subtended, θ, by each of these pixels at the viewer's retina. We thus use the term *spatial* here to indicate things that are not moving as opposed to implying that there is some sense of spatial constancy, regardless of viewing distance.

2.6.1 Just noticeable difference, contrast, and Weber's law

Contrast is the visual property that makes an object distinguishable from other objects and from a background. It is related to the difference in the color and brightness of the object and other objects within the same field of view. Because the HVS is more sensitive to contrast than absolute luminance, we perceive the world similarly regardless of changes in illumination. It is well known that the perception of a constant increment in illumination is not uniform, but instead varies with illumination level. This is captured by Weber's law which identifies that, over a reasonable amount of our visible range, the following expression for contrast, C, holds true, for a target T and a background B:

$$\triangle C = \frac{\triangle Y}{Y} = \frac{\text{JND}(Y_T - Y_B)}{Y_B} = \text{constant} \tag{2.4}$$

Weber's law implies that the just-noticeable difference (JND) between two stimuli is proportional to the magnitude of the stimuli. This is sometimes confused with Fechner's law that states that a subjective sensation is proportional to the logarithm of the stimulus magnitude. Figure 2.18 shows an example test chart for JND testing. For most observers the JND value is between 0.01 and 0.03.

Weber's law fits well over 2–3 decades of $\log(Y)$ and is illustrated in Figure 2.19. As background illumination is increased above or decreased below this region the slope will change. As we will see in later chapters, this subjective assessment of luminance sensitivity can be important in image and video quantization.

2.6.2 Frequency-dependent contrast sensitivity

One of the most important issues with HVS models concerns the relationship between contrast sensitivity and spatial frequency. This phenomenon is described by the

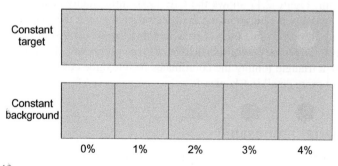

FIGURE 2.18

Just-noticeable differences at different contrast increments.

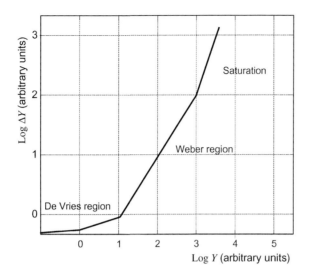

FIGURE 2.19

JND curve for human vision.

contrast sensitivity function (CSF). The CSF model describes the capacity of the HVS to recognize differences in luminance and chrominance as a function of contrast and spatial frequency.

A spatial contrast sensitivity chart or sine wave grating is shown in Figure 2.20. Typical contrast sensitivity responses for luminance and chrominance are shown in

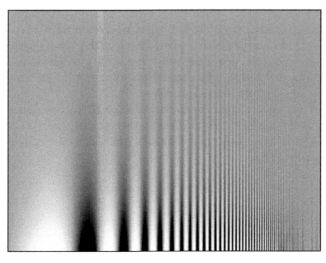

FIGURE 2.20

Contrast sensitivity chart.

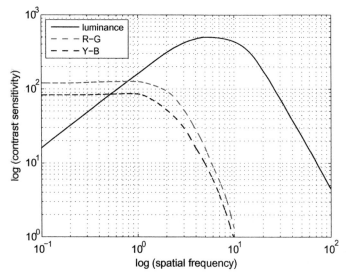

FIGURE 2.21

Luminance and chrominance CSF responses.

Figure 2.21 where contrast sensitivity is defined as:

$$C(f) = \frac{Y_B(f)}{\text{JND}(Y_T(f) - Y_B(f))} \tag{2.5}$$

This illustrates the sensitivity of the visual system to spatial sinusoidal patterns at various frequencies and contrast levels. The bandpass characteristic is primarily due to the optical transfer function of the eye together with the retinal (lateral inhibition) processing and some properties of region V1. Its shape is influenced not only by the spacing and arrangement of photoreceptors, but also possibly by the limitation on the number of spatial patterns representable in V1.

Essentially, the HVS is more sensitive to lower spatial frequencies and less sensitive to high spatial frequencies. Mannos and Sakrison [22] presented a CSF model for grayscale images as a non-linear transformation followed by a Modulation Transfer Function (equation (2.6)):

$$C(f) = 2.6 \left(0.0192 + 0.114 f \right) e^{-(0.114 f)^{1.1}} \tag{2.6}$$

where spatial frequency is:

$$f = \sqrt{f_h^2 + f_v^2}$$

f_h is the horizontal frequency component and f_v is the vertical frequency component. It is often preferable to work in terms of a spatial frequency normalized to the display

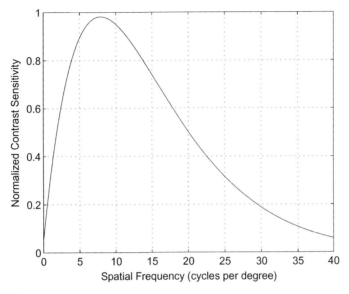

FIGURE 2.22

Luminance contrast sensitivity function.

characteristics in terms of cycles per pixel, thus:

$$f_d = \frac{f}{f_e}$$

where f_e is the number of pixels per degree for the experiment being conducted, which depends on the resolution of the display, the dimensions of the display and the viewing distance. This fits well with measured results and its CSF response function (normalized) is shown in Figure 2.22.

The contrast sensitivity of human vision is plotted here against spatial frequency. We can observe that:

1. The sensitivity to luminance information peaks at around 5–8 cycles/deg. This corresponds to a contrast grid with a stripe width of 1.8 mm at a distance of 1 m.
2. Luminance sensitivity falls off either side of this peak and has little sensitivity above 50 cycles/deg.
3. The peak of the chrominance sensitivity curves occurs at a lower spatial frequency than that for luminance and the response falls off rapidly beyond about 2 cycles/ deg. It should also be noted that our sensitivity to luminance information is about three times that for *R-G* and that the *R-G* sensitivity is about twice that of *B-Y*.

The contrast sensitivities of the HVS lead us to our first basic means of compressing images—The perceptual CSF model can be used for reduction of imperceptible information and, if we use luminance and color difference signals as the basis of our representation, then we can sample the chrominance signals at around half the rate of the luminance signal, without any loss of perceived quality. Furthermore, both luminance and chrominance signals can be more coarsely quantized at higher spatial frequencies due to their reduced contrast sensitivity. These mechanisms for achieving this will be explored in Chapter 4.

2.6.3 Multiscale edges

Edge localization models of early vision have generally featured some arrangement of tuned spatial filters at multiple image scales followed by a feature extraction process and finally integration of these features into a map (such as Marr's primal sketch [4]). The orientation and frequency selectivity of the visual system in many ways resembles multiscale transforms such as the wavelet or complex wavelet transform and these have been employed in various HVS models. As we will see later in the book, these tools are also used extensively in image and video compression, where the transformed image data in frequency, orientation, and scale can be exploited to remove invisible content by, for example, filtering with a CSF characteristic.

2.6.4 Perception of textures

Texture is an important visual cue that we exploit in tasks such as edge localization, depth perception, and general object recognition. The visual system appears to operate under the assumption of high entropy in a scene. For example, the distribution of pebbles on a beach provides good depth cues because they appear to get smaller the further they are from the viewer. It has been proposed that texture perception employs a filter–rectify–filter (FRF) process, which first applies a non-linear transform to the output of the initial multiscale spatial filters, followed by a second filtering stage to provide a low frequency surface on which to do edge localization. This model implies that humans should be less accurate at determining texture defined edges compared to those associated with luminance gradients and this has been confirmed experimentally.

This also suggests that the HVS would experience some degree of change blindness associated with textures and again this has been confirmed from subjective trials. Consider the two images in Figure 2.23. They look identical but on closer inspection are actually significantly different. Again this observation reinforces the notion that a texture analysis–synthesis approach to compression may offer potential.

2.6.5 Shape and object recognition

There is evidence from lesion studies that visual processing can be divided into three broad stages:

1. Extracting local features.
2. Building shapes and surfaces from these features.
3. Creating representations of objects in the scene.

There is also some evidence that the HVS uses other properties from a scene to help group features, for example proximity, color and size similarity, common fate (regions

FIGURE 2.23

Texture change blindness. (Images courtesy of Tom Troscianko.)

with similar direction and speed properties are likely to be part of the same object) and continuity. Marr [4] proposed that features are grouped using average local intensity, average size, local density, local orientation, local distances between neighboring pairs of similar items, and local orientation of the line joining neighboring pairs of items.

To actually recognize objects, further integration of features over larger spatial distances must take place. A number of explanations for this exist including feedback from higher functional areas of the visual cortex that modulate edge detector responses so as to enhance contours [2]. Surfaces are also likely to be important grouping cues as features on the same surface are likely to be part of the same object.

An object representation is thus likely to be formed from a combination of low level image analysis with more symbolic higher level abstract representations coupled through significant feedback between layers and regions. Object recognition is complex because objects have both intrinsic and extrinsic properties. The extrinsic factors, such as change of viewpoint, occlusion or illumination changes, can have a major impact on recognizability. These extrinsic properties can provide context but can also confound recognition.

2.6.6 The importance of phase information

It has been widely reported that the phase spectrum is highly significant in determining the visual appearance of an image; distortions to phase information, especially over larger areas, can result in poor image quality. See for example Figure 2.24. Equally, it has been shown that it is possible to obtain the phase indirectly from local magnitude information. There is evidence that complex cells in the human visual cortex V1 respond to local magnitude information rather than phase. It has thus been suggested that the HVS might use local magnitude information and local phase to determine an image's appearance. Vilankar et al. [14] conducted an experiment to quantify the contributions of local magnitude and local phase toward image appearance as a function of spatial frequency using images distorted using the dual tree complex wavelet transform. They confirmed that both local magnitude and local phase do

FIGURE 2.24

The importance of phase information in visual perception. Left: original. Right: phase distorted version using the complex wavelet transform (Reproduced with permission from Vilankar et al. [14]).

indeed play equally important roles and, in some cases, local phase can dominate the image's appearance. While we are still at the early stages of understanding these processes, there is no doubt that they may in future provide a better understanding of the effects of image compression on the human visual system.

For the above reasons, it is widely accepted that signal processing operations on images should preserve phase information. Linear phase FIR digital filters are therefore (almost) exclusively used in compression operations, both for pre- and post-processing and for interpolation.

2.7 Perception of scale and depth

2.7.1 Size or scale

Size and scale are complex topics in vision. When a spatial pattern is viewed at increasing distance, the image it projects on the retina moves further from the retina; the image it projects stimulates a response reflecting the higher spatial frequency. Thus, at this level changes in scale simply relate to changes in spatial frequency. It has been shown that adaptation to a certain frequency at one distance does affect the perception of a different frequency at a different distance, even if both project the same spatial frequency at the retina. So, even though distance is not integrated into the frequency information from the scene, depth information must be incorporated at higher levels to compensate and allow humans to assess the size of an object at different distances. This ability to judge the size of objects regardless of distance is called size constancy and can lead to misleading and entertaining results, especially when combined with other depth cues such as perspective (Figure 2.25).

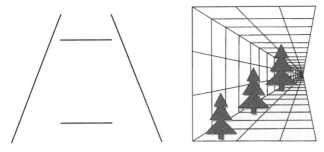

FIGURE 2.25

Perspective-based depth cues can be very compelling and misleading.

2.7.2 Depth cues

There is no doubt that depth assessment is a dominant factor in our ability to interpret a scene; it also serves to increase our sense of engagement in displayed image and video content. Stereopsis, created through binocular human vision, is often credited as being the dominant depth cue, and has indeed been exploited, with varying degrees of success, in creating new entertainment formats in recent years. It is however only one of the many depth cues used by humans—and arguably it is not the strongest. A list of depth cues used in the human visual system is given below:

- **Our model of the 3-D world:** Top-down familiarity with our environment enables us to make relative judgements about depth.
- **Motion parallax:** As an observer moves laterally, nearer objects move more quickly than distant ones.
- **Motion:** Rigid objects change size as they move away or toward the observer.
- **Perspective:** Parallel lines will converge at infinity—allowing us to assess relative depths of oriented planes.
- **Occlusion:** If one object partially blocks the view of another it appears closer.
- **Stereopsis:** Humans have two eyes and binocular disparity information, obtained from the different projections of an object onto each retina, enables us to judge depth.
- **Lighting, shading, and shadows:** The way that light falls on an object or scene tells us a lot about depth and orientation. See Figure 2.26.
- **Elevation:** We perceive objects that are closer to the horizon as further away.
- **Texture gradients:** As objects recede into the distance, consistent texture detail will appear to be finer scale and will eventually disappear due to limits on visual acuity.
- **Accommodation:** When we focus on object our ciliary muscles will either stretch (a thinner lens for distant objects) or relax (a fatter lens for closer objects). This provides an oculomotor cue for depth perception.
- **Convergence:** For nearer objects, our eyes will converge as they focus. This stretches the extraocular muscles, giving rise to depth perception.

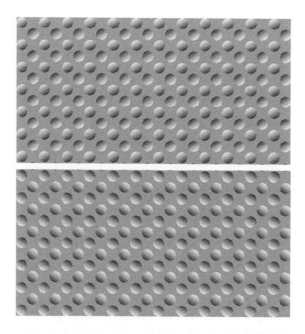

FIGURE 2.26

Pits and bumps—deceptive depth from lighting.

It should be noted that certain depth cues can be confusing and can cause interesting illusions. For example, we have an in-built model that tells us that light comes from above (i.e. the sun is above the horizon) and our expectation of shadows reflects this top-down knowledge. Figure 2.26 shows exactly this. The top diagram clearly shows alternating rows of bumps and pits starting with a row of bumps at the top. The bottom diagram is similar except that it starts with a row of pits at the top. In reality, the only difference with these diagrams is that the bottom one is the top one flipped by 180°. Try this by turning the page upside down. This provides an excellent example of how hard wired certain visual cues are and how these can lead to interesting illusions. Another example of an illusion driven by top-down processes is the *hollow mask*. In Figure 2.27, we can see a normal picture of Albert Einstein. In reality this is a photograph of a concave mask. Our visual system is so highly tuned to faces that, even when our disparity cues conflict with this, we cannot help but see this as a convex face. The right picture shows that, even when we rotate the mask to an angle where the features are distorted and you can see it is clearly concave, it still looks like a convex face!

2.7.3 Depth cues and 3-D entertainment

3-D video technology exploits stereopsis generated through binocular vision. Stereopsis is produced using two cameras spaced at the interocular distance, to create two views which can be displayed in a manner so that the left view is only fed to the left eye and the right view is fed to the right eye. While this does indeed create a sensation

FIGURE 2.27

The hollow mask illusion.

of depth, and can be effective if used well for certain types of content, there are several issues that cause problems. These relate to conflict between the accommodation and the convergence of the eyes.

When we watch a 3-D movie we view it on a 2-D screen so that is where we naturally focus. However, 3-D objects projecting out of the screen cause our eyes to converge because they appear nearer. This can lead to confusing depth cues causing fatigue and, in some cases, nausea. In addition, it has been reported that up to 20% of the population do not see stereoscopically. Of the remainder, a large proportion do not like to watch movies in 3-D and choose to see them in 2-D. There are a range of reasons given for this, including the wearing of glasses, the reduced light levels (up to 50%) due to current 3-D glasses or the general feeling of forced, over-emphasized or inconsistent depth cues.

Equally, directors, cinematographers and technologists are only gradually beginning to understand the complex interactions between their domains. For example, directors and cinematographers use shots to support narrative and the average shot length in a movie is surprisingly short (around 4 s). It is now known that the HVS takes longer to adapt to the transitions between shots in 3-D than in 2-D, so shot lengths should be correspondingly longer in the 3-D case (more of this later). The rules of cinematography have had to change to cope with 3-D.

2.8 Temporal and spatio-temporal response

The temporal response of the HVS is dependent on a number of internal and external factors. Essentially the impulse response of photoreceptors defines the maximum temporal sampling of the retinal image. However, external influences such as ambient lighting, display brightness, and viewing distance also cause variations. If an object is moving across the visual field and not tracked, its image will blur (motion blur) due to the fact that photoreceptors take more that about 100 ms to reach their peak response. These observations have significant impact, not only on camera and display design but also on cinematography (e.g. rules are often imposed to limit camera pan speeds) [16, 17].

2.8.1 Temporal CSF

The visual system is sensitive to temporal variations in illumination. This character-istic is well known and exploited in the design of camera and displays in determining their temporal update rates. The temporal limit is imposed by the rate of response of the photoreceptors, the retinal circuitry and the response times in the visual cortex [2]. The eye retains the sensation of the incident image for a short time after it has been removed (persistence of vision). It has until recently been thought that an update rate of 60 Hz is sufficient to convey the perception of smooth motion. However, recent investigations are questioning this in the context of higher spatial resolution and larger displays. This is discussed in more detail in Chapter 13.

Kelly [19] showed that the visual sensitivity response to a temporally varying pattern at different frequencies was band-pass in nature, highly dependent on display brightness and peaking at around 7 or 8 Hz for dimmer displays, becoming diminish-ingly small at around 24 Hz. However, for brighter displays, the peak rises to around 20 Hz, disappearing at around 80 Hz. The frequency at which the temporal variation becomes unnoticeable is important as it relates to the flickering we experience due to the refresh of a display. The variation with display brightness explains why we can (just about) get away with 24 fps sampling for cinema but need higher rates for TV screens and even higher rates for computer monitors (as we sit much closer to them).

2.8.2 Spatio-temporal CSF

A diagram showing the spatio-temporal characteristics of the HVS is shown in Figure 2.28 [18]. This shows the band-pass temporal characteristic described above for low spatial frequencies. However, as spatial frequency increases, the temporal response becomes more low-pass in nature and similarly, with higher temporal fre-quencies, the spatial response tends to a low-pass characteristic. Thus for faster mov-ing objects, we cannot easily assimilate their spatial textures. This trade-off between spatial and temporal update rates has been exploited in interlaced scanning as we will examine in Chapter 4.

2.8.3 Flicker and peripheral vision

The flicker fusion threshold is the frequency at which an intermittent light stim-ulus appears steady to the observer. This depends on a range of factors including

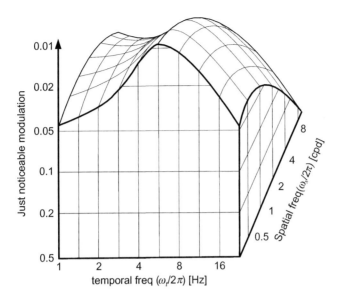

FIGURE 2.28

Spatio-temporal CSF. (Adapted from Kelly [18].)

the frequency and depth of the modulation, the wavelength of the illumination, the position on the retina where the stimulation occurs, and the degree of light or dark adaptation. Other factors such as fatigue can also have an influence.

Tyler [23] demonstrated that the flicker threshold (or critical flicker frequency (CFF)) varies with brightness (higher for a brighter source) and with retinal location (rods respond faster than cones). As a consequence flicker can be sensed in peripheral vision at higher frequencies than in foveal vision. This is important as all of our assumptions about flicker are based on central vision and small experimental screens. This explains why, as screens have become brighter and their sizes have become larger, flicker has become more noticeable. This is why TV manufacturers have had to upsample to 300 or even 600 Hz.

Figure 2.29 [23] shows the visual-field CFF contours for square-wave modulation (in cycles per second) as a function of eccentricity and meridian, with the field sizes scaled to stimulate a constant number of cones at each eccentricity, for two observers. This demonstrates flicker frequencies up to 90 Hz in the periphery.

It is also important to note that CFF generally refers to unstructured stimuli. If the interaction with spatial characteristics is of interest, the temporal contrast sensitivity function (CSF) is more informative. Interestingly the temporal CSF has been observed to be independent of eccentricity up to 30° on the nasal horizontal meridian [24]. In reality of course with natural scenes, complex interactions exist across different interacting patterns and motions in a scene.

FIGURE 2.29

Variation of critical flicker frequency (Reproduced with permission from Tyler [23]).

2.9 Attention and eye movements

The HVS creates for us an illusion of fully perceiving a scene while in reality only a relatively small amount of information is passed to the higher brain levels. In this way we perceive only what is considered absolutely necessary, ignoring all other information, in order to compress the huge amount of visual data that continuously enters the visual system. The process that achieves this is referred to as attention and is supported at all cortical levels, but perhaps foremost through foveation and eye movements.

2.9.1 Saliency and attention

Attention modulates the responses of other parts of the visual system to optimize the viewing experience and to minimize the amount of additional information needed to achieve this. Attention can be bottom-up (exogenous) or top-down (endogenous) or a combination of both. In the first case this happens when lower-level responses attract attention (e.g. through detection of motion in the periphery). This process is often referred to as saliency and is based on the strength of the low level HVS response to the specific feature. Top-down or endogenous attention is guided (more) consciously by the task at hand.

There has been a substantial amount of research into models of saliency, perhaps most notably by Itti et al. [25] where top-down attention modulates bottom-up features. Eye movements are used to enable us to foveate on rapidly changing areas of interest and we will consider these briefly next. Attention and foveation have been proposed as a means of allocating bits optimally according to viewing in certain video compression scenarios [26,27].

2.9.2 **Eye movements**

Apart from when we compensate for head movements, our eyes move in two distinct ways:

1. **Saccades:** As rapid involuntary movements between fixation points. The visual system is blanked during saccading, so we do not experience rapid motion effects due to scanning. Saccades can take two forms:

   a. **Microsaccades:** Small involuntarily movements used to refresh cell responses that can fall off due to adaptation.
   b. **Foveation:** The eyes also move rapidly in photopic vision, to allow exploration of new areas of a scene providing an impression of increased acuity across the visual field.

2. **Smooth pursuit:** These are attention-guided smoother voluntary movements which happen, for example, when tracking moving objects. This process keeps the object of interest centered on the fovea so as to maintain high spatial acuity and reduce motion blur. Under the conditions of smooth pursuit, the spatio-temporal limits discussed earlier change dramatically. Girod [29] recomputed these characteristics and demonstrated that this type of eye movement has the effect of extending the temporal limit of our visual system, showing that frequencies of several hundred hertz can be perceived.

Eye tracking is used extensively in vision research [32] to assess the gaze and fixations of an observer in response to a stimulus. Some of the earliest and most famous research on this was published by Yarbus [28]. The results of Yarbus's experiment, where observers were provided with a range of viewing tasks associated with the Visitor painting, are shown in Figure 2.30. It is interesting to observe how the saccade patterns relate to the specified task.

2.10 **Visual masking**

Visual masking is the reduction or elimination of the visibility of one brief stimulus, called the "target," by the presentation of a second stimulus, called the "mask." An overview of the influences of masking and other psychovisual effects is presented by Girod [30a].

2.10.1 **Texture masking**

The visibility threshold for a target increases when the background is textured rather than plain. Spatial (or texture) masking causes the contrast threshold for a given spatial frequency to rise when a high contrast mask is present. The inhibitory nature of interacting receptive fields causes the HVS sensitivity to decrease to certain spatial patterns when they are viewed in the context of other patterns. The influence of the mask is related to the similarity between the spatial frequency content of the mask and the target. This effect can be explained by recognizing that the mask pattern elicits

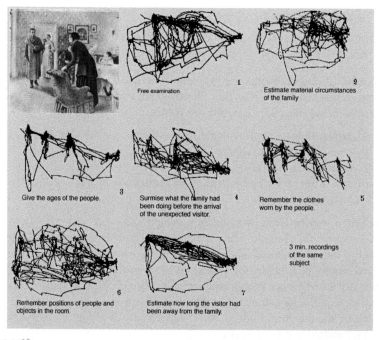

FIGURE 2.30

Eye movements in response to task (Public domain image from: http://commons. wikimedia.org/wiki/File:Yarbus_The_Visitor.jpg).

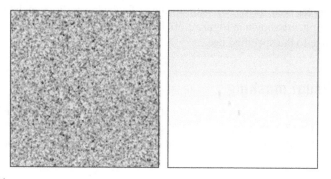

FIGURE 2.31

Example of texture masking.

responses in the same spatial filters as stimulated by the target test pattern, making them indistinguishable at later visual processing stages [2]. An example of texture masking is shown in Figure 2.31. This shows an identical target in both sub-figures (three rectangles). The target is highly visible on the right with a plain background but is very difficult to detect on the left.

2.10.2 Edge masking

It has been observed that there is a distinct masking effect in the vicinity of a spatial edge [30a]. This has also been demonstrated by Zhang et al. [31] for the case of higher dynamic range content. An example of this, showing the experimental setup of the target and edge and the noise visibility variations according to edge contrast and distance of the target from the edge, is shown in Figure 2.32.

2.10.3 Temporal masking

Girod [30a] demonstrated the masking effects of video in the presence of temporal discontinuities (such as those that occur at shot cuts or scene changes in a movie).

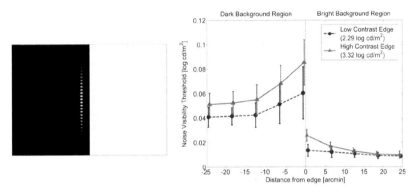

FIGURE 2.32

Edge masking for high and low dynamic range content.

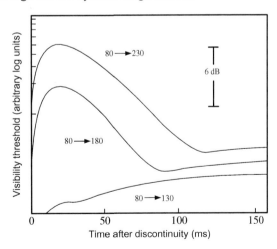

FIGURE 2.33

Temporal masking effects for various edge step sizes. (Reproduced with permission from Girod [30b].)

This effect is shown in Figure 2.33 for a gamma predistorted video signal. It can be observed that, as the extent of temporal discontinuity increases (shown here for an 8 bit edge), the temporal masking effect extends in duration over approximately 100 ms.

2.11   Summary: a perceptual basis for image and video compression

2.11.1   Influential factors

As we have seen throughout this chapter, many factors influence the way we see things: ambient light, scene or display brightness, spatio-temporal content of the scene, viewing distance, attention, task, expectation, physiological variations between subjects and many other subtle environmental factors.

Many researchers have tried to understand these influences on perception, using various abstract psychophysics experiments. While these provide very valuable insights into the limits and variations of our visual responses, few deal with real scenes and the complex interactions within them. However, we can take away several important guidelines and some of these have been used in the design of acquisition, display devices, and for optimizing signal representations and compression methods. Others hold potential for the future. These are summarized in the next subsection.

2.11.2   What have we learnt?

A summary of the observed characteristics of our visual system alongside ways in which these characteristics can be exploited in the design of vision-based technology are provided below.

Color processing

HVS characteristics	Implications
The HVS perceives color using three color channels in the visible part of the electromagnetic spectrum	Acquisition and display technology and visual signal representations for transmission and storage can be based on a tristimulus framework
Based on opponent theory, the HVS processes visual signals as a luminance channel and two chrominance channels	Luminance-only content is readily accepted as natural. Visual signal representations can be based on color spaces that split luminance and chrominance
The HVS has higher acuity in its luminance channel than its chrominance channels	Chrominance channels can be allocated lower bandwidth than the luminance channel

Spatial response

HVS characteristics	Implications
The HVS response to spatial frequencies typically peaks at 5–8 cpd and then falls off, becoming diminishingly small beyond 40–60 cpd	Frequencies beyond the upper range need not be processed. Higher frequencies can be coded more coarsely with implications for quantization. The lower frequency luminance roll-off is not normally exploited
Edges are an important aspect in human vision system processing. Sensitivity increases with contrast level	Edges should be preserved. Artificial edge artifacts introduced by quantization are very noticeable and should be avoided
The HVS can detect noise more readily in plain areas than in textured areas	Spatial (contrast) masking of quantization noise can be exploited in textured regions

Temporal response

HVS characteristics	Implications
The HVS temporal contrast sensitivity peaks at around 5–20 Hz and diminishes above 20–80 Hz (brightness dependent)	Visual processing technology should be capable of sampling at twice the visual threshold and should take account of viewing conditions
Our ability to see individual frames as a continuum (without flicker) occurs at approximately 50 Hz	Frame rates for acquisition and display should normally be at least 50 Hz
Critical flicker frequency is higher in the periphery than in the fovea	Higher frame refresh rates are required for larger screens and closer viewing

Spatio-temporal response

HVS characteristics	Implications
The eye can resolve spatial frequencies better at lower temporal frequencies while at higher spatial frequencies the temporal response becomes low-pass	This trade-off between temporal and spatial acuity can be exploited in interlaced scanning to provide the perception of increased frame rates

Processing textures

HVS characteristics	Implications
There is evidence that texture in-filling occurs in the HVS	Analysis–synthesis approaches to image and video coding hold potential
Texture variations and changes are more difficult to detect than shape changes	As above

Depth cues

HVS characteristics	Implications
The HVS exploits many and varied depth cues at all stages of processing	Stereoscopic 3-D content may not be the only way to achieve more immersive content

Dynamic range

HVS characteristics	Implications
Our visual response depends on the mean brightness of the display	Cinema frame rates can be lower than those for TV or computer monitors. Higher dynamic range displays offer better perceptions of reality and depth
There is a linear relationship between JND and background brightness over a wide range of brightness levels, with further increases at low and high values	This can be exploited in applying intensity-dependent quantizations, allowing step-sizes to increase more rapidly for low and high level signals
There is a non-linear relationship between luminance and perceived brightness	Implications for signal coding in that a non-linear (gamma) function is applied to camera outputs prior to coding

Attention, foveation, and eye movements

HVS characteristics	Implications
Our high acuity vision is limited to a 2 degree arc	Bit allocation strategies could target those regions of an image where attention is focused (if known!)
The spatio-temporal frequency response of the HVS is altered significantly by eye movements. Under smooth pursuit conditions, much higher temporal acuity is possible	Especially with larger and brighter screens, displays must be able to cope with temporal frequencies of many hundreds of hertz

Physiological variations between subjects and environments

HVS characteristics	Implications
The responses of individuals will vary naturally and with age	Subjective quality assessment experiments should be based on results from a large number of subjects (typically > 20)
Human responses vary significantly according to viewing conditions	Subjective tests should be conducted under tightly controlled viewing conditions, to enable cross referencing of results

The implications and exploitation of these characteristics are explored in much more detail in the remainder of this book.

References

[1] R. Snowden, P. Thompson, T. Troscianko, Basic Vision, Oxford University Press, 2006.

[2] G. Mather, Foundations of Sensation and Perception, second ed., Psychology Press, 2009.

[3] B. Wandell Foundations of Vision, Sinauer Assoc, 1995.

[4] D. Marr, Vision, Freeman, 1982. Reprinted MIT Press (2010).

[5] H. von Helmholtz, Handbook of Physiological Optics, first ed., Hamburg and Leipzig, Voss, 1896.

[6] B. Julesz, Binocular depth perception of computer-generated patterns, Bell System Technical Journal 39 (1960) 1125–1162.

[7] D. Hubel, T. Wiesel, Receptive fields of single neurones in the cat's striate cortex, Journal of Physiology 148 (1959) 574–591.

[8] R. Gregory, Eye and Brain, fifth ed., Princeton University Press, 1997.

[9] M. Clowes, Transformational grammars and the organization of pictures, in: A. Grasselli (Ed.), Automatic Interpretation and Classification of Images, Academic Press, 1969.

[10] H. Kolb, How the retina works, American Scientist 91 (2003) 28–35.

[11] R. Lotto, D. Purves, The effects of color on brightness, Nature Neuroscience 2 (11) (1999) 1010–1014.

[12] D. Mustafia, A. Engela, K. Palczewskia, Structure of cone photoreceptors, Progress in Retinal and Eye Research 28 (4) (2009) 289–302.

[13] A. Netravali, B. Haskell, Digital Pictures: Representation, Compression and Standards, second ed., Plenum Press, 1995.

[14] K. Vilankar, L. Vasu, D. Chandler, On the perception of band-limited phase distortion in natural scenes, in: Human Vision and Electronic Imaging XVI. Proceedings of the SPIE, vol. 7865, 2011, p. 78650C.

[15] L. Cormack, Computational models of early human vision, in: A. Bovic (Ed.), Handbook of Image and Video Processing, Academic Press, 2000.

[16] A. Watson (Ed.), Digital Images and Human Vision, MIT Press, 1998.

[17] A. Watson, A. Allhumada, J. Farell, Windows of visibility: psychophysical theory of fidelity in time sampled visual motion displays, Journal of the Optical Society of America A 3 (3) (1986) 300–307.

[18] D. Kelly, Adaptation effects on spatio-temporal sine wave thresholds, Vision Research 12 (1972) 89–101.

[19] D. Kelly, Visual responses to time-dependent stimuli, Journal of the Optical Society of America 51 (1961) 422–429.

[20] R. Shapley, M. Hawken, Neural mechanisms for color perception in the primary visual cortex, Current Opinion in Neurobiology 12 (4) (2002) 426–432.

[21] <http://en.wikipedia.org/wiki/File:CIE1931xy_blank.svg>.

[22] J. Mannos, D. Sakrison, The effects of visual error criteria on the encoding of images, IEEE Transactions on Information Theory IT-20 (1974) 525–536.

[23] C. Tyler, Analysis of visual modulation sensitivity. III. Meridional variations in peripheral flicker sensitivity, Journal of the Optical Society of America A4 (8) (1987) 1612–1619.

[24] V. Virsu, J. Rovamo, P. Laurinen, R. Nasanen, Temporal contrast sensitivity and cortical magnification, Vision Research 22 (1982) 1211–1217.

[25] L. Itti, C. Koch, E. Niebur, A model of saliency-based visual attention for rapid scene analysis, IEEE Transactions on Pattern Analysis and Machine Intelligence 20 (11) (1998) 1254–1259.

[26] D. Agrafiotis, C. Canagarajah, D. Bull, J. Kyle, H. Seers, M. Dye, A perceptually optimised video coding system for sign language communication at low bit rates, Signal Processing: Image Communication 21 (7) (2006) 531–549.

[27] S. Davies, D. Agrafiotis, C. Canagarajah, D. Bull, A multicue bayesian state estimator for gaze prediction in open signed video, IEEE Transactions on Multimedia 11 (1) (2009) 39–48.

[28] A. Yarbus, Eye Movements and Vision, Plenum, New York, 1967.

[29] B. Girod, Motion compensation; visual aspects, accuracy and fundamental limits, in: M. Sezan, R. Lagendijk (Eds.), Motion Analysis and Image Sequence Processing, Kluwer, 1993, pp. 126–152.

[30a] B. Girod, Psychovisual aspects of image communication, Signal Processing 28 (3) (1992) 239–251.

[30b] B. Girod, The information theoretical significance of spatial and temporal masking in video signals, in: Proceedings SPIE/SPSE Conference on Human Vision, Visual Processing and Digital Display, Los Angeles, CA, USA, 1989, pp. 178–187.

[31] Y. Zhang, D. Agrafiotis, M. Naccari, M. Mrak, D. Bull, Visual masking phenomena with high dynamic range content, in: Proceedings of the IEEE International Conference on Image Processing, 2013, pp. 2284–2288.

[32] S. Liversedge, I. Gilchrist, S. Everling, Oxford Handbook of Eye Movements, Oxford University Press, 2011.

[33] J. Bowmaker, H. Dartnall, Visual pigments of rods and cones in a human retina, Journal of Physiology 298 (1980) 501–511.

Discrete-Time Analysis for Images and Video

CHAPTER OUTLINE

3.1 **Signal and picture sampling** **64**
   3.1.1 The sampling theorem 65
   3.1.2 Multidimensional sampling lattices 69
3.2 **Statistics of images** **70**
   3.2.1 Histograms and distributions 70
   3.2.2 Mean values 70
   3.2.3 Correlation in natural images 71
3.3 **Filtering and transforms** **76**
   3.3.1 Discrete-time linear systems 76
   3.3.2 Convolution 76
   3.3.3 Linear filters 77
   3.3.4 Filter frequency response 79
   3.3.5 Examples of practical filters 80
   3.3.6 Non-linear filters 81
   3.3.7 Linear transforms and the DFT 82
3.4 **Quantization** **84**
   3.4.1 Basic theory of quantization 84
   3.4.2 Adaptation to signal statistics 86
   3.4.3 HVS weighting 87
   3.4.4 Vector quantization 87
3.5 **Linear prediction** **87**
   3.5.1 Basic feedforward linear predictive coding 87
   3.5.2 Linear prediction with the predictor in the feedback loop 90
   3.5.3 Wiener Hopf equations and the Wiener filter 92
3.6 **Information and entropy** **93**
   3.6.1 Self-information 93
   3.6.2 Entropy 94
   3.6.3 Symbols and statistics 96
3.7 **Summary** **97**
References **97**

Communicating Pictures. http://dx.doi.org/10.1016/B978-0-12-405906-1.00003-9
© 2014 Elsevier Ltd. All rights reserved.

The first thing to emphasize about this chapter is that it is not intended to be a course in discrete-time signal analysis. Many excellent texts on this topic already exist covering both 1-D [1] and 2-D [2] signals. We assume here that the reader is familiar with the basics of sampling, linear systems analysis, digital filters, transforms, information theory, and statistical signal processing. The purpose of this chapter is to provide an overview of these important topics as they form the basis of many of the compression techniques described in this book.

We firstly review the sampling theorem and consider the cases of 2-D (image) and 2-D + time (video) sampling. This provides the discrete-time samples that represent our image or video and that are processed during the various stages of compression and coding. Next in Section 3.2, we introduce the means of describing signals using statistical representations. We focus on second order statistics and we use these to characterize the redundancy contained in an image or video signal. It is this redundancy that is exploited during the compression process.

Filters and transforms form the basic work-horses of most compression systems and these are introduced in Section 3.3. The signal distortions introduced in most compression systems can be attributed to coefficient quantization after filtering or transformation; this important aspect is covered in Section 3.4. Prediction operators are also in common usage across many aspects of compression, from intra-prediction to motion estimation. The basic concepts of linear prediction are introduced in Section 3.5 and we pay particular attention to the mitigation of decoder drift. This architecture is important as it provides a framework for all current video compression systems. Finally, as a precursor to Chapter 7, we review the basics of information theory; this topic is key to all symbol encoding methods in use today.

3.1 Signal and picture sampling

The manner in which a signal is sampled has a profound effect on subsequent processing steps. Particularly in the context of sensory signals such as images or video, the samples must provide a suitable and realistic representation of the underlying (continuous) natural scene, when viewed on a screen. Sampling must thus be performed in a manner that avoids the introduction of perceptual spatio-temporal artifacts due to aliasing. The sampling process addresses three primary questions:

1. What spatial and temporal sampling rates are needed to avoid aliasing and to produce the desired signal fidelity?
2. How should the samples be distributed in space and time?
3. What pre-processing is required prior to sampling and what post-processing is needed prior to reconstruction?

These issues are influenced by a number of factors including: the frequency content of the signal being sampled, the perceptual thresholds of the observer as related to the prevailing viewing conditions, the characteristics of the capture and display devices, and the costs associated with storage and transmission. This section attempts to answer these questions.

3.1.1   The sampling theorem

In one dimension

Consider a continuous (let us assume time domain) signal $x_a(t)$ sampled using a sequence of delta functions $\delta(t - nT)$ to yield:

$$x_s(t) = x_a(t)s(t) \tag{3.1}$$

where:

$$s(t) = \sum_{n=-\infty}^{\infty} \delta(t - nT)$$

Using the Fourier Transform we can obtain the frequency domain representation of these signals, thus:

$$X_a(\omega) = \int_{-\infty}^{\infty} x_a(t)e^{-j\omega t} dt \tag{3.2}$$

and

$$S(\omega) = \frac{2\pi}{T} \sum_{k=-\infty}^{\infty} \delta(\omega - k\omega_s) \tag{3.3}$$

Invoking the modulation property of the Fourier Transform gives:

$$X_s(\omega) = \frac{1}{T} \sum_{k=-\infty}^{\infty} X_a(\omega - k\omega_s) \tag{3.4}$$

where ω_s represents the continuous frequency variable.

The spectrum of a sampled signal thus comprises scaled replicated versions of the original (continuous time) spectrum. In this context, Shannon's sampling theorem captures the requirements for sampling in order to avoid aliasing [3]:

> *If a signal $x_a(t)$ is bandlimited with $X_a(\omega) = 0$ for $|\omega| > \omega_m$, then $x_a(t)$ is uniquely determined by its samples provided that $\omega_s \geq 2\omega_m$. The original signal may then be completely recovered by passing $x_s(t)$ through an ideal low-pass filter.*

The spectral characteristics of the sampling process as described by Shannon's theorem are shown in Figure 3.1. Paying particular attention to the bottom two sub-figures, the penultimate sub-figure shows the case where perfect reconstruction is possible with an ideal reconstruction filter. In contrast, the bottom sub-figure shows the case where ω_m exceeds the frequency $\omega_s/2$ and aliasing distortion occurs. The frequency $\omega_s/2$ is often referred to as the Nyquist frequency.

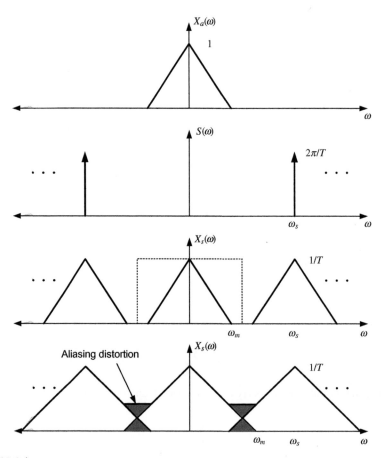

FIGURE 3.1

Spectral characteristics of sampling and aliasing.

Example 3.1 (1-D signal aliasing)

Consider a sinusoidal signal $x(t) = \cos(3\pi t/2)$ sampled at $\omega_s = \frac{4}{3}\omega_m$ with $T = 1$. Show that this sampling violates the sampling theorem and characterize the impact of the alias in the time and frequency domains.

Solution.   The result is shown in Figure 3.2. The sinusoid $x(t) = \cos(3\pi t/2)$ is sampled at $\omega_s = \frac{4}{3}\omega_m$ with $T = 1$. As can be observed, an alias is created at a frequency of $\frac{\omega_s}{4}$ (to be accurate, this is actually an alias of the negative frequency component of the sinusoid at $-\omega_m$). This can be observed in both the signal domain plot (top) and the frequency domain plot (bottom). Here, as is conventional with discrete-time systems, we plot normalized frequency $\Omega = \omega T$, where the sampling is given by $\Omega = 2\pi$.

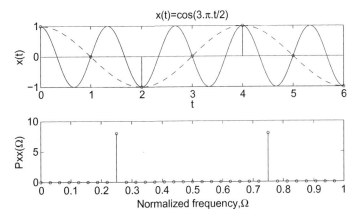

FIGURE 3.2

Demonstration of aliasing for a 1-D signal. Top: sinusoid sampled below Nyquist frequency. Bottom: Fourier plot showing spectral aliasing.

Extension to 2-D

Assuming a rectangular (orthogonal) sampling pattern in 2-D, we can extend our 1-D approach to the case of sampling of a 2-D signal. This gives:

$$X_s(\omega_1, \omega_2) = \frac{1}{T_1 T_2} \sum_{k_1, k_2} X_a(\omega_1 - k_1 \omega_s, \omega_2 - k_2 \omega_s) \qquad (3.5)$$

The effect of sampling and aliasing on the 2-D spectrum is shown in Figure 3.3. In 2-D images, aliasing errors will take the appearance of ringing in a direction perpendicular to high frequencies or sharp edges. It should be emphasized that while Figure 3.3 shows a circularly symmetric spectrum, in practice the spectrum could be skewed according to the prevailing horizontal and vertical frequencies or textures in the image.

Extension to 3-D

In the case of spatio-temporal sampling, it is conventional to sample at a fixed frame rate using either progressive or interlaced sampling (see Chapter 4). As we saw in Chapter 2, the visual sensitivities in the spatial and temporal dimensions are quite different. For example, a conventional 1080p/50 4:4:4 format will capture 103,680,000 samples per second. This however comprises 2,073,600 samples per picture but only 50 pictures per second. Thus there is a large difference in the number of samples per unit dimension between the spatial and temporal domains. It appears that the temporal domain gets a raw deal and to some extent this is true. We will come back to this in Chapter 13.

As described in Chapter 2, visual sensitivity depends on the mean brightness of the display. For a bright TV signal, it is generally considered, based on the temporal contrast sensitivity function and critical flicker frequencies, that a temporal update

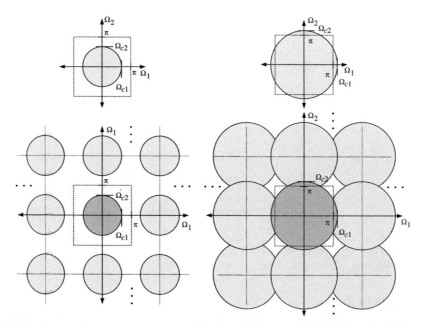

FIGURE 3.3

2-D spectral characteristics of sampling and aliasing. Left: Top—original signal spectrum; Bottom—sampled signal spectrum with no aliasing. Right: Top—original signal spectrum; Bottom—sampled signal spectrum with aliasing due to sub-Nyquist sampling.

rate of 50–70 Hz is sufficient, whereas for computer monitors which are viewed more closely in brighter environments, a higher update rate is normally specified.

Example 3.2 (How many spatial samples do we need?)
Consider the case of a 16:9 screen with a width of 1 m, viewed at a distance of $3H$ (where H is the height of the screen). How many horizontal samples are required to satisfy the contrast sensitivity limits of the human visual system?

Solution. For the case of a 16:9 screen with a width of 1 m, viewed at $3H$, we can compute the number of horizontal samples required from simple trigonometry. The height of the screen is 0.5625 m and the half viewing angle is then:

$$\theta = \tan^{-1}\left(\frac{0.5}{1.6875}\right)$$

Assuming a spatial resolution limit of 30 cpd (see Chapter 2), this yields 1980 horizontal samples per line, which is pretty close to the 1920 that are used in practice.

Although this calculation is useful, as screen sizes increase and viewing distances decrease, there will be a demand to take a fresh look at spatio-temporal sampling, but more of that later in Chapter 13.

3.1.2 **Multidimensional sampling lattices**

A sampling lattice, in real-valued K-dimensional space, is the set of all possible vectors that can be formed from linear combinations of a set of K linearly independent basis vectors, $\mathbf{v}_k \in \mathcal{R}^K$, $k = \{1, 2, 3, \ldots, K\}$. So the new vectors are formed from a weighted combination as follows:

$$\mathbf{w} = \sum_{k=1}^{K} c_k \mathbf{v}_k; \quad \forall c_k \in \mathcal{Z} \tag{3.6}$$

The matrix $\mathbf{V} = [\mathbf{v}_1, \mathbf{v}_2, \ldots, \mathbf{v}_K]$ that combines the K basis vectors is called the *sampling matrix* or the *generating matrix*. For example, the generating matrix for our familiar rectangular lattice is:

$$\mathbf{V}_1 = \begin{bmatrix} 1 & 0 \\ 0 & 1 \end{bmatrix} \tag{3.7}$$

However, we can explore more exotic sampling grids and their associated regions of support. Consider for example the generating matrix:

$$\mathbf{V}_2 = \begin{bmatrix} \sqrt{3}/2 & 0 \\ 1/2 & 1 \end{bmatrix} \tag{3.8}$$

The sampling lattice for this generating matrix is shown in Figure 3.4. As can be seen, this represents an hexagonal sampling grid. The lattice structure represents a simple and convenient tool for generating and analyzing sampling structures that are not necessarily hypercubic in structure. All the theorems for 1-D and 2-D sampling still apply and a generalized form of the Shannon sampling theory exists which dictates the structure and density of the sampling grid dependent on the signal spectrum. There is

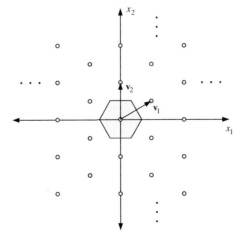

FIGURE 3.4

Hexagonal sampling lattice and its reciprocal as defined by equation (3.8).

a significant body of work relating to this more general approach to multidimensional sampling, which is relevant to image and video applications. The reader is referred to the work of Dubois [4] or Wang et al. [5] for excellent introductions to this topic.

3.2 Statistics of images

3.2.1 Histograms and distributions

A histogram is a discrete approximation of a probability density function based on the actual occurrences of quantized pixel values. It can provide important information about the distribution of sample amplitudes within the image, about the dynamic range of the signal as opposed to the wordlength of its representation, and about whether the image is under- or overexposed. An example histogram for the 256×256 image Stampe_SV4 is given in Figure 3.5.

Spatial and subband distributions

An understanding of spatial domain and frequency domain distributions can be of significant benefit in compression systems. For example, the Pyramid Vector Quantization method described in Chapter 11 exploits the Laplacian-like subband coefficient distribution while the Parametric Video Coding approaches outlined in Chapter 13 exploit the subband structure of Dual-Tree Complex Wavelet coefficients in providing segmentation into coherent texture regions.

3.2.2 Mean values

Mean and correlation provide a partial statistical characterization of a random process in terms of its averages and moments. They are useful as they offer tractable

FIGURE 3.5

Example image histogram for 256×256 image Stampe_SV4.

mathematical analysis, they are amenable to experimental evaluation and they are well suited to the characterization of linear operations on random processes.

The mean for a stochastic process is given by:

$$\mu_n = E\{x[n]\}$$
$$= \int_{-\infty}^{\infty} x[n] f(x[n]) dx[n] \tag{3.9}$$

where $f(\cdot)$ is the probability density function (pdf) of the process and $E\{\cdot\}$ is the expectation operator. In the case of a stationary process the pdf of the random variable is the same for all n and hence the mean is constant. Expectation can be interpreted as an average value obtained by repeating an experiment a number of times. Hence:

$$\mu_n = E\{x[n]\} = \lim_{N \to \infty} \left[\frac{1}{N} \sum_{i=1}^{N} x_i[n] \right] \tag{3.10}$$

3.2.3 Correlation in natural images

The autocorrelation (or autocovariance) of a sequence expresses the linear statistical dependencies between its samples. It is defined for a real-valued signal with a lag of m samples as:

$$R_{xx}[m] = E\{x[n]x[n+m]\} \tag{3.11}$$

where $x[n]$ is a stationary random process. In particular:

$$R_{xx}[0] = E\left\{|x[n]|^2\right\} \tag{3.12}$$

is the average power of the signal. In practice, the correlation is estimated based on a finite length sequence **x**, of N samples as follows:

$$R_{xx}[m] = \frac{1}{N} \sum_{n=0}^{N-|m|-1} x[n]x[n+m] \tag{3.13}$$

Equation (3.13) produces a biased estimate because of the attenuation for large lags. An unbiased estimate can also be used which is noisier, but preferable in certain situations:

$$R_{xx}[m] = \frac{1}{N-|m|} \sum_{n=0}^{N-|m|-1} x[n]x[n+m] \tag{3.14}$$

The autocovariance is computed in the same manner as the autocorrelation, but with the signal means removed. When the autocorrelation or autocovariance functions are normalized by their maximum value, they are generally referred to as autocorrelation coefficients or autocovariance coefficients respectively. These have values

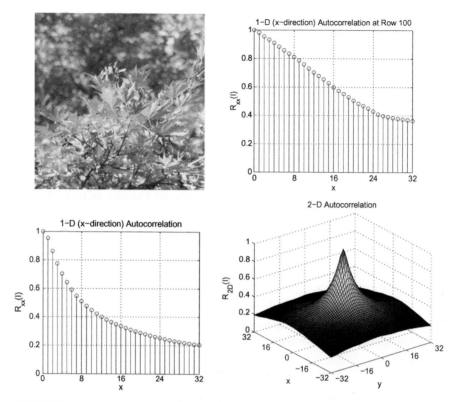

FIGURE 3.6

Autocorrelation plots for Acer image (512×512). Top left to bottom right: original image; autocorrelation function for row 100; autocorrelation function for whole image; 2-D autocorrelation surface.

between -1 and $+1$. For example, the autocorrelation coefficient is given by:[1]

$$\rho_x(k) = \frac{r_x(k)}{r_x(0)} \tag{3.15}$$

The autocorrelation matrix comprises autocorrelation values at various lags. The correlation matrix for a *wide sense stationary* (WSS) real-valued signal is the expectation of the outer product of the signal vector with itself. Thus:

$$\mathbf{R}_x = E\left\{\mathbf{x}\mathbf{x}^\mathrm{T}\right\} \tag{3.16}$$

---

[1] *Note:* We have dropped the double subscript as it is redundant and where no confusion results we will use lower case r to denote autocorrelation.

$$\mathbf{R}_x = \begin{bmatrix} r_x[0] & r_x[1] & \cdots & r_x[M-1] \\ r_x[1] & r_x[0] & & r_x[M-2] \\ \vdots & & & \vdots \\ r_x[M-1] & r_x[M-2] & \cdots & r_x[0] \end{bmatrix} \qquad (3.17)$$

The correlation matrix is Toeplitz and Hermitian. For the case of real-valued signals this means that it is transpose-invariant.

Spatial autocorrelation in natural images

Let us now consider the correlation properties of some images. Figures 3.6 and 3.7 show the autocorrelation properties of two 512×512 images—*Acer* and *Stampe_SV4*. The first has a highly textured, spatially active foreground with a blurred background and the second is highly structured with several plain areas. As well as the original

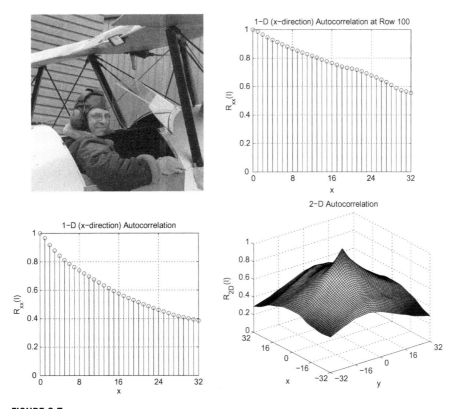

FIGURE 3.7

Autocorrelation plots for Stampe_SV4 image (512×512). Top left to bottom right: original image; autocorrelation function for row 100; autocorrelation function for whole image; 2-D autocorrelation surface.

image, each figure includes (i) an autocorrelation plot (unbiased with zero mean) for lags up to 32 pixels for a single row of pixels, (ii) a similar plot based on an average of all rows in the image, and (iii) a two-dimensional version. It can be observed that in both cases the correlation falls off gradually with increasing lag and that the values for the Acer image fall off more quickly that those for the Stampe_SV4 image. This is because of the structure in the latter compared to the more stochastic nature of the former. Figure 3.8 shows a similar series of plots, but this time for a 256×256 image, with only values up to a lag of 16 shown. As one would expect, the autocorrelation values fall off in a similar fashion to Figure 3.7 except that the corresponding lag values are halved.

Exploitation of this inter-sample correlation is the basis of prediction, transform, and filterbank compression methods. In each case, the purpose of the transform or predictor is to decorrelate the spatial information to make it more amenable to frequency-related perceptual quantization. As we can see from the figures, the autocorrelation

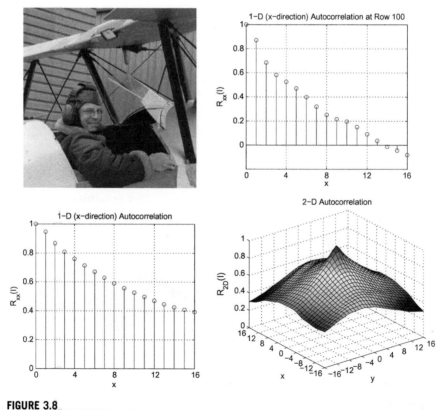

FIGURE 3.8

Autocorrelation plots for Stampe_SV4 image (256×256). Top left to bottom right: original image; autocorrelation function for row 100; autocorrelation function for whole image; 2-D autocorrelation surface.

values for large lags become increasingly small, while those for smaller lags are high. Because decorrelating transforms and predictors work better on regions with higher correlation, they are normally applied to small regions (typically 8×8 blocks) rather than to the whole image. An 8×8 block has been conventionally chosen as a good compromise between computational efficiency and coding gain. It is important to note, however, that the image resolution will have a significant impact on this—as lower resolution images will exhibit, like for like, lower correlations across the same size block.

Temporal autocorrelation in natural image sequences

Let us now consider the time axis. An example of temporal correlation, for the Foreman sequence, is shown in Figure 3.9. This indicates a similar characteristic to the spatial case considered above. Two plots are shown. The first is the time correlation for a single pixel computed using 300 frames of the Foreman sequence. The second is the same calculation, but this time averaged over the 16×16 block of pixels indicated in the top sub-figure. Again we can observe that significant temporal

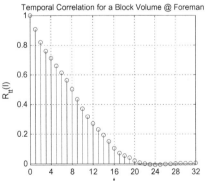

FIGURE 3.9

Temporal autocorrelation plots for Foreman (30 fps). Top to bottom right: sample frame showing selected 16×16 block used; temporal correlation for a single pixel; temporal correlation for a 16×16 block.

correlation exists between the adjacent video frames and indeed over 8 or so frames the correlation remains relatively high. This indicates that there is significant temporal redundancy present in most image sequences. As we will see later this is exploited in most compression methods, by applying motion compensated prediction, prior to transformation.

3.3 Filtering and transforms

We assume here that the reader is familiar with basic linear system theory, filters and transforms. For background reading, the reader is referred to Refs. [1,2,8,9].

3.3.1 Discrete-time linear systems

Many of the systems that we encounter in this book exhibit the properties of linearity and shift invariance.[2] Let us briefly describe these properties.

Shift invariance

A shift invariant system is one where a shift in the independent variable of the input signal causes a corresponding shift in the output signal. So if the response of a system to an input $x_0[n]$ is $y_0[n]$ then the response to an input $x_0[n - n_0]$ is $y_0[n - n_0]$.

Linearity

If the response of a system to an input $x_0[n]$ is $y_0[n]$ and the response to an input $x_1[n]$ is $y_1[n]$ then a system is referred to as linear if:

1. The response to $x_0[n] + x_1[n]$ is $y_0[n] + y_1[n]$ (additivity).
2. The response to $ax_0[n]$ is $ay_0[n]$, where a is any complex constant (scaling).

A corollary of this is the principle of superposition which states that if $x[n] = ax_0[n] + bx_1[n]$ then the output of a linear shift invariant system will be $y[n] = ay_0[n] + by_1[n]$.

3.3.2 Convolution

If the response of a linear time-invariant (LTI) system to a unit impulse, $\delta[n]$, is $h[n]$ then $h[n]$ is termed the *impulse response* of the system. The shift invariance, linearity and sifting properties of LTI systems give:

$$y[n] = \sum_{k=-\infty}^{\infty} x[k]h[n - k] \tag{3.18}$$

---

[2]Linear systems that are shift invariant are often referred to as linear time invariant (LTI). Although the independent variable is not always time, we will use this terminology here for convenience.

or:
$$y[n] = x[n] \star h[n]$$

This is known as the convolution sum and can be used to determine the response of a discrete-time LTI system to an arbitrary input sequence.

Example 3.3 (Convolution)
Using the convolution equation (3.18), determine the response of a linear system with impulse response $h[n] = \{1, 2, 1\}$ to an input $x[n] = \{1, 2, 0, 1, -1, 2\}$.

Solution. The convolution sum embodies the principle of superposition. Hence we can form the overall system output from the sum of the responses to each input sample as follows:

n	0	1	2	3	4	5	6	7
$y_0[n]$	1	2	1					
$y_1[n]$		2	4	2				
$y_2[n]$			0	0	0			
$y_3[n]$				1	2	1		
$y_4[n]$					-1	-2	-1	
$y_5[n]$						2	4	2
$y[n]$	1	4	5	3	1	1	3	2

3.3.3 Linear filters

The transfer function of a discrete-time linear system is given by:

$$H(z) = \frac{a_0 + a_1 z^{-1} + a_2 z^{-2} + \cdots + a_M z^{-M}}{b_0 + b_1 z^{-1} + b_2 z^{-2} + \cdots + b_N z^{-N}} \qquad (3.19)$$

Since $x[n-k] \xleftrightarrow{Z} z^{-k} X(z)$, the equivalent difference equation is (with b_0 scaled to unity):

$$y[n] = a_0 x[n] + a_1 x[n-1] + a_2 x[n-2] + \cdots + a_M x[n-M] \qquad (3.20)$$

$$- b_1 y[n-1] - b_2 y[n-2] - \cdots - b_N y[n-N]$$

or:

$$y[n] = \sum_{k=0}^{M} a_k x[n-k] - \sum_{k=1}^{N} b_k y[n-k] \qquad (3.21)$$

In cases where $\{b_k\} = 0$ then the filter is described as having a *finite impulse response* (FIR). In other cases it is known as *infinite impulse response* (IIR). FIR filters are much more common in image and video processing than their IIR counterparts. The reason for this is primarily due to their phase response. We saw in Chapter 2 that phase distortion has an important influence on perceptual quality, and it is straightforward to design FIR filters with a linear phase characteristic. Such a characteristic only introduces a simple shift in the filter response with no phase distortion. We will provide some examples of practical FIR filters that are used in video compression a little later, but first let us examine the filter frequency response in a little more detail.

Extension to 2-D

The 2-D convolution of a signal $x[m, n]$ with an impulse response $h[m, n]$ is given by:

$$y[m, n] = \sum_{k=-\infty}^{\infty} \sum_{l=-\infty}^{\infty} x\left[m - k, n - 1\right] h[k, l] = x[m, n] \star h[m, n] \qquad (3.22)$$

2-D digital filters are used extensively in image and video coding, for example combined with sample-rate changing, in the analysis or synthesis stages of filter banks (Chapter 6), in interpolation filters for sub-pixel motion estimation (Chapters 8 and 9) or in more general pre- and post-processing operations. However, for reasons of flexibility and complexity these are, wherever possible, implemented separably; i.e. as a cascade of 1-D filtering stages.

Separability

Separability is an important property of digital filters and transforms that is exploited extensively in image and video processing. A system is separable when (for real-valued coefficients):

$$\mathbf{h} = \mathbf{h}_1 \mathbf{h}_2^{\mathrm{T}} \qquad (3.23)$$

where $h_1[m]$ is a 1-D impulse response that operates across the rows of the signal and $h_2[n]$ is the corresponding impulse response that operates down the columns. We can now rewrite equation (3.22) as follows:

$$y[m, n] = \sum_{k=-\infty}^{\infty} h_1[k] \sum_{l=-\infty}^{\infty} x\left[m - k, n - 1\right] h_2[l] \qquad (3.24)$$

If we define:

$$y_2[m, n] = \sum_{l=-\infty}^{\infty} x\left[m, n - 1\right] h_2[l]$$

then we can rewrite equation (3.24) as:

$$y[m, n] = \sum_{k=-\infty}^{\infty} y_2\left[m - k, n\right] h_1[k] \qquad (3.25)$$

So this implies that if equation (3.23) holds, then we can achieve 2-D filtering by first filtering the columns of the 2-D signal with $h_2[n]$, followed by filtering the rows of the output from the first stage using $h_1[m]$. Due to the fact we are cascading linear operations, these two stages can be interchanged if necessary.

It should be noted that not all filters are separable and that separability does impose some orientation constraints on the filter characteristic. In general, a filter is separable if all its rows and all its columns are linearly dependent. This property can be observed in the solution of Example 3.4.

Example 3.4 (Filter separability)

Consider the two filter impulse responses below:

$$\mathbf{h}_1 = \begin{bmatrix} 0.5 & 0.3 & 0.2 \end{bmatrix}^T$$
$$\mathbf{h}_2 = \begin{bmatrix} 0.6 & 0.1 & 0.3 \end{bmatrix}^T$$

Form the equivalent 2-D digital filter.

Solution. The 2-D filter is given by:

$$\mathbf{h} = \mathbf{h}_1\mathbf{h}_2^T = \begin{bmatrix} 0.30 & 0.05 & 0.15 \\ 0.18 & 0.03 & 0.09 \\ 0.12 & 0.02 & 0.06 \end{bmatrix}$$

As a further exercise, satisfy yourself that the 2-D filtering operation is identical to the use of separable filters.

3.3.4 Filter frequency response

From our understanding of convolution and digital filtering, we know that the response of a discrete linear system to a sampled sinusoidal input signal will be a number of weighted and shifted versions of the input signal. The output is thus also a sinusoidal signal at the same frequency, but with different amplitude and phase. We can therefore think of our filter in the frequency domain as a frequency-dependent modifier of amplitude and phase. If we generalize this with an input signal that is a complex exponential:

$$x[n] = e^{j\Omega n}$$

then from the convolution equation (3.18) we have:

$$y[n] = \sum_k h[k]e^{j\Omega(n-k)} = e^{j\Omega n} \sum_k h[k]e^{-j\Omega k}$$

Thus:

$$y[n] = x[n]H(\Omega)$$

where:

$$H(\Omega) = \sum_{k=-\infty}^{\infty} h[k]e^{-j\Omega k} \tag{3.26}$$

The term $H(\Omega)$ describes the change in amplitude and phase experienced by the complex exponential as it passes through the system and it is referred to as the *frequency response* of the system.

3.3.5 Examples of practical filters

LeGall wavelet analysis filters

A common set of separable filters, that are employed to analyze a signal into low- and high-pass subbands prior to sub-sampling and quantization, are the 5/3 LeGall filters. The impulse responses of the low- and high-pass filters are:

$$\mathbf{h}_0 = [-1, 2, 6, 2, -1]/4$$

$$\mathbf{g}_0 = [1, -2, 1]/4$$

The frequency responses of these filters are shown in Figure 3.10. It is interesting to note that the low-pass and high-pass filters are not symmetrical and that there appears to be significant spectral leakage between them. However, as we will see in Chapter 6, these filters do perform well and offer the property of perfect reconstruction when combined in a filterbank.

H.264/AVC half-pixel interpolation filter

A second important filtering operation is interpolation, for example where a region of a picture is upsampled prior to performing sub-pixel motion estimation (see Chapters 8 and 9). In the currently deployed video coding standard, H.264/AVC, a six-tap FIR

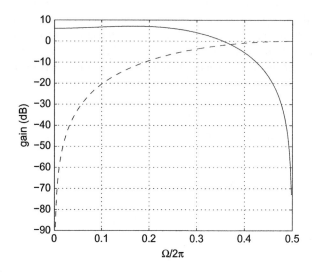

FIGURE 3.10

Filterbank responses for the LeGall low-pass and high-pass analysis filters.

FIGURE 3.11

Filter response for H.264 half-pixel interpolation filter.

filter is employed:

$$\mathbf{h} = [1, -5, 20, 20, -5, 1]/32 \qquad (3.27)$$

The frequency response of this filter is shown in Figure 3.11. It can be observed that the characteristic is not dissimilar to that of the LeGall low-pass analysis filter. However, in this case, the filter has no effect on the original integer-pixel resolution samples, it is simply used to generate the interpolated values. The filter characteristic is therefore a trade-off between obtaining a high cut-off frequency with rapid roll-off and low-pass band ripple. Such trade-offs are common in filter design [1].

3.3.6 Non-linear filters

Non-linear filters are not based on linear relationships between an input and an output via a system function. Instead they represent a much broader class of operations that do not have any explicit frequency domain transfer function. Non-linear filters find application in a range of image processing and coding applications such as: denoising, edge preserving operations and in some forms of prediction. For example, the deblocking filters used in the HEVC and H.264/AVC compression standards are content-adaptive non-linear filters as described in Chapter 9.

A common class of filter that is used in some motion vector prediction applications is the rank-order filter, in particular the median filter. We introduce these briefly below.

Rank-order and median filters

A rank-order filter is an operator that ranks the input samples and selects an output according to its position in the ranked list. The following are examples of rank-order filters:

$$y[m, n] = \min_{\mathcal{N}(m,n)} (x(i, j)) \tag{3.28}$$

$$y[m, n] = \max_{\mathcal{N}(m,n)} (x(i, j)) \tag{3.29}$$

where $\mathcal{N}$ represents the neighborhood, related to location (m, n) over which the operation is applied. The min operator will attenuate isolated peaks or ridges in an image, whereas the max operator will fill isolated troughs or holes. It can be seen that repeated application of this type of filter will have a continued effect of eroding the isolated regions described.

The median operator will rank all the N samples within the region and output the ranked value in position[3] $(N + 1)/2$, thus:

$$y[m, n] = \underset{\mathcal{N}(m,n)}{\mathrm{med}} (x(i, j)) \tag{3.30}$$

The median operator is better than a linear filter in removing outliers (e.g. salt and pepper noise). It will preserve straight edges, but will round corners and distort texture features. The extent of feature flattening will depend on the window size used.

We can of course combine these operations to produce non-linear difference filters such as that below:

$$y[m, n] = \min_{\mathcal{N}(m,n)} (x(i, j)) - \max_{\mathcal{N}(m,n)} (x(i, j)) \tag{3.31}$$

which produces a non-linear gradient of the image.

Morphological filters

The erosion and dilation operations referred to above can be generalized and combined to form a powerful toolbox of operators known as morphological filters. These shape-based operators support region erosion and dilation and geodesic reconstruction and have found applications in segmentation [6], contrast enhancement and in feature prioritization for low bit rate video coding [7].

3.3.7 Linear transforms and the DFT

Linear transforms, in particular the *Discrete Cosine Transform* (DCT), form the basis of most image intra-frame and residual coding and are covered in detail in Chapter 5. We provide an introduction to these here, focusing on the *Discrete Fourier Transform* (DFT). Fourier Transforms are used extensively across signal processing applications to perform spectral analysis based on discrete representations in both the signal domain and its associated frequency domain. An excellent introduction to Fourier Transforms is provided in Ref. [8].

---

[3]*Note:* Median filters are normally defined with an odd number of taps. In cases where the number of samples is even, it is conventional to take the average of the middle two terms.

While the Discrete Time Fourier Series and the Discrete Time Fourier Transform provide a basis for analyzing periodic and aperiodic signals respectively, the latter produces a continuous frequency domain function so is not well suited to digital implementation. The ubiquitous Discrete Fourier Transform overcomes this and enables discrete-time processing in both the signal and its associated frequency domain.

The Discrete Fourier Transform

Let us consider the 1-D version of the DFT initially. If $x[n]$ is a finite duration signal defined over the range $0 \leq n \leq N_1 - 1$, a periodic signal $\tilde{x}[n]$ can be constructed which equals $x[n]$ over one cycle. Let $N \geq N_1$ and let $\tilde{x}[n]$ be periodic in N. If we compute the Discrete Time Fourier Series (DTFS) of this new sequence:

$$c_k = \frac{1}{N} \sum_{n=\langle N \rangle} \tilde{x}[n] e^{-jk(2\pi/N)n}$$

Letting $X(k) = N \cdot c_k$, this becomes the analysis equation of the DFT:

$$X(k) = \sum_{n=0}^{N-1} x[n] e^{-jk(2\pi/N)n}; \quad k = 0, 1, 2, \ldots, N - 1 \tag{3.32}$$

The inverse DFT is given by:

$$x[n] = \frac{1}{N} \sum_{k=0}^{N-1} X(k) e^{jk(2\pi/N)n}; \quad n = 0, 1, 2, \ldots, N - 1 \tag{3.33}$$

The N terms of the DFT represent samples of the continuous function $X(\Omega)$, equally spaced over a 2π interval. These can be easily shown to equal the samples of the z-transform on the unit circle in the z-plane.

Normally we use the DFT in matrix vector form:

$$\mathbf{X} = \mathbf{W}\mathbf{x} \tag{3.34}$$

where

$$W_N = e^{-j(2\pi/N)}$$

Hence we have in matrix form:

$$
\begin{bmatrix} X(0) \\ X(1) \\ X(2) \\ \vdots \\ X(N-1) \end{bmatrix} = \begin{bmatrix} W^0 & W^0 & W^0 & \cdots & W^0 \\ W^0 & W^1 & W^2 & \cdots & W^{N-1} \\ W^0 & W^2 & W^4 & & W^{2(N-1)} \\ \vdots & & & & \vdots \\ W^0 & W^{N-1} & W^{2(N-1)} & \cdots & W^{(N-1)(N-1)} \end{bmatrix} \begin{bmatrix} x[0] \\ x[1] \\ x[2] \\ \vdots \\ x[N-1] \end{bmatrix}
$$

$$\tag{3.35}$$

Similarly the inverse DFT is given by:

$$\mathbf{x} = \frac{1}{N} \mathbf{W}^* \mathbf{X} \tag{3.36}$$

The 2-D DFT

In a similar manner to that for filters in Section 3.3.3, transform operations can be separable and this is the conventional way of applying 2-D transforms, largely because of the associated complexity savings. We cover this in more detail in Chapter 5.

The DFT and compression

It may seem logical to attempt to exploit the ability of the DFT to analyze a signal in terms of its frequency domain components. There are however a number of significant issues with this which mean that this is not done in practice. These are:

1. Although images are invariably real-valued, the DFT coefficients are complex-valued so we are doubling the amount of information.
2. For short sequences or small 2-D regions, like those typical in compression, the discontinuities produced by the underlying assumptions of periodicity can have a major impact on performance as they produce ringing or spectral leakage in the frequency domain.
3. Ringing is normally addressed in spectral analysis applications by applying a non-rectangular window function (such as a Hamming window) to the input data prior to extension. These reduce spectral ringing but smear sharp frequency domain edges, distorting the characteristics of the underlying signal. Again, this is exactly what we don't want in a compression system!

Comparisons between the DFT and more appropriate transforms such as the DCT are made in Chapter 5.

3.4 Quantization

3.4.1 Basic theory of quantization

A quantizer maps an input value (or values in the context of a vector quantizer) to an output value. It can either take a continuous amplitude discrete-time signal and produce a quantized digital signal (with finite wordlength) or it can take a previously digitized sample and quantize it further to reduce its dynamic range. The former is common in acquisition devices such as A to D converters and the latter in compression systems where, for example, the coefficients of a transformed input signal might be mapped to a different range, with small values going to zero.

Once an image or video signal has been sampled to produce a discrete set of values with a given wordlength, the compression process normally involves some form of signal analysis (typically transformation) to decorrelate the content, followed by quantization to approximate the original signal in a way that retains as much perceptual quality as possible. Ignoring numerical imprecision, quantization is the stage in any compression process where loss is introduced. It is thus crucial to understand the impact of quantization on the perceptual quality of an image or video signal.

Uniform quantization

Uniform quantizers adopt a constant step size $\triangle$, between reconstruction levels. If the input to the quantizer is x, this will map to a reconstruction value, y_i, where y_i represents the center of the range as dictated by the step size parameter. We define a quantization error, $q[n] = x[n] - y[n]$, which is modeled as a zero mean, uniformly distributed signal within the range $[-\triangle/2, \triangle/2]$. This signal has energy or variance given by:

$$\sigma_q^2 = \frac{1}{\triangle} \int_{-\triangle/2}^{\triangle/2} q^2 dq = \frac{\triangle^2}{12} \tag{3.37}$$

Two common uniform quantizer profiles are shown in Figure 3.12. On the left is the conventional midtread quantizer for positive valued signals, and on the right is a similar profile, but for two-sided input signals. Decision levels for x are indicated along with the corresponding reconstruction values for y.

For the case of the conventional midtread quantizer, the quantization operation on an input x, to produce a reconstruction level y_i, in terms of the decision levels x_i, is defined as:

$$\dot{x} = Q(x) = y_i; \quad x_{i-1} < x \leq x_i \tag{3.38}$$

This is typically achieved using a rounding operation, for example:

$$\dot{x} = \text{NINT}\left(\frac{x}{\triangle}\right) \tag{3.39}$$

where NINT is the operation of rounding to the nearest integer.

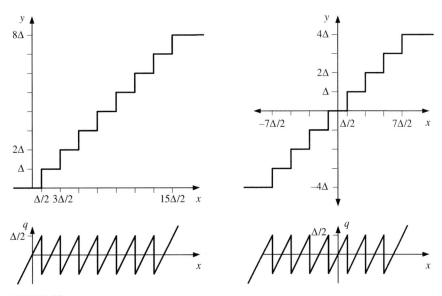

FIGURE 3.12

Common uniform quantizer characteristics.

In order to reconstruct a quantized signal, it will normally have to be rescaled to its original range. This rescaling operation is defined as:

$$\tilde{x} = \Delta \dot{x} \tag{3.40}$$

3.4.2 Adaptation to signal statistics

Quantizers with non-uniform step sizes can be beneficial in cases when:

- The signal being quantized does not exhibit a uniform PDF.
- Low level signals or other signal ranges are known to contain excess noise.
- There is a desire to map a range of low level signals to zero.
- There is a need to incorporate a non-linear amplitude mapping as part of the quantization process.

Deadzone quantizer

The most common form of non-uniform quantizer is the *Deadzone* quantizer. This has a broader decision range for the band of inputs close to zero. It has the benefit during compression of ensuring that noisy low level signals are not allocated bits unnecessarily. The characteristic of a deadzone quantizer is shown in Figure 3.13 (left).

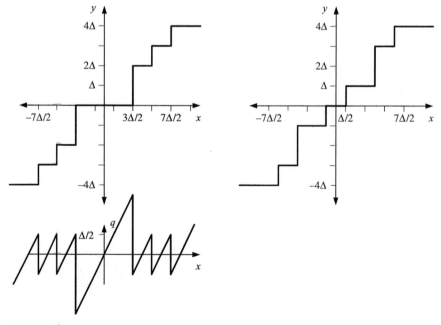

FIGURE 3.13

Common non-uniform quantizers. Left: center deadzone. Right: Lloyd Max quantizer.

Lloyd Max quantizer

The Lloyd Max algorithm [10] is a well-known approach to designing non-uniform quantizers optimized according to the prevailing PDF of the input signal. Rather than allocate a uniform step size, as would be optimum for a uniform PDF, the Lloyd Max approach identifies decision boundaries according to the mean values of equal area partitions of the PDF curve. The characteristic of a Lloyd Max quantizer is shown in Figure 3.13 (right).

3.4.3 HVS weighting

In practice for many compression applications, the quantizer step size is weighted according to some perceptual criterion, such as the spatial frequency of a coefficient produced by a decorrelating transform operation. This is considered in more detail in Chapter 5.

3.4.4 Vector quantization

Vector quantization (VQ) is a method which maps a group of samples to a single quantization index. It is a method that has been adapted in highly robust codecs such as those based on Pyramid Vector Quantization (PVQ) as discussed in Chapter 11. We will not cover this in detail here but the interested reader is referred to Refs. [9,11].

3.5 Linear prediction

3.5.1 Basic feedforward linear predictive coding

The idea of linear prediction is to exploit statistical relationships in random data sequences in order to decorrelate them prior to coding.

A feedforward analysis filter for linear prediction is shown in Figure 3.14 (top) and the corresponding synthesis filter is shown in Figure 3.14 (bottom). Ignoring the quantizer block for the time being, the transfer function of the analysis filter is given by:

$$\begin{aligned} E(z) &= X(z) - X_p(z) \\ &= X(z)(1 - P(z)) \end{aligned} \tag{3.41}$$

$$H(z) = \frac{E(z)}{X(z)} = 1 - P(z) \tag{3.42}$$

where the notation used is as shown in the figure. Assuming the predictor block in Figure 3.14 is an FIR filter, then in the sample domain we have:

$$e[n] = x[n] - \sum_{i=1}^{N} a_i x[n-i] \tag{3.43}$$

If we now consider the decoder or synthesis filter:

$$Y(z) = E(z) + Y(z)P(z) \tag{3.44}$$

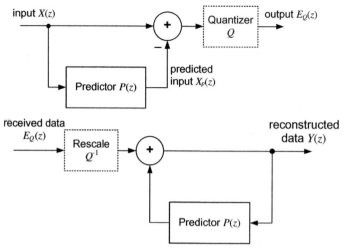

FIGURE 3.14

Feedforward linear prediction. Top: encoder. Bottom: decoder.

or:

$$Y(z) = \frac{E(z)}{1 - P(z)} = X(z) \tag{3.45}$$

Example 3.5 (Linear predictive coding)

Consider the case in Figure 3.14 when $P(z) = z^{-1}$ and the input to the predictive coder is $x[n] = \{1, 2, 3, 2, 5, 4, 2, 4, 5, 6\}$. Assuming that no quantization takes place, confirm that the decoder output will be identical to the input sequence.

Solution. We have $e[n] = x[n] - x[n-1]$ and $y[n] = e[n] + y[n-1]$. So, assuming zero initial conditions and working through for each input sample:

$x[n]$	1	2	3	2	5	4	2	5	6
$e[n]$	1	1	1	-1	3	-1	-2	3	1
$y[n]$	1	2	3	2	5	4	2	5	6

Hence $y[n] = x[n]$.

Predictor dynamic range

The benefit of prediction filtering as a basis for coding is that although the dynamic range of the transmitted error signal can be double that of the original signal, the variance of the error signal is significantly reduced compared to the original.

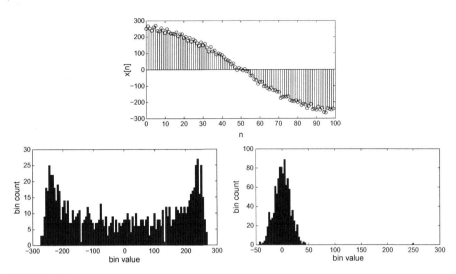

FIGURE 3.15

Prediction signal dynamic range. Top: input signal. Bottom left: distribution of 1000 samples of input signal. Bottom right: distribution of 1000 samples of prediction residual.

This is demonstrated in Figure 3.15, where a sinusoidal signal with additive Gaussian noise ($\sigma_v^2 = 0.05$):

$$x[n] = \cos(0.01\pi n) + v[n] \tag{3.46}$$

is filtered using a predictor $P(z) = z^{-1}$.

Linear predictive coding with quantization

The problem with the feedforward architecture arises when quantization is introduced in the process. In this case we have:

$$E_Q(z) = Q\{X(z) - X_p(z)\} \tag{3.47}$$

and

$$Y(z) = Q^{-1}\{E_Q(z)\} + Y(z)P(z) \tag{3.48}$$

or:

$$Y(z) = \frac{Q^{-1}\{E_Q(z)\}}{1 - P(z)} \neq X(z) \tag{3.49}$$

As one would expect, the quantization process introduces errors in the reconstructed output. This is to be expected since the operation is a lossy one. However, the situation is worse because there is the potential for drift between the encoder and the decoder. This is illustrated in Example 3.6.

Example 3.6 (Linear predictive coding with quantization)

Consider the case in Figure 3.14 where, as before, $P(z) = z^{-1}$ and the input to the predictive coder is $x[n] = \{1, 2, 3, 2, 5, 4, 2, 4, 5, 6\}$. Assuming that quantization takes place, where the quantization and rescaling operations take the form:

$$e_Q[n] = \text{rnd}\left(\frac{e[n]}{2}\right); \quad e_R[n] = 2e_Q[n]$$

with rounding of 0.5 values toward 0, characterize the decoder output relative to the input signal.

Solution. Assuming 0 initial conditions and following the same approach as in Example 3.5, but this time including the quantization and rescaling operations:

$x[n]$	1	2	3	2	5	4	2	5	6
$e[n]$	1	1	1	−1	3	−1	−2	3	1
$e_Q[n]$	0	0	0	0	1	0	−1	1	0
$e_R[n]$	0	0	0	0	2	0	−2	2	0
$y[n]$	0	0	0	0	2	2	0	2	2

Hence $y[n]$ is not the same as $x[n]$. Furthermore significant drift has occurred due to the accumulation of rounding errors at the decoder.

3.5.2 Linear prediction with the predictor in the feedback loop

In order to avoid problems with drift, we must replicate the effects of the decoding operation within the encoder. This ensures that, in the absence of data corruption during transmission, both the encoder and the decoder process exactly the same data and operate in unison. This modified architecture is shown in Figure 3.16. The expression

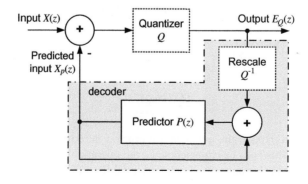

FIGURE 3.16

Feedback-based linear prediction.

for our encoded output (without quantization) is now given by:

$$E(z) = X(z) - X_p(z)$$
$$= X(z) - \frac{E(z)P(z)}{1 - P(z)} \tag{3.50}$$

Therefore:

$$E(z) = X(z)(1 - P(z))$$

and, as before, at the decoder:

$$Y(z) = E(z) + Y(z)P(z) \tag{3.51}$$

or:

$$Y(z) = \frac{E(z)}{1 - P(z)} = X(z) \tag{3.52}$$

If we now include quantization in the process, the difference between the feedback and feedforward architectures will become clear. To ease analysis we model the quantization process as additive quantization noise, $V(z)$ with a variance of σ_N^2. Then we have, as illustrated in Figure 3.17:

$$E(z) = X(z) - X_p(z)$$
$$= X(z) - \frac{[E(z) + V(z)]P(z)}{1 - P(z)} \tag{3.53}$$

Therefore:

$$E(z) = X(z)(1 - P(z)) - V(z)P(z) \tag{3.54}$$

and at the decoder:

$$Y(z) = E(z) + Y(z)P(z) + V(z) \tag{3.55}$$

or:

$$Y(z) = \frac{E(z) + V(z)}{1 - P(z)} \tag{3.56}$$

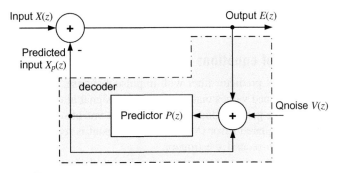

FIGURE 3.17
Feedback-based linear predictor with quantization noise modeling.

Substituting from equation (3.54) for $E(z)$ in equation (3.56), it is straightforward to see that:

$$Y(z) = X(z) + V(z) \qquad (3.57)$$

So we no longer have any drift between encoder and decoder, we simply have a mismatch due to the additive noise of the quantization process. This is a well known but very important result as this prediction architecture is used extensively in video compression.

Example 3.7 (Feedback-based linear predictive coding with quantization)

Consider the case in Figure 3.16 where, as before, $P(z) = z^{-1}$ and the input to the predictive coder is $x[n] = \{1, 2, 3, 2, 5, 4, 2, 4, 5, 6\}$. Assuming that quantization takes place, where the quantization and rescaling operations take the form:

$$e_Q[n] = \text{rnd}\left(\frac{e[n]}{2}\right); \quad e_R[n] = 2e_Q[n]$$

with rounding of 0.5 values toward 0, characterize the decoder output relative to the input signal.

Solution. Assuming 0 initial conditions and following the same approach as in Example 3.6, but this time based on the architecture in Figure 3.16:

$x[n]$	1	2	3	2	5	4	2	5	6
$e[n]$	1	2	1	0	3	−1	−2	3	2
$e_Q[n]$	0	1	0	0	1	1	−1	1	1
$e_R[n]$	0	2	0	0	2	2	−2	2	2
$y[n]$	0	2	2	2	4	4	2	4	6

As expected, $y[n]$ is not identical to $x[n]$ because of the lossy quantization process. However, the variance of the predicted signal is significantly reduced, compared to the original, and also there is no drift between encoder and decoder.

3.5.3 Wiener Hopf equations and the Wiener filter

If we have a linear FIR predictor filter with impulse response, **h**, we wish the filter coefficients to be assigned values that force the error signal after prediction to be as small as possible. If we perform this optimization in terms of second order statistics (i.e. based on Mean Squared Error (MSE)) then the result is referred to as a Wiener filter—an optimum non-recursive estimator.

If we have a desired sequence (training sequence) **d**, then:

$$e[n] = d[n] - y[n] \qquad (3.58)$$

and we can define a cost function as:

$$J = E\left\{|e[n]|^2\right\} \tag{3.59}$$

where, when the gradient $\nabla(J) = 0$ with respect to all filter coefficients, then the filter is said to be optimum in the MSE sense. It can be shown that a necessary and sufficient condition for J to attain a minimum is that the corresponding value of the estimation error is orthogonal to each sample that enters the estimation process at time n. This is known as the *principle of orthogonality*. Based on this observation, it is straightforward to show that the optimum filter coefficients can be computed from the autocorrelation of the filter input sequence $r[k]$ and the cross correlation of the filter input and the desired response $p[k]$ for a lag of k samples. This is known as the Wiener Hopf equation:

$$\sum_{i=0}^{\infty} h_o[i]r[i-k] = p[k] \tag{3.60}$$

or in matrix vector form:

$$\mathbf{h}_o = \mathbf{R}^{-1}\mathbf{p} \tag{3.61}$$

It is also possible to show, assuming stationarity, that the error surface is exactly a second order function of the filter coefficients. It will thus possess a unique minimum point corresponding to the optimum filter weights. For further details, the reader is referred to Ref. [12].

3.6 Information and entropy

The concept of generating a quantitative measure of information was originated by Shannon in 1948 [13]. He used entropy, a measure of the uncertainty in a random variable, to quantify the value of the information contained in a message.

Understanding the basics of information theory is important in the context of image and video compression, since it is used extensively for lossless symbol encoding after transformation and quantization. Typically, quantized transform coefficients are scanned and run-length encoded. Symbols representing runs of zeros and values of non-zero coefficients are coded, using variable length codewords, in a manner that reflects their information content. In this way a lossless reduction in bit rate can be achieved.

Before we look at symbol statistics, let us first review the basics of information theory. For more in-depth coverage, please refer to Ref. [14].

3.6.1 Self-information

Information is used to reduce the uncertainty about an event or signal. Shannon [13] defined the *self-information* of an event A, which could for example comprise a set of outcomes from an experiment, as:

$$i(A) = \log_2 \left(\frac{1}{P(A)} \right) = -\log_2 \left(P(A) \right) \qquad (3.62)$$

The use of logarithms is intuitive, since $\log(1) = 0$ and $-\log(x)$ increase as x decreases from 1 to 0. Hence, if the probability of an event is low, then the information associated with it is high. For example, if $P(A) = 0.5$ then $-\log_2(0.5) = 1$ and if $P(A) = 0.25$ then $-\log_2(0.25) = 2$. Other logarithm bases can be used, but base 2 is the most common as it conveniently provides a measure of information in bits.

Independent events

The information obtained from two independent events is the sum of the information from each event. This can be seen as follows:

$$i(AB) = \log_2 \left(\frac{1}{P(AB)} \right)$$

and, if A and B are independent, then:

$$P(AB) = P(A)P(B) \qquad (3.63)$$

Therefore:

$$i(AB) = \log_2 \left(\frac{1}{P(A)P(B)} \right) = \log_2 \left(\frac{1}{P(A)} \right) + \log_2 \left(\frac{1}{P(B)} \right)$$

or

$$i(AB) = i(A) + i(B) \qquad (3.64)$$

3.6.2 Entropy

If we define a source that generates a series of random and independent events with outcomes from an alphabet $S = \{s_1, s_2, s_3, \ldots, s_N\}$, then the average self-information associated with an event is:

$$H = \sum_{i=1}^{N} P(s_i)i(s_i) = -\sum_{i=1}^{N} P(s_i) \log_2 \left(P(s_i) \right)$$

The quantity H, the average self-information, is referred to as the entropy of the source and this tells us how many bits per symbol are required to code that source. Shannon demonstrated that no coding scheme can code the source losslessly at a rate lower than its entropy.

Entropy and first order entropy

The entropy of a source is given by:

$$H(S) = \lim_{N \to \infty} \frac{1}{N} G_N \tag{3.65}$$

where

$$G_N = -\sum_{i_1=1}^{N} \sum_{i_2=1}^{N} \cdots \sum_{i_M=1}^{N} P(d_1 = s_{i_1}, d_2 = s_{i_2}, \dots, d_M = s_{i_M})$$
$$\times \log P(d_1 = s_{i_1}, d_2 = s_{i_2}, \dots, d_M = s_{i_M})$$

and $D = \{d_1, d_2, d_3, \dots, d_M\}$ and M is the length of the sequence. However, if each symbol is independent and identically distributed (*iid*), then it can be shown that:

$$G_N = -M \sum_{i_1=1}^{N} P(d_1 = s_{i_1}) \log \left(P(d_1 = s_{i_1}) \right) \tag{3.66}$$

and the entropy is then:

$$H(S) = -\sum_{i=1}^{N} P(s_i) \log_2 \left(P(s_i) \right) \tag{3.67}$$

In most realistic cases equations (3.65) and (3.67) do not give the same result. Hence we differentiate them by referring to equation (3.67) as the first order entropy.

The relationship between symbol probability and the self-information of that symbol is given in Figure 3.18. The figure also illustrates the relationship between the probability of a symbol and the contribution that symbol makes to the overall entropy

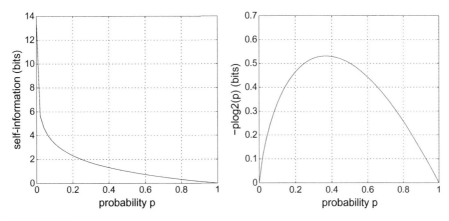

FIGURE 3.18

Self-information and probability. Left: plot of self-information vs probability for a single event. Right: plot of the self-information of an event weighted by its probability.

of the source. Figure 3.18 indicates that symbols with probabilities at either end of the scale contribute little to the source entropy while those in the middle of the range contribute values around 0.5.

Example 3.8 (Entropy)
An alphabet $S = \{s_1, s_2, s_3, s_4, s_5, s_6, s_7, s_8\}$. If the corresponding probabilities of occurrence of these symbols during transmission is $P(S) = \{0.06, 0.23, 0.3, 0.15, 0.08, 0.06, 0.06, 0.06\}$, calculate the first order entropy for the source.

Solution. First order entropy is given by:

$$H = -\sum_S P(s_i) \log_2\left(P(s_i)\right)$$

$$= -(0.06 \times \log_2 0.06 + 0.23 \times \log_2 0.23 + \cdots + 0.06 \times \log_2 0.06)$$

$$= 2.6849 \text{ bits/symbol}$$

3.6.3 Symbols and statistics

In practice, since equation (3.65) is impossible to compute, first order entropy is almost always used instead. We therefore have to estimate the entropy of the source and this depends heavily on the statistics of the source symbols. This point is emphasized in Example 3.9.

Example 3.9 (Source statistics and entropy)
Consider the sequence of symbols: $D = \{1, 2, 3, 2, 1, 2, 3, 4, 5, 6, 5, 6, 7, 8, 9, 10\}$. Compute the first order entropy for this sequence.

Solution. This seems like a straightforward question. We are given no information about the statistics of the source so let us estimate the probabilities of each symbol, based on relative frequencies of occurrence. In this way we have:

$$P(4) = P(7) = P(8) = P(9) = P(10) = \frac{1}{16}$$

$$P(1) = P(3) = P(5) = P(6) = \frac{2}{16}$$

$$P(2) = \frac{3}{16}$$

Thus the first order entropy is

$$H = -\sum_S P(s_i) \log_2\left(P(s_i)\right) = 3.2028 \text{ bits}$$

So this says that, on average, we need just over 3.2 bits per symbol for this source.
However, if we now look closer at the sequence we can observe that the symbols are correlated. So what happens if we process this sequence with a simple first order linear predictor? Our residual sequence, $E = \{1, 1, 1, -1, -1, 1, 1, 1, 1, 1, 1, -1, 1, 1, 1, 1, 1\}$.

So now we have a new set of probability estimates:

$$P(1) = \frac{13}{16}$$
$$P(-1) = \frac{3}{16}$$

In this case, we recalculate our entropy to be 0.6962 bits/symbol. So which is correct? In fact neither is right or wrong, it's just that in the latter case, we have better understood the structure of the source and provided a better estimate of its entropy.

For example, we can estimate the probabilities of occurrence of pixel values in an image, or a region of an image, through computing a histogram of its quantization levels and dividing the content of each quantization bin by the total number of pixels. Alternatively we can pre-compute them using a histogram of one or more similar training images. The latter is more common, although we rarely use actual pixel values as the input symbols in practice. More often we use some more efficient means based on prediction or run length coding of quantized spatial or transform domain data. More of this later in Chapter 7.

3.7 Summary

This chapter has reviewed some preliminary topics in signal processing and information theory that help to place the remainder of this book in context. We have not burdened the reader with series of proofs and theorems but have instead focused on the use of examples to aid understanding of some important topics. We examined some basic statistical measures and used these to characterize picture redundancy. Filters and transforms were introduced as they form the basis of most compression methods and these will be covered further in Chapters 5 and 6. We also revisit the area of quantization in the same chapters.

We spent some time here on prediction structures as these are used frequently in compression and can provide substantial coding gains over non-predictive methods, both for spatial and temporal processing. Intra-prediction will be covered in Chapters 9 and 12 and motion prediction in Chapters 8 and 9.

Finally, we reviewed the basics of information theory, explaining how its performance depends heavily on assumed source models and context. This forms the basis of symbol encoding as used in most compression standards. We specifically focus on entropy coding methods in Chapter 7 and follow this up in Chapter 12 in the context of standards.

References

[1] S. Mitra, Digital Signal Processing: A Computer Based Approach, fourth ed., McGraw Hill, 2011.
[2] J. Woods, Multidimensional Signal, Image and Video Processing and Coding, second ed., Academic Press, 2012.

[3] C. Shannon, Communication in the presence of noise, Proceedings of Institute of Radio Engineers 37 (1) (1949) 10–21.

[4] E. Dubois, The sampling and reconstruction of time-varying imagery with application in video systems, Proceedings of the IEEE 73 (1985) 502–522.

[5] Y. Wang, J. Ostermann, Y.-Q. Zhang, Video Processing and Communications, Prentice Hall, 2001.

[6] R. O'Callaghan, D. Bull, Combined morphological-spectral unsupervised image segmentation, IEEE Transactions on Image Processing 14 (1) (2005) 49–62.

[7] P. Czerepiński, D. Bull, Enhanced interframe coding based on morphological segmentation, IEE Proceedings — Vision, Image and Signal Processing 144 (4) (1997) 220–226.

[8] A. Oppenheim, A. Willsky, H. Nawab, Signals and Systems, second ed., Prentice Hall, 1999; A. Netravali, B. Haskell, Digital Pictures: Representation, Compression and Standards, second ed., Plenum Press, 1995.

[9] J.-R. Ohm, Multimedia Communication Technology, Springer, 2004.

[10] S. Lloyd, Least squares quantisation in PCM (1957), Reprint in IEEE Trans Communications 30 (1982) 129–137.

[11] Y. Linde, A. Buzo, R. Gray, An algorithm for vector quantiser design, IEEE Transactions on Communications 28 (1980) 84–95.

[12] M. Hayes, Statistical Digital Signal Processing and Modeling, Wiley, 1999.

[13] C. Shannon, A mathematical theory of communication, Bell System Technical Journal 27 (3) (1948) 379–423.

[14] T. Cover, J. Thomas, Elements of Information Theory, Wiley, 2006.

Digital Picture Formats and Representations

CHAPTER OUTLINE

4.1 Pixels, blocks, and pictures **100**
   4.1.1 Pixels, samples, or pels 100
   4.1.2 Moving pictures 102
   4.1.3 Coding units and macroblocks 103
   4.1.4 Picture types and groups of pictures 104
4.2 Formats and aspect ratios **106**
   4.2.1 Aspect ratios 106
   4.2.2 Displaying different formats 110
4.3 Picture scanning **111**
   4.3.1 Interlaced vs progressive scanning 111
   4.3.2 Standards conversion 112
4.4 Gamma correction **114**
4.5 Color spaces and color transformations **116**
   4.5.1 Color descriptions and the HVS 116
   4.5.2 Sub-sampled color spaces 121
   4.5.3 Bayer filtering 123
4.6 Measuring and comparing picture quality **124**
   4.6.1 Compression ratio and bit rate 124
   4.6.2 Objective distortion and quality metrics 125
   4.6.3 Subjective assessment 128
4.7 Rates and distortions **129**
   4.7.1 Rate–distortion characteristics 129
   4.7.2 Rate–distortion optimization 131
4.8 Summary **132**
References **132**

Since the origins of digital video communications, there has been a need to understand the links between the human visual system and picture representations. Building on Chapters 1, 2, and 3, this chapter describes the picture formats, processing techniques,

For colour versions of Figures 4.10, 4.11, 4.14, 4.15, 4.17 and 4.18 please refer to the electronic version or the website.

Communicating Pictures. http://dx.doi.org/10.1016/B978-0-12-405906-1.00004-0

and assessment methods that underpin the coding process; it introduces the processing methods and sampling structures that enable compression to be so effective.

Building on the mathematical preliminaries from Chapter 3, we first show how samples are represented in digital images and video signals, emphasizing the important aspects of color and spatio-temporal representations. We examine how samples are grouped to form pictures and how pictures are grouped to form sequences of pictures or videos.

The mapping of light levels to brightness as perceived by the human visual system is considered in Section 4.4, showing how the signal from the camera must be transformed using a non-linear mapping (gamma correction) to ensure best use of the bits available. Next, developing further the trichromacy theory of vision introduced in Chapter 2, we introduce the area of color spaces in Section 4.5; these are important since, with relatively little effort, we can create representations that reduce the bit rate for a digital video by 50%. Finally, we introduce the important topic of image and video quality assessment—both as a means of comparing the performance of different compression systems and as a means of controlling encoder modes during compression.

4.1   Pixels, blocks, and pictures

4.1.1   Pixels, samples, or pels

A still image or picture is a spatial distribution of sample values that is constant with respect to time. These are often referred to as pixels or pels.

Monochrome image

A monochrome image is a two-dimensional matrix of luma samples, **S,** with spatial dimensions (often called resolution) $X \times Y$. This is shown in Figure 4.1. If we zoom in on an actual image, such as that shown in Figure 4.2, where the pixel resolution (in

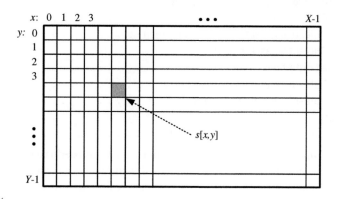

FIGURE 4.1

Image sample array.

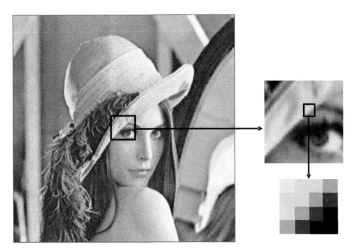

FIGURE 4.2

Image samples.

terms of the angle it subtends at the retina) starts to be within the resolving power of the HVS, we begin to see pixelation effects. As we zoom in still further, we lose the structure of the underlying image and the pixel structure dominates.

Figure 4.3 presents this in a slightly different way, through the use of four different sample rates. If you move away from this figure you will see (at around 5 m) that the top two images start to look very similar. If you can get even further away (at around 10 m), all four images will take on a similar appearance. This emphasizes the fact that resolution is not about the number of horizontal and vertical samples, but about the angle that each pixel subtends at the viewer's retina.

Color image

In the case of a color picture, each element of $\mathbf{S}$ is itself a vector $\mathbf{s}$, with three dimensions representing the intensities in a given color space (e.g. RGB, HSI, or YC_bC_r).

So we have in matrix form:

$$\mathbf{S} = \begin{bmatrix} \mathbf{s}[0,0] & \mathbf{s}[0,1] & \cdots & \mathbf{s}[0,X-1] \\ \mathbf{s}[1,0] & \mathbf{s}[1,1] & \cdots & \mathbf{s}[1,X-1] \\ \vdots & \vdots & & \vdots \\ \mathbf{s}[Y-1,0] & \mathbf{s}[Y-1,1] & \cdots & \mathbf{s}[Y-1,X-1] \end{bmatrix} \tag{4.1}$$

where (e.g. for an RGB signal):

$$\mathbf{s}[x,y] = \begin{bmatrix} s_R[x,y] & s_G[x,y] & s_B[x,y] \end{bmatrix}$$

FIGURE 4.3

Pixelation at varying resolutions. Top left to bottom right: 256×256; 64×64; 32×32; 16×16.

4.1.2 Moving pictures

A video signal is a spatial intensity pattern that changes with time. Pictures are captured and displayed at a constant frame rate, $1/T$ where T is the sampling period. In order to trade off temporal artifacts (flicker, judder etc.) against available bandwidth, a frame rate of 24, 25, or 30 Hz has conventionally been chosen. More recently, as screens have become larger and spatial resolutions have increased, higher frame rates have been introduced at 50 and 60 Hz. The new UHDTV standard Rec.2020 specifies a maximum frame rate of 120 Hz. The reader is referred to Chapter 12 for a more in-depth discussion of frame rate requirements.

In terms of our discrete signal representation, we add an extra dimension (time or frame index), z, to the two spatial dimensions associated with a still image, thus:

$$\mathbf{S} = \begin{bmatrix} \mathbf{s}[0,0,z] & \mathbf{s}[0,1,z] & \cdots & \mathbf{s}[0,X-1,z] \\ \mathbf{s}[1,0,z] & \mathbf{s}[1,1,z] & \cdots & \mathbf{s}[1,X-1,z] \\ \vdots & \vdots & & \vdots \\ \mathbf{s}[Y-1,0,z] & \mathbf{s}[Y-1,1,z] & \cdots & \mathbf{s}[Y-1,X-1,z] \end{bmatrix} \tag{4.2}$$

where (again for an RGB signal):

$$\mathbf{s}[x, y, z] = \begin{bmatrix} s_R[x, y, z] & s_G[x, y, z] & s_B[x, y, z] \end{bmatrix}$$

4.1.3 Coding units and macroblocks

As we will see later, video compression algorithms rarely process information at the scale of a picture or a pixel. Instead the coding unit is normally a square block of pixels. In standards up to and including H.264, this took the form of a 16×16 block, comprising luma and chroma information, called a macroblock. We will examine the nature of color spaces and chroma sub-sampling in Section 4.5. So for now let us simply examine the high level organization of this macroblock.

Macroblocks

A typical macroblock structure is illustrated in Figure 4.4. The macroblock shown corresponds to what is known as a 4:2:0 format (see Section 4.5) and comprises a

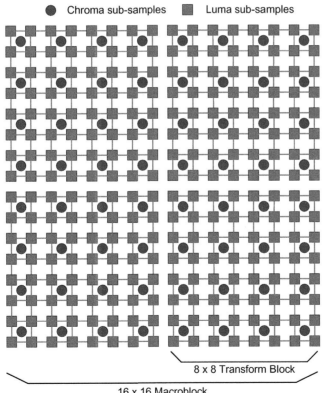

● Chroma sub-samples   ■ Luma sub-samples

8 x 8 Transform Block

16 x 16 Macroblock

FIGURE 4.4

Typical macroblock structure.

16×16 array of luma samples and two sub-sampled 8×8 arrays of chroma (color difference) samples. This macroblock structure, when coded, must include all of the information needed to reconstruct the spatial detail. For example, this might include transform coefficients, motion vectors, quantizer information, and other information relating to further block partitioning for prediction purposes. A 16×16 block size is normally the base size used for motion estimation; within this, the decorrelating transforms are normally applied at either 8×8 or 4×4 levels.

Coding tree units

HEVC has extended the size of a macroblock up to 64×64 samples to support higher spatial resolutions, with transform sizes up to 32×32. It also provides much more flexibility in terms of block partitioning to support its various prediction modes. Further details on this are provided in Chapter 12.

4.1.4 Picture types and groups of pictures

Frame types

Most video coding formats support three types of picture (or frame[1]), classified in terms of their predictive properties. These are described below:

- **Intra-coded (I) pictures:** I pictures (or frames) are coded without reference to any other picture, but can be used as a reference for other pictures.

  - They provide anchor pictures and support random access.
  - They offer the highest resilience to errors as they do not propagate errors from previous pictures.
  - Because they do not exploit temporal correlation, they have the lowest compression ratio of all frame types.

- **Predicted (P) pictures:** P pictures are inter-coded, i.e. they contain the compressed difference between the current frame and a prediction of it based on a previous I or P frame.

  - Like I frames, they can be used as a reference for other frames.
  - In order to decode a P frame, the transformed residual must be accompanied by a set of motion vectors, that provide information on the prediction reference location.
  - P pictures are more efficient than I pictures as they exploit both spatial and temporal correlations.
  - Unlike I frames, they will propagate errors due to predictive coding.

- **Bi-predicted (B) pictures:** B pictures comprise the compressed difference between the current frame and a prediction of it based on up to two I, P, or B frames.

---

[1] These terms can generally be used interchangeably.

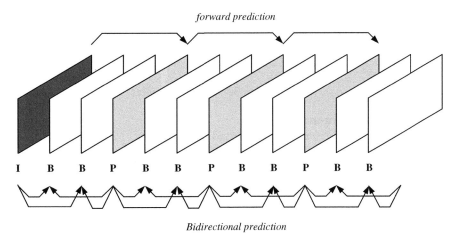

forward prediction

I B B P B B P B B P B B

Bidirectional prediction

FIGURE 4.5

Typical group of pictures structure and prediction modes.

- In standards up until H.264/AVC, B pictures were not used as a reference for predicting other pictures. However, H.264/AVC and HEVC have generalized their definition and in these cases they can be used as a reference, although normally in a hierarchical fashion (see Chapter 12).
- They require accompanying motion vectors.
- They provide efficient coding due to exploitation of temporal and spatial redundancies and can be quantized more heavily if not used as a reference.
- They will not propagate errors if not used as a reference for other pictures.

Groups of pictures (GOPs)

All video coding methods and standards impose some bitstream syntax in order for the bitstream to be decodable. Part of this relates to the order of encoding, decoding, and display of pictures and the way in which pictures of different types are combined. A typical GOP structure that illustrates this (typical of standards such as MPEG-2) is shown in Figure 4.5.

Example 4.1 (Intra- vs inter-frame coding—the benefits of prediction)
Consider a 512×512 @30 fps color video with 8 bit samples for each color channel. Compare the intra-compressed, inter-compressed, and uncompressed bit rates if the average number of bits per pixel after compression are as follows:

- **Intra-frame mode:** 0.2 bpp (assume this is an average value that applies to all color channels).
- **Inter-frame mode:** 0.02 bpp (assume a GOP length of 100 pictures and that this value includes any overheads for motion vectors).

Solution.

(a) Uncompressed

The uncompressed bit rate can be calculated as:

$$512 \times 512 \times 30 \times 24 \simeq 189 \text{ Mb/s}$$

(b) Compressed intra-mode

In intra-mode, all frames are compressed independently to give a bit rate of:

$$512 \times 512 \times 30 \times 0.2 \times 3 \simeq 4.7 \text{ Mb/s}$$

(c) Compressed inter-mode

In inter-mode, 1 in every 100 frames are intra-frames and the other 99 are inter-frames. Hence the bit rate is:

$$512 \times 512 \times 30 \times \left(\frac{0.6 + 99 \times 0.06}{100}\right) \simeq 514 \text{ kb/s}$$

This indicates that we can achieve significant savings (the figures given here are not unrealistic in practice) through the use of inter-frame coding to exploit temporal as well as spatial correlations.

4.2 Formats and aspect ratios

We examined the range of commonly used video formats in Chapter 1, in the context of their raw bit rates and the limitations of available transmission bandwidths. The video industry has come a long way since it started coding QCIF (176×144 luma samples) images for low bit rate conferencing applications around 1989. We are now poised, with the introduction of HEVC, to be able to acquire, process, store, deliver, and display UHDTV formats, initially at a resolution of 3840×2160, but soon rising to 8k (7680×4320) for some applications. The relative sizes of these formats are illustrated in Figure 4.6, where the size drawn is normalized to the viewing angle subtended for a single pixel at the retina. The top subfigure is particularly startling in showing the relative dimensions of QCIF and UHDTV formats.

4.2.1 Aspect ratios

Table 4.1 illustrates some of the more common video format aspect ratios. The choice of aspect ratio has been the subject of much debate; choosing one closest to the ratio of viewing angles of the retina would seem the most obvious choice. These are approximately 95° temporally and roughly 70° above and below the horizontal meridian. It is not difficult to see, however, that this causes a problem—it argues that the aspect ratio should be infinite! So let's have another go—what about the foveal visual field? This approximately circular region of our retina covers 2° and 3° of our visual

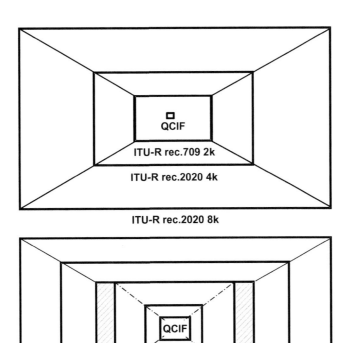

FIGURE 4.6

Aspect ratios of common formats, normalized according to resolution.

Table 4.1 Aspect ratios of film and TV.

Format	Aspect ratio
SDTV	1.33 (4:3)
WSTV/HDTV	1.78 (16:9)
Widescreen 35 mm film	1.85
70 mm film	2.10
Cinemascope anamorphic 35 mm film	2.35
Cinemascope anamorphic 35 mm (1970+)	2.39

field and visual acuity rapidly decreases away from the fovea, typically 20/20 in the fovea and 20/200 at a field angle of 20°. So because this is roughly circular, should we assume that a square aspect ratio is best or that the screen should be circular? The answer is clearly somewhere between 1:1 and infinity and, maybe, we do not need rectangular screens! Some argue that the *golden ratio* of 1.61:1 is the most aesthetically pleasing, although the reason for this is not particularly clear.

A practical solution, certainly for larger and closer screens, is to create the impression that the screen fills our peripheral vision, without completely doing so. This provides a better sense of immersion in the content while not requiring overly wide screens—hence the convergence on typical aspect ratios of 1.77–2.39. However, none of this accounts for the changes in viewing pattern with vary large screens, where we exhibit a greater tendency to track objects with eye and head movements. This introduces all kinds of new problems that we will touch on again in Chapter 13.

In practice the 16:9 format, selected for most consumer formats, is a compromise that optimizes cropping of both film and 4:3 content. The 16:9 ratio (1.78) is very close to the geometric mean of 1.33 and 2.39, so all formats occupy the same image circle. At the end of the day, the best aspect ratio will depend a lot on composition and the relationships between the main subject, the edges of the frame, and the amount of uninteresting space. The art of the cinematographer is to compose the content in a manner that is compelling, regardless of frame aspect ratio.

Figure 4.7 gives an example of some common aspect ratios superimposed on a scene. The preference, of course, is also dictated by the viewing conditions. Within the constraints of comfortable viewing, comparing 4:3 and 16:9 formats based on identical width, height, or area, most subjects prefer the format that provides the greatest viewing area.

For the case of a sensor or display with N_p pixels and an aspect ratio, r, it is straightforward to compute the horizontal and vertical pixel counts, X and Y, as follows:

$$X = \sqrt{r N_p}$$

$$Y = \sqrt{\frac{N_p}{r}} \tag{4.3}$$

| 1.85:1 | 4:3 | 16:9 | 2.39:1 |

FIGURE 4.7

Widescreen formats.

Variation of screen viewing angle with viewing distance

FIGURE 4.8

Variation of field of view with viewing distance (aspect ratio = 16:9 here).

Field of view ratio

Let us consider the viewing angle ratio (or field of view ratio) rather than the aspect ratio. It can easily be shown that this is independent of screen size, s, but will vary non-linearly with viewing distance, d. The commonly adopted TV and computer monitor format, 16:9, is used as an example in Figure 4.8, but other aspect ratios exhibit similar characteristics. This figure plots the viewing angle ratio against the viewing distance in multiples of screen height H. A number of important things can be taken from this graph. Firstly, as d increases, the viewing angle ratio $\theta_w/\theta_h \to r$ where r is the aspect ratio of the screen. Secondly, as θ_w/θ_h decreases below $3H$ there is a rapid fall-off in this ratio down to unity at $d = 0$. For a 16:9 aspect ratio, the viewing distance of $1.5H$ (specified for UHDTV) corresponds to a ratio of approximately 1.65, actually quite close to the golden ratio mentioned above! Example 4.2 examines this in more detail.

Example 4.2 (Field of view ratio)

Derive an expression for the field of view ratio for a screen of size s (diagonal dimension), with an aspect ratio, r, viewed at a distance, d.

Solution. Frequently the dimensions of a screen are given in terms of its diagonal dimension, s. It is a simple matter to calculate the screen height H and width W, using equation (4.4):

$$H = \frac{s}{\sqrt{1+r^2}}; \quad W = \frac{s}{\sqrt{1+1/r^2}} \tag{4.4}$$

The horizontal and vertical viewing angles that correspond to a screen of aspect ratio $r(W \times H)$ viewed at a distance d are then:

$$\theta_w = 2 \arctan \left(\frac{W}{2d} \right) \tag{4.5}$$

$$\theta_h = 2 \arctan \left(\frac{H}{2d} \right) \tag{4.6}$$

and their ratio, defined here as the field of view ratio, Ψ, is thus:

$$\Psi = \frac{\arctan \left(\frac{W}{2d} \right)}{\arctan \left(\frac{H}{2d} \right)} \tag{4.7}$$

This is plotted for a 16:9 aspect ratio screen for various viewing distances in Figure 4.8.

4.2.2 Displaying different formats

The existence of multiple formats (not to mention 3-D, 2-D, etc.) creates significant extra work for content producers, especially film makers. It is quite common for films or older TV programmes to be cropped or distorted in some way and in general this is annoying for the consumer who feels that they are not getting the most out of their screen!

For example, displaying a 2.39:1 format on a 16:9 screen produces a *letter box* appearance. This can be avoided by distorting the original to fit or by zooming, but both have their problems. The other way around, for example a 4:3 format displayed on a 16:9 screen, produces a *pillar box* appearance. Again this can be stretched to produce short and fat subjects or zoomed to lose edge content.

Pan and scan and Active Format Description

Active Format Description (AFD) is a mechanism used in digital broadcasting to avoid the problems described above. This is a means of signaling the active content of a picture to enable intelligent processing of the picture at the receiver, in order to map the format to the display in the best way. Considering, for example, Figure 4.7, if we wish to watch the 2.39:1 version on a 16:9 display, the blue box (representing the 16:9 picture) does not need to stay in the center of the 2.39:1 picture. It could instead move left or right to frame the action in the best way for the smaller screen.

This capability was introduced as a non-normative feature in MPEG-2 on a GOP basis and is also supported by newer standards (e.g. through SEI data in H.264/AVC). AFD codes include information about the location of the active video and about any protected area which must be displayed. The picture edges outside of this area can be discarded without major impact on the viewing experience.

4.3 **Picture scanning**

4.3.1 **Interlaced vs progressive scanning**

Interlaced scanning is an attempt to trade-off bandwidth, flicker, and resolution, by constructing each frame from two consecutive fields. Fields are sampled at different times such that consecutive lines belong to alternate fields; in the US a 60 Hz format is adopted, whereas in Europe a 50 Hz format is employed.

Early analog television systems needed to minimize bandwidth while maintaining flickerless viewing. Interlacing was devised as a compromise between a high temporal update rate (a field rate at 50 or 60 Hz) which reduces flicker for most content, and lower effective bandwidth (due to the 25 or 30 Hz frame rate). The persistence of the phosphor in early CRT displays, coupled with the persistence of human vision, meant that, provided the vertical dimension of a picture is scanned rapidly enough, the HVS does not see the effect of alternating lines and experiences continuous motion. An illustration of the difference between interlaced and progressive scanning is given in Figure 4.9. In modern digital cameras and displays the scanning, as shown in Figure 4.9, is no longer performed by deflecting an electron beam. The row-wise and field-wise scanning pattern is however the same as indicated in the figure.

Notation. The notation we will adopt here is that used by the European Broadcasting Union (EBU) and SMPTE:

$$Y\pi/f$$

where Y = the number of vertical lines in the frame; π = p (progressive) or i (interlaced); and f = frame rate.

For example: 1080p/50 represents an HD progressive format with spatial resolution 1920×1080 and temporal sampling at 50 fps. Similarly 1080i/30 represents the same spatial resolution, but this time interlaced at 30 and 60 fps.

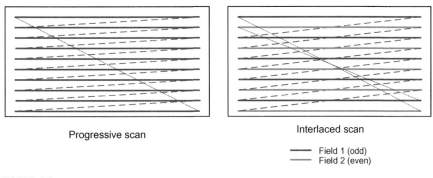

Progressive scan

Interlaced scan

—— Field 1 (odd)
—— Field 2 (even)

FIGURE 4.9

Interlaced vs progressive frame scanning.

FIGURE 4.10

Example of effects of interlaced scanning with poor deinterlacing.

Problems caused by interlacing

The effects of interlacing become apparent in the presence of fast motion as illustrated by the frame shown in Figure 4.10. The inset shows the *combing* artifacts that characterize this format, due to the temporal offset between the two constituent fields. Interlace also introduces a second problem of aliasing, commonly referred to as *twitter*. This occurs where information changes vertically at a rate similar to the field scanning rate. This becomes particularly prevalent when the vertical resolution of the content is similar to the resolution of the format. For example, the clothes of a person wearing fine horizontal stripes will appear to twitter. Filtering is often applied to reduce this effect. An example of twitter can be observed in Ref. [2].

It is finally worth noting that the bandwidth advantages of interlacing associated with analog systems is reduced in the presence of digital video compression, especially if a frame with two fields is processed as one. MPEG-2 and subsequent standards have introduced an adaptive field-frame mode in an attempt to improve this situation.

The artifacts associated with interlacing can be minimized if the receiver and display employ good deinterlacing methods [5]. These employ locally adaptive filtering to the two fields as part of the combining process.

4.3.2 Standards conversion

3:2 pull-down

The need to convert between film and TV formats has been a requirement since the beginnings of public television broadcasting. Film and digital cinema content is traditionally captured at 24 fps using progressive scanning. This needs to be converted to 29.97 fps in the US based on their 60 Hz interlaced scanning and 25 fps in Europe based on 50 Hz. For the case of 25 fps it is normal for the film to be simply speeded up by a factor of 25/24. This produces some audiovisual changes but these are not generally noticeable. For motion to be accurately rendered at 29.97 fps, however, a telecine device must be employed to resample the 24 fps signal. A technique known as 3:2 pull-down is normally employed.

3:2 pull-down uses the approach described in Algorithm 4.1 and it is illustrated in Figure 4.11.

Some streaming sites, such as Netflix, have strict requirements for their content and require an inverse telecine operation to detect and remove 3:2 pull-down from telecine video sources, thereby reconstructing the original 24 frames per second film format. This improves the quality for compatibility with non-interlaced displays and eliminates some of the redundant data prior to compression.

Algorithm 4.1 3:2 pull-down.

---

1. Play the film at 23.976 frames/s;
2. Split each cinema frame into video fields. At 23.976 frames/s, there are four frames of film for every five frames of 60 Hz video: i.e. 23.976/29.97 = 4/5;
3. Convert these four frames to five by alternately placing the first film frame across two fields, the next across three, and so on.

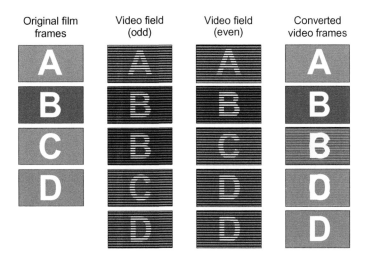

FIGURE 4.11

3:2 pull-down example.

Motion compensated standards conversion

More general standards or format conversion is provided by motion compensation methods such as those based on phase correlation (see Chapter 6). These are required to convert content between interlaced and progressive formats, between HD and SD and other formats, and between formats of different frame rates and color spaces. Phase correlation is a Fourier-based method that exploits duality between displacement in the time (or space) domain and phase shift in the frequency domain. It offers better immunity to noise and luminance changes than conventional block matching methods. Subpixel accuracy can easily be achieved, as is required for the generation of new frames in the target format.

4.4 Gamma correction

Gamma correction [5] is used to correct the differences between the way a camera captures content, the way a display displays content and the way our visual system processes light. Our eyes do not respond to light in the same way that a camera captures it. In basic terms, if twice the number of photons hits a digital image sensor, then the output voltage will be twice as big. For the case of older CRT-based cameras this was not, however, the case and there was a highly non-linear relationship between light intensity and output voltage. The HVS also has a very non-linear response to luminance levels, being more sensitive to small changes in darker areas and much less sensitive in light areas (recall Weber's law from Chapter 2). Without correction, to avoid banding effects due to perceptible jumps between coding levels, around 11 bits

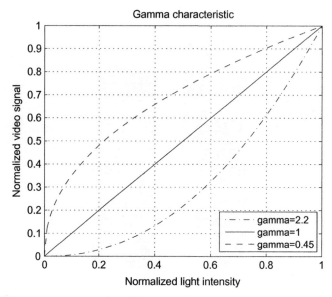

FIGURE 4.12

Gamma curves for $\gamma = \{0.45, 1, 2.2\}$.

would be needed. However, over most of the higher end of the scale, the coding would be highly redundant since coding levels would be imperceptible. With gamma pre-correction we can get an acceptable representation with only 8 bits.

Photopic human vision characteristics can be approximated by a power law function. Correction of the signal output from the camera using this type of function ensures that not too many bits are allocated to brighter regions where humans are less discriminating and more to darker regions, where they are more so. Video cameras therefore normally perform a non-linear mapping of illumination intensity to output value, providing a more uniform perceptual scale with finer increments for darker values. This is known as the Gamma Characteristic of the device as is described by the transfer function:

$$V = c_1 \Phi^\gamma + c_2 \tag{4.8}$$

where Φ is the luminous flux (normalized) and V is the value (voltage) of the signal, c_1 is the camera sensitivity and c_2 is an offset value. Figure 4.12 shows transfer characteristic plots for three gamma values, $\gamma = \{0.45, 1, 2.2\}$. These values are typical of those used in modern TV systems. Figure 4.13 shows some corrected examples of an image at various gamma levels.

FIGURE 4.13

Examples of gamma correction.

It should be noted that modern formats such as Rec.709 use a piecewise transfer characteristic from the luminance to the output. This is linear at the lower region and then changes to the power law for the rest of the range. For Rec.709 this is:

$$V = \begin{cases} 4.5\Phi & \Phi < 0.018 \\ 1.099\Phi^{0.45} - 0.099 & \Phi \geq 0.018 \end{cases} \tag{4.9}$$

Although a topic of some confusion, it is becoming accepted to refer to the gamma-corrected luminance signal as *luma*.

4.5 Color spaces and color transformations

4.5.1 Color descriptions and the HVS

Trichromacy theory

As we have examined in Chapter 2, the cone cells in the retina are the main photore-ceptors for normal photopic conditions and have sensitivities peaking around short (S: 420–440 nm), middle (M: 530–540 nm), and long (L: 560–580 nm) wavelengths. (The low light sensors (rods) peak around 490–495 nm.) The sensation of color in humans is the result of electromagnetic radiation in the range 400–700 nm, incident on the retina.

One of the challenges when color television was introduced was the trade-off between delivering realistic colors to the screen vs conserving bandwidth. In simple terms, the transmission of color information (assuming the use of three color channels (RGB)) occupies three times the bandwidth of a monochrome transmission. The solution to this for analog TV was (inspired by the trichromatic theory of vision (see Chapter 2)) to encode the color information separately from the luma information, with significantly reduced resolution in order to conserve bandwidth. This had the advantage that the luma signal was backwards compatible with older monochrome TV sets while, for color sets, the higher resolution luma and the lower resolution color signals combined in the HVS to produce a perceptually high-resolution color image. As technology has progressed since the introduction of analog color television, there has been substantial effort invested in developing new color space representations appropriate for modern digital encodings, formats and displays. Some of these are reviewed below. For further information, the reader is referred to Ref. [1] and the very good introduction to color spaces presented by Poynton [3,5].

Much of colorimetry is based on the tri stimulus theory of color—i.e. that any color can be formed from three primaries so long as they are orthogonal, but also on Grassman's laws which reflect the linearity and additivity rules of color. These state that the color match between any two stimuli is constant when the intensities of the two stimuli are increased or decreased by the same factor. For example, if we have two sources, C_1 and C_2, then if:

$$\begin{aligned} C_1 &= R_{C1}(R) + G_{C1}(G) + B_{C1}(B) \\ C_2 &= R_{C2}(R) + G_{C2}(G) + B_{C2}(B) \quad \text{and} \\ C_3 &= C_1 + C_2 \end{aligned} \tag{4.10}$$

then:

$$C_3 = [R_{C1} + R_{C2}](R) + [G_{C1} + G_{C2}](G) + [B_{C1} + B_{C2}](B) \qquad (4.11)$$

Color spaces

We saw in Chapter 2 that the CIE has defined the coordinates of color spaces, based on a trichromatic representation, which describes the set of color sensations experienced by the human observer. The first of these, CIE 1931 (developed in 1931), was based on the concept of a standard observer, matching the observations from a large number of (normal) subjects to controlled stimuli. These stimuli were based on the primaries:

$$[\, R_0 \quad G_0 \quad B_0 \,] = [\, 700 \quad 546.1 \quad 435.8 \,] \text{ nm}$$

Color space transformations

According to the basic principles of colorimetry, various other linear combinations of the RGB stimulus values can be derived that might offer better properties than the original values. CIE introduced an alternative reference system (or color space), referred to as XYZ, which could provide a more flexible basis for representing tristimulus values. A color space maps a range of physically produced colors (e.g. from mixed lights) to an objective description of those color sensations based on a trichromatic additive color model, although not usually the LMS space. CIE XYZ can represent all the color sensations that an average person can experience and is a transformation of the $[R_0, G_0, B_0]$ space. The CIE have defined this transform as:

$$\begin{bmatrix} X \\ Y \\ Z \end{bmatrix} = \begin{bmatrix} 2.365 & -0.515 & 0.005 \\ -0.897 & 1.426 & -0.014 \\ -0.468 & 0.089 & 1.009 \end{bmatrix} \begin{bmatrix} R_0 \\ G_0 \\ B_0 \end{bmatrix} \qquad (4.12)$$

CIE defined XYZ such that the Y component corresponds to the luminance signal, normalized to equal energy white. Also, unlike some RGB systems where contributions have to be negative to match a color, the XYZ system is wholly positive. However, the problem with this is that it causes the XYZ values to be complex numbers, i.e. they cannot be physically realized by an actual stimulus.

Chromaticity values can also be defined in the XYZ system as follows:

$$x = \frac{X}{X + Y + Z}$$
$$y = \frac{Y}{X + Y + Z} \qquad (4.13)$$
$$z = \frac{Z}{X + Y + Z}$$

Large x values correspond to red or orange hues, y values correspond to green, blue-green or yellow-green, and large z corresponds to large blue, violet or purple hues. Because $x + y + z = 1$, only two of these values are actually required, so normally this is expressed as an Yxy system.

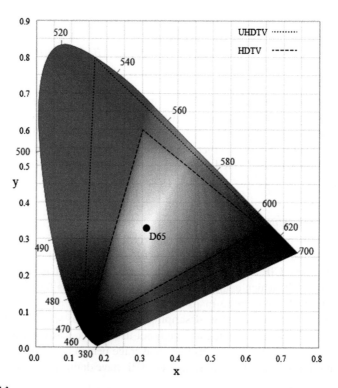

FIGURE 4.14

CIE 1931 chromaticity diagram. (Public domain image from: http://commons.wikimedia.org/wiki/File:CIExy1931_Rec_2020_and_Rec_709.svg.)

Chromaticity diagrams

The chromaticity coordinates can be expressed on a 2-D chromaticity diagram like that shown in Figure 4.14. This diagram shows all the hues perceivable by the standard observer for various (x, y) pairs and indicates the spectral wavelengths of the dominant single frequency colors. The D65 point shown on this diagram is the white point, which in this case corresponds to a color temperature of 6500K (representative of average daylight).

CIE XYZ or CIE 1931 is a standard reference which has been used as the basis for defining most other color spaces. This maps human color perception based on the two CIE parameters x and y. Although revised in 1960 and 1976, CIE 1931 is still the most commonly used reference.

If we consider again the CIE 1931 color space in Figure 4.14, this also shows two triangles that correspond to the colors representable by the Rec.709 (HDTV) and the Rec.2020 (UHDTV) color spaces. It is interesting to note that the latter can represent many colors that cannot be shown with Rec.709. The Rec.2020 primaries correspond to 630 nm (R), 532 nm (G), and 467 nm (B). The Rec.2020 space covers approximately 76% of the CIE 1931 space whereas Rec.709 covers only 36%, thus offering better coverage of more saturated colors.

Color spaces for analog TV

In almost all cases, a format YC_1C_2 is used for analog and digital formats, where Y represents signal luminance. This is done for the two reasons provided above:

- Compatibility with monochrome displays.
- Possible bandwidth reduction on the chrominance signals with no loss of perceptual quality.

For example, in the case of US NTSC systems for color TV, the primaries are defined in terms of CIE xyz coordinates as follows:

$$\mathbf{m} = \begin{matrix} & x & y & z \\ R: & 0.67 & 0.33 & 0.00 \\ G: & 0.21 & 0.71 & 0.08 \\ B: & 0.14 & 0.08 & 0.78 \end{matrix} \tag{4.14}$$

The actual values used in the TV transmission format are those after gamma correction. Consequently, after various normalizations and rotations [1] the resulting color space, defined as YIQ, is given by the following transformation (where R', G', B represent the gamma corrected camera output signals):

$$\begin{bmatrix} Y \\ I \\ Q \end{bmatrix} = \begin{bmatrix} 0.299 & 0.587 & 0.114 \\ 0.596 & -0.274 & -0.322 \\ 0.211 & -0.523 & 0.311 \end{bmatrix} \begin{bmatrix} R' \\ G' \\ B' \end{bmatrix} \tag{4.15}$$

A similar mapping for PAL systems in Europe is given by:

$$\begin{bmatrix} Y \\ U_t \\ V_t \end{bmatrix} = \begin{bmatrix} 0.299 & 0.587 & 0.114 \\ -0.147 & -0.289 & -0.436 \\ 0.615 & -0.515 & -0.100 \end{bmatrix} \begin{bmatrix} R' \\ G' \\ B' \end{bmatrix} \tag{4.16}$$

Color spaces for digital formats

With the introduction of digital formats, starting with SDTV (Rec.601), there was a need to adjust the color space transformation to deal with these digital formats and finite wordlengths. The components again are referenced against the gamma corrected camera outputs and are quantized to 8 bits. The representation for the ITU-R Rec.601 system is known as YC_bC_r and is given by the following transformation from corrected 8-bit RGB signals in the range 0 to 255 (N.B. Studio $R'G'B'$ signals are in the range 17 to 235 and require a modified conversion matrix):

$$\begin{bmatrix} Y \\ C_b \\ C_r \end{bmatrix} = \begin{bmatrix} 0.257 & 0.504 & 0.098 \\ -0.148 & -0.291 & 0.439 \\ 0.439 & -0.368 & -0.071 \end{bmatrix} \begin{bmatrix} R' \\ G' \\ B' \end{bmatrix} + \begin{bmatrix} 16 \\ 128 \\ 128 \end{bmatrix} \tag{4.17}$$

In order to provide coding headroom, the full range of 256 values is not used. Instead a range of 220 levels from 16 to 235 are employed. It should be noted that some authors add a prime to the Y (luma) component above to differentiate it from the luminance signal, because it results after non-linear encoding based on gamma corrected RGB primaries. C_b and C_r are the blue-difference and red-difference chroma components.

Similarly the ITU-R Rec.709 system for HDTV formats is given by:

$$\begin{bmatrix} Y \\ C_b \\ C_r \end{bmatrix} = \begin{bmatrix} 0.183 & 0.614 & 0.062 \\ -0.101 & -0.338 & 0.439 \\ 0.439 & -0.399 & -0.040 \end{bmatrix} \begin{bmatrix} R' \\ G' \\ B' \end{bmatrix} + \begin{bmatrix} 16 \\ 128 \\ 128 \end{bmatrix} \tag{4.18}$$

Consider as an example the image of Thala Beach shown in Figure 4.15. This displays the Y, C_b, and C_r components. Several things can be observed from this—for example, the white clouds are represented as middle values in both chroma components, the blue sky is strong in the blue channel; the sand and browner aspects of the tree show strongly in the red channel. Furthermore, the two chroma components appear fuzzy to the human eye relative to the luma channel and this provides some evidence that we could downsample them without loss of overall perceived detail.

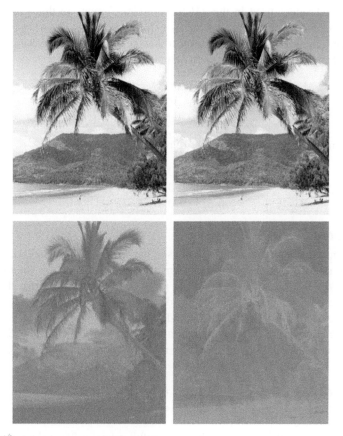

FIGURE 4.15

Color difference images using the YC_bC_r decomposition. Top left: original. Right: luma channel. Bottom left: C_b channel. Right: C_r channel.

4.5.2 **Sub-sampled color spaces**

Most capture devices produce an RGB output, where each component requires high precision and high resolution to provide perceptually high quality images. As discussed above, this format is rarely used for transmission as it requires three times the bandwidth of a monochrome channel. Hence, as we have seen, a color space transformation is applied to produce a luminance signal Y and two color difference signals, C_b and C_r. The motivations for this are:

1. **Chroma sub-sampling:** As we have discussed in Chapter 2 the human visual system has a relatively poor response to color information. The effective resolution of our color vision is about half that of luminance information. Separation of luma and chroma components thus means that we can sub-sample the chroma by a factor of 2 without affecting the luma resolution or the perceived image quality. This provides an immediate compression benefit as we are reducing the bit rate by 50% before we start compression proper!
2. **Luma processing:** As well as having compatibility with monochrome displays, the availability of an independent luma signal is useful during compression. For example, we normally perform motion estimation only on the luma component and use the resulting vectors to compensate both luma and chroma components. We also often base objective quality assessments only on the luma component.

Chroma sub-sampling

The human eye has a relatively poor response to chroma spatial detail. Chroma components are thus normally sub-sampled (usually by a factor of 2) to reduce bandwidth requirements. The most commonly used sub-sampling patterns are shown in Figure 4.16. This shows the relationships between luma and chroma samples for a range of common sub-sampling formats. These are:

- **4:4:4**—This is the full resolution image with luma and chroma samples at all pixel locations. It is used when there is a requirement for the highest quality studio work.
- **4:2:2**—This is the conventional format for studio production. Most professional cameras output video in a 4:2:2 format. It provides sub-sampling horizontally but not vertically. This is advantageous in the case of interlaced scanning as it eliminates problems with color degradation.

FIGURE 4.16

Common chroma sub-sampling formats.

- **4:2:0**—This system is the most common format for broadcast delivery, internet streaming and consumer devices. It sub-samples the chroma signals by a factor of 2 in both horizontal and vertical directions, maximizing the exploitation of perceptual color redundancy.
- **4:1:1**—This system was used in DV format cameras. It sub-samples by a factor of 4 horizontally but not vertically. Its quality is inferior to 4:2:0 formats, so it is rarely used now.
- **4:0:0**—This is not shown in the figure but represents the simple case of a luma only or monochrome image.

The question that arises is how are the chroma signals downsampled? The answer is by pre-processing with a linear phase FIR filter designed to avoid aliasing. Similarly signals are upsampled for reconstruction on an image to display using an appropriate linear phase FIR interpolation filter. It should be noted that the sub-sampling filter will dictate where the chroma samples are sited relative to the luma samples.

Notation 4:X:Y. This notation is confusing to most people and there have been several attempts to fit rules that explain its meaning, all of which fail under some circumstances. The reason for this is that, historically, the 4:X:Y format was devised at the time of PAL and NTSC systems when subcarrier-locked sampling was being considered for component video. The first digit represented the luma horizontal sampling reference as a multiple of $3\frac{3}{8}$ MHz. The second number X represented the horizontal sampling factor for chroma information relative to the first digit. The third number, Y, was similar to X, but in the vertical direction.

It does not take a lot of thinking to realize that the 4:2:0 and 4:1:1 formats do not comply with this definition. So people have started thinking up other rules that fit. This author's view is that we should just use them as labels for the relevant format. However, if you need a rule, then the best one is as follows (not very elegant but it sort of works for most formats):

1. Assume a 4×2 block of pixels, with 4 luma samples in the top row.
2. X is the number of chroma samples in the top row.
3. Y is the number of additional chroma samples in the second row.

An issue of course is that with some 4:2:0 formats (as shown in Figure 4.16), the chroma samples are not actually co-sited with the luma samples.

Example 4.3 (Color space transformations in ITU-R Rec.601)
Consider the YC_bC_r format used with ITU-R Rec.601. Compute the component values for the following cases:

(a) $R'G'B' = [0\,0\,0]$.
(b) $R'G'B' = [220\,0\,0]$.
(c) $R'G'B' = [0\,220\,0]$.

Solution. Using equation (4.17), we can compute these values as follows:

(a) $YC_bC_r = [16\ 128\ 128]$

A signal at the base black level in terms of luminance and at central and equal chroma levels:

(b) $YC_bC_r = [73\ 95\ 224]$

A signal that is higher in the red color difference channel with mid levels of luminance and blue channel:

(c) $YC_bC_r = [127\ 64\ 47]$

Notice the difference between this and (b). The green channel is weighted more heavily as it contributes more to our sense of brightness.

Example 4.4 (Chroma sub-sampling and compression)
If a color movie of 120 min duration is represented using ITU-R.601 (720×576 @30 fps@8 bits, 4:2:0 format):

(a) What hard disk capacity would be required to store the whole movie?
(b) If the movie is MPEG-2 encoded at a compression ratio CR = 50:1 and transmitted over a satellite link with 50% channel coding overhead, what is the total bit rate required for the video signal?

Solution.

(a) The bit rate is given by:

$$720 \times 576 \times 30 \times 8 \times \frac{3}{2} = 149{,}299{,}200 \text{ b/s} = 149 \text{ Mb/s}$$

For a 120 min movie, the total storage space required is:

$$120 \times 60 \times 149{,}299{,}200 = 1{,}074{,}954{,}240{,}000 \text{ bits} = 134 \text{ GB}$$

(b) With a compression ratio of 50:1, the bit rate would be approximately 3 Mbps. With 50% coding overhead, this represents a total bit rate of 4.5 Mbps.

4.5.3 Bayer filtering

Professional cameras often have three CCD or CMOS sensors, one for each color channel. An alternative approach is to use a single sensor in conjunction with a Color Filter Array. The color filter array (often referred to as a Bayer filter or Bayer mosaic) is a pattern of color filters applied to a single image sensor in order to enable it

FIGURE 4.17

Bayer filter mosaic.

to produce R, G, and B samples. This approach is used in most single-chip digital image sensors and increasingly in professional equipment. A Bayer pattern array [4] is shown in Figure 4.17. Half of the filter elements are green and the remainder are split between blue and red. This approximates human photopic vision where the M and L cones combine to produce a bias in the green spectral region.

Obviously the effective sensor resolution of the Bayer filtered array for color signals is reduced compared to a three or four sensor array, and the degree of degradation is largely influenced by the demosaicing algorithm used. As a rule of thumb, according to tests on actual resolutions, you can roughly divide the number of pixels in the array by two, so a 10 Mpixel array effectively becomes a 5 Mpixel array after demosaicing.

There are many classes of demosaicing algorithm and we will not consider them in detail here. These range from relatively simple bilinear interpolation, to bicubic, or spline interpolation. These methods are however prone to artifacts and more sophisticated methods, for example based on the exploitation of local pixel statistics, or gradient estimation, can provide improvements. More advanced methods based on superresolution techniques have also been proposed to address aliasing issues.

4.6 Measuring and comparing picture quality

We introduce some basic topics in image and video quality assessment here. An excellent introduction to video quality metrics and subjective assessment is provided by Winkler [6]. We also discuss these topics in much more detail in Chapter 10.

4.6.1 Compression ratio and bit rate

A very basic measure of compression performance is the compression ratio. This is simply defined as:

$$C_1 = \frac{\text{No. bits in original video}}{\text{No. bits in compressed video}} \quad \text{(unitless)} \quad (4.19)$$

This provides an absolute ratio of the compression performance and only if used in the context of similar content, similar sized images, and for similar target qualities, can it be used as a comparator—otherwise it should not be. A commonly employed alternative, used for still images, is the normalized bit rate:

$$C_2 = \frac{\text{No. bits in compressed video}}{\text{No. pixels in original video}} \quad \text{(bits per pixel)} \quad (4.20)$$

Again this will vary according to content type and image resolution but also in terms of the original signal wordlength. It is just a rate indicator and provides no assessment of quality. A commonly used parameter for video is the actual bit rate:

$$C_3 = \frac{\text{No. bits in compressed video}}{\text{No. frames in original video}} \times \text{frame rate} \quad \text{(bits per second)} \quad (4.21)$$

Again this gives no information on the quality of the compressed signal, but it does provide us with an absolute value of how much channel capacity is taken up by the signal coded in a given manner.

In order for all of these measures to be useful they need to be combined with an associated measure of image or video quality. These are considered below.

4.6.2 Objective distortion and quality metrics

Mean square error (MSE)

Consider an image $\mathbf{S} = s[x, y]$: $x = 0, \ldots, X - 1$; $y = 0, \ldots, Y - 1$ and its reconstructed version $\tilde{\mathbf{S}}$. If the area of the image (or equivalently for a monochrome image, the number of samples in it) is $A = XY$ then the MSE is given by:

$$\text{MSE} = \frac{1}{A} \left(\sum_{x=0}^{X-1} \sum_{y=0}^{Y-1} (s[x, y] - \tilde{s}[x, y])^2 \right) = \frac{1}{A} \left(\sum_{x=0}^{X-1} \sum_{y=0}^{Y-1} (e[x, y])^2 \right) \quad (4.22)$$

A signal to noise ratio is also common in many communications applications:

$$\text{SNR} = 10 \cdot \log \left(\frac{E\left\{ (s[x, y] - \mu_s)^2 \right\}}{E\left\{ (s[x, y] - \tilde{s}[x, y])^2 \right\}} \right) = 20 \cdot \log \left(\frac{\sigma_s}{\sigma_e} \right) \text{ dB} \quad (4.23)$$

If we extend the MSE calculation to a video sequence with K frames then the MSE is given by:

$$\text{MSE} = \frac{1}{KA} \left(\sum_{z=0}^{K-1} \sum_{x=0}^{X-1} \sum_{y=0}^{Y-1} (s[x, y, z] - \tilde{s}[x, y, z])^2 \right) \quad (4.24)$$

Peak signal to noise ratio (PSNR)

Mean Square Error-based metrics are in common use as objective measures of distortion, due mainly to their ease of calculation. An image with a higher MSE will generally express more visible distortions than one with a low MSE. In image and video compression, it is however common practice to use Peak Signal to Noise Ratio (PSNR) rather than MSE to characterize reconstructed image quality. PSNR is of particular use if images are being compared with different dynamic ranges and it is employed for a number of different reasons:

1. MSE values will take on different meanings if the wordlength of the signal samples changes.
2. Unlike many natural signals, the mean of an image or video frame is not normally zero and indeed will vary from frame to frame.
3. PSNR normalizes MSE with respect to the peak signal value rather than signal variance, and in doing so enables direct comparison between the results from different codecs or systems.
4. PSNR can never be less than zero.

The PSNR for our image **S** is given by:

$$PSNR = 10 \cdot \log_{10} \left(\frac{A s_{max}^2}{\sum_{x=0}^{X-1} \sum_{y=0}^{Y-1} (e[x, y])^2} \right) \qquad (4.25)$$

where for a wordlength B: $s_{max} = 2^B - 1$. We can also include a binary mask, $b[x, y]$, so that the PSNR value for a specific arbitrary image region can be calculated. Thus:

$$PSNR = 10 \cdot \log_{10} \left(\frac{s_{max}^2 \sum_{x=0}^{X-1} \sum_{y=0}^{Y-1} b[x, y]}{\sum_{x=0}^{X-1} \sum_{y=0}^{Y-1} (e[x, y])^2} \right) \qquad (4.26)$$

Although there is no perceptual basis for PSNR, it does fit reasonably well with subjective assessments, especially in cases where algorithms are compared that produce similar types of artifact. It remains the most commonly used objective distortion metric, because of its mathematical tractability and because of the lack of any widely accepted alternative. It is worth noting, however, that PSNR can be deceptive, for example when:

1. There is a phase shift in the reconstructed signal. Even tiny phase shifts that the human observer would not notice will produce significant changes.
2. There is visual masking in the coding process that provides perceptually high quality by hiding distortions in regions where they are less noticeable.
3. Errors persist over time. A single small error in a single frame may not be noticeable, yet it could be annoying if it persists over many frames.
4. Comparisons are being made between different coding strategies (e.g. synthesis based vs block transform based (see Chapters 10 and 13)).

Consider the images in Figure 4.18. All of these have the same PSNR value (16.5 dB), but most people would agree that they do not exhibit the same perceptual qualities. In particular, the bottom right image with a small spatial shift is practically

FIGURE 4.18
Quality comparisons for the same PSNR (16.5 dB). Top left to bottom right: original; AWGN (variance 0.24); grid lines; salt and pepper noise; spatial shift by 5 pixels vertically and horizontally.

indistinguishable from the original. In cases such as this, perception-based metrics can offer closer correlation with subjective opinions and these are considered in more detail in Chapter 10.

Example 4.5 (Calculating PSNR)
Consider a 3×3 image **S** of 8-bit values and its approximation after compression and reconstruction, $\tilde{\mathbf{S}}$. Calculate the MSE and the PSNR of the reconstructed block.

1	2	3
4	5	6
7	8	9

S

1	2	2
4	4	8
7	8	8

$\tilde{\mathbf{S}}$

Solution. We can see that $X = 3$, $Y = 3$, $s_{\max} = 255$, and $A = 9$. Using equation (4.22) the MSE is given by:

$$\text{MSE} = \frac{7}{9}$$

and from equation (4.25), it can be seen that the PSNR for this block is given by:

$$\text{PSNR} = 10 \cdot \log_{10}\left(\frac{9 \times 255^2}{7}\right) = 49.2 \text{ dB}$$

PSNR for color images and for video

For color images with three *RGB* values per pixel, PSNR can be calculated as above, except that the MSE is now the sum over all squared value differences divided by image size and by three. Alternatively, the PSNR could be computed separately for each color component. However, because the human eye is not equally sensitive to all channels, it is common to just compute the luma PSNR. This can easily be performed when using a format such as YC_bC_r, where the luma component represents a weighted average of the color channels.

Furthermore, it is important to note that the PSNR calculation for a video signal, based on averaging PSNR results for individual frames, is not the same as computing the average MSE for all frames in the sequence and then computing the PSNR. The former biases algorithms that produce SNR fluctuations, so normally the latter is used.

Typical values for the PSNR in lossy image and video compression are between 30 and 50 dB, where higher is better. Values over 40 dB are normally considered very good and those below 20 dB are normally unacceptable. When the two images are identical, the MSE will be zero, resulting in an infinite PSNR.

Mean absolute difference (MAD)

Mean square error-based metrics, although relatively simple to implement, still require one multiplication and two additions per sample. In order to reduce complexity further, especially in some highly search-intensive operations such as motion estimation, other simpler metrics have been employed. One such metric is Mean Absolute Difference (MAD). The MAD between two images or regions, s_1 and s_2, is defined as:[2]

$$\text{MAD} = \frac{1}{A} \sum_{x=0}^{X-1} \sum_{y=0}^{Y-1} |s_1[x, y] - s_2[x, y]| \tag{4.27}$$

where A is defined as previously.

4.6.3 Subjective assessment

Mathematical, distortion-based metrics such as MSE and PSNR can in some cases be poor indicators of subjective quality. Therefore, it has been necessary to develop perception-based criteria to achieve better (subjectively meaningful) quality assessments. One example of a perception-based quality assessment tool is the Mean Opinion Score (MOS). With MOS measurements, a number of observers view an image

---

[2]*Note:* In search applications it is normal to use Sum of Absolute Difference (SAD), omitting the division operation to reduce complexity.

Table 4.2 ITU-R Rec.BT.500 subjective evaluation impairment scale.

Score	Impairment	Quality rating
5	Imperceptible	Excellent
4	Perceptible but not annoying	Good
3	Slightly annoying	Fair
2	Annoying	Poor
1	Very annoying	Bad

and assess its quality on a five-point scale from bad to excellent. The subjective quality of the image (the MOS) is then characterized by performing statistical analysis on the ratings across a representative number of subjects.

Probably the most common testing procedure still is the Double Stimulus Continuous Quality Scale (DSCQS) approach, which is described in ITU-R Rec.BT.500 [7], although several multimedia quality evaluations will be based on single stimulus tests. DSCQS testing employs a five-point impairment scale as shown in Table 4.2. ITU-R Rec.BT.500 sets out a wide range of testing criteria and methods, covering viewing conditions, choice of test material, characteristics of the observer ensemble, methods for conducting the test, and methods for analyzing the results. Further details are provided in Chapter 10.

The problem with subjective tests is that they are time consuming and can be expensive to run. Hence this provides additional impetus for the development of robust perceptual quality metrics that can be used, with confidence, in their place.

4.7 Rates and distortions

4.7.1 Rate–distortion characteristics

The bit rate for any compressed sequence will depend on:

- The encoding algorithm used (intra- vs inter-frame, integer or subpixel motion estimation, coding modes, or block sizes available).
- The content (high spatio-temporal activity will in general require more bits to code).
- The encoding parameters selected. At the coarsest level this includes things such as spatial resolution and frame rate. At a finer granularity, issues such as quantizer control, intra- vs inter-modes, and block size choices will be key. The difference between an encoding based on good parameter choices vs one based on poor choices is huge.

Following the argument in Section 4.6.1 we need a description of coder performance that captures how it trades rate against distortion. This is achieved using a graph that plots rate vs distortion (or rate vs quality) for a given codec and a given video file. This allows us to compare codec performances and to assess how parameter

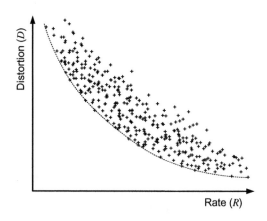

FIGURE 4.19

Rate–distortion plot for various coding parameter choices.

selections within a single codec can influence performance. However, if we randomly select coding parameters, some of these combinations might produce good results and some will inevitably produce poor results. The plot of points might, for many different parameter choices, look something like that in Figure 4.19. The aim is, of course, to ensure that the operating points produced by the codec in practice are those on the dotted curve—the Pareto curve that produces the lowest possible distortion for any given coding rate. To achieve this we must perform Rate–Distortion (RDO) or Rate–Quality Optimization (RQO).

Take for example a simple codec where the only control over RD performance is quantizer step size. In this case, adjusting the quantizer would produce different codec operating points as shown in Figure 4.20.

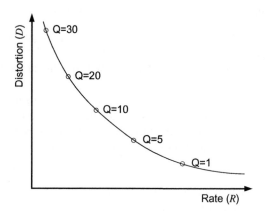

FIGURE 4.20

Rate–distortion plot showing quantizer controlled operating points.

FIGURE 4.21

Example rate–quality comparison curves for typical standard definition entertainment content.

An example Rate–Quality curve (based on PSNR rather than distortion) is shown in Figure 4.21. This compares the performance of two codecs (MPEG-2 and H.264/AVC).

4.7.2 Rate–distortion optimization

Rate–Distortion Optimization (RDO) aims to maximize image or video quality, subject to bit rate constraints. RDO requirements and methods are non-normative in all video coding standards, but nonetheless are a key differentiator in terms of encoder performance. Many of the advanced features (e.g. block size selections) included in modern standards such as H.264/AVC and HEVC will not deliver savings without optimization included as part of the coder control process. The RDO must select the best modes and parameter sets for each region of the video. Often a Lagrangian optimization approach is adopted where the distortion measure, D, is based on sum-of-squared difference and the rate, R, includes all bits associated with the decision including header, motion, side information, and transform data.

For all possible parameter vectors, $\mathbf{p}$, the aim is to solve the constrained problem:

$$\min_{\mathbf{p}} D(\mathbf{p}) \quad \text{s.t.} \quad R(\mathbf{p}) \leq R_T \tag{4.28}$$

where R_T is the target bit rate, or to solve the unconstrained Lagrangian formulation:

$$\mathbf{p}_{opt} = \arg\min_{\mathbf{p}} \{D(\mathbf{p}) + \lambda R(\mathbf{p})\} \tag{4.29}$$

where λ controls the rate–distortion trade-off. Further details on RDO methods can be found in Chapter 10.

4.8 Summary

This chapter has introduced the digital representations, formats, processing techniques, and assessment methods that underpin the coding process. Techniques such as gamma correction and color space conversion have been shown to be essential in providing a digital description of a video signal that is best suited to further compression. Indeed, by exploiting the color processing and perception attributes of the HVS, we have shown that appropriate luma/chroma formats can be designed that reduce the bit rate for a digital video by 50%, even before we apply the conventional compression methods that are described in the following chapters. Finally, we have emphasized that assessment methods are an essential element in video codec design, both as a means of comparing the performance of different compression systems and as a basis for optimizing a codec's rate–distortion performance.

References

[1] A. Netravali, B. Haskell, Digital Pictures: Representation, Compression and Standards, second ed., Plenum Press, 1995.
[2] <http://en.wikipedia.org/wiki/Interlaced_video>, September 2013.
[3] C. Poynton, A Guided Tour of Color Space, August 2013. <http://www.poynton.com/PDFs/Guided_tour.pdf>.
[4] B. Bayer, Color Imaging Array, US Patent 3,971,065, 1976.
[5] C. Poynton, Digital Video and HD, second ed., Morgan Kaufmann, 2012.
[6] S. Winkler, Digital Video Quality, John Wiley, 2005.
[7] Recommendation ITU-R BT.500-13, Methodology for the Subjective Assessment of the Quality of Television Pictures, ITU-R, 2012.

Transforms for Image and Video Coding

CHAPTER OUTLINE

5.1 **The principles of decorrelating transforms** **134**
    5.1.1 The basic building blocks 134
    5.1.2 Principal components and axis rotation 135
5.2 **Unitary transforms** **137**
    5.2.1 Basis functions and linear combinations 137
    5.2.2 Orthogonality and normalization 138
    5.2.3 Extension to 2-D 139
5.3 **Basic transforms** **140**
    5.3.1 The Haar transform 140
    5.3.2 The Walsh–Hadamard transform 141
    5.3.3 So why not use the discrete Fourier transform? 144
5.4 **Optimum transforms** **146**
    5.4.1 Discarding coefficients 146
    5.4.2 The Karhunen–Loeve transform (KLT) 147
    5.4.3 The KLT in practice 148
5.5 **Discrete cosine transform (DCT)** **149**
    5.5.1 Derivation of the DCT 149
    5.5.2 DCT basis functions 151
    5.5.3 Extension to 2-D: separability 153
5.6 **Quantization of DCT coefficients** **155**
5.7 **Performance comparisons** **160**
    5.7.1 DCT vs DFT revisited 160
    5.7.2 Comparison of transforms 160
    5.7.3 Rate–distortion performance of the DCT 161
5.8 **DCT implementation** **163**
    5.8.1 Choice of transform block size 163
    5.8.2 DCT complexity reduction 164
    5.8.3 Field vs frame encoding for interlaced sequences 166
    5.8.4 Integer transforms 167
    5.8.5 DCT demo 167
5.9 **JPEG** **167**

Communicating Pictures. http://dx.doi.org/10.1016/B978-0-12-405906-1.00005-2

5.10 Summary **169**

References **169**

As we saw in Chapter 3, most natural scenes captured using regular sampling will exhibit high levels of inter-pixel correlation. This means that the actual entropy of the pixels in a region of an image is likely to be significantly lower than that given by the first order entropy or, in other words, that the spatial representation is redundant. If our aim is to compress the image, it is therefore likely that an alternative representation may be available that is more amenable to reducing redundancy. The aim of image transformation is to create such a representation by decorrelating the pixels in an image or image region. This is achieved by mapping from the image domain to an alternative domain where the pixel energy is redistributed and concentrated into a small number of coefficients. This process is sometimes referred to as energy compaction.

As we will see, this redistribution of energy does not result in data compression as the transform itself is a lossless operation. What we can achieve, however, is a concentration of energy into a small number of high-valued coefficients. These larger coefficients are likely to have a higher psychovisual significance than their low-valued counterparts and can be coded accordingly.

This chapter firstly introduces the principles and properties of decorrelating transforms and explains their relationship to principle component analysis and eigen-analysis. We introduce the optimum Karhunen–Loeve Transform (KLT) in Section 5.4.2 and, after discussing its limitations, focus the remainder of the chapter on deriving and characterizing the discrete cosine transform (DCT). The DCT is introduced in Section 5.5 and its extension to 2-D in Section 5.5.3. Quantization of DCT coefficients is presented in Section 5.6 and performance comparisons are provided in Section 5.7. We conclude with a discussion of DCT implementation and complexity in Section 5.8 and with a brief overview of JPEG in Section 5.9.

5.1   The principles of decorrelating transforms

5.1.1   The basic building blocks

Transformation presents a convenient basis for compression and this comes about through three mechanisms:

1. It provides data decorrelation and creates a frequency-related distribution of energy allowing low energy coefficients to be discarded.
2. Retained coefficients can be quantized, using a scalar quantizer, according to their perceptual importance.
3. The sparse matrix of all remaining quantized coefficients exhibits symbol redundancy which can be exploited using variable length coding.

The components in a typical transform-based image coding system are shown in Figure 5.1. As we will explain later, for the purposes of transform coding an input

For colour versions of Figures 5.14, 5.15 and 5.16 please refer to the electronic version or the website.

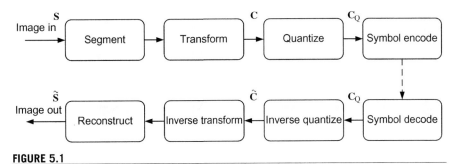

FIGURE 5.1

Typical transform-based image compression architecture.

image is normally segmented into small $N \times N$ blocks where the value of N is chosen to provide a compromise between complexity and decorrelation performance. The components shown in this figure perform the following functions:

1. **Segmentation:** This divides the image into $N \times N$ blocks—where typically $N = 8$. This is a reversible one-to-one mapping.
2. **Transformation (map or decorrelate):** This transforms the raw input data into a representation more amenable to compression. Again this is normally a reversible one-to-one mapping.
3. **Quantization:** This reduces the dynamic range of the transformed output, according to a fidelity and/or bit rate criterion, to reduce psychovisual redundancy. For correlated spatial data, the resulting block of coefficients will be sparse. This is a many-to-one mapping and is not reversible, hence once quantized, the original signal cannot be perfectly reconstructed. This is thus the basis of lossy compression.
4. **Symbol encoding (codeword assigner):** The sparsity of the quantized coefficient matrix is exploited (typically by run-length coding) to produce a string of symbols. The symbol encoder assigns a codeword (a binary string) to each symbol. The code is designed to reduce coding redundancy and it normally uses variable length codewords. This operation is reversible.

5.1.2 Principal components and axis rotation

The purpose of transformation in the context of data compression is energy compaction. This can be achieved through identification of trends in the data and then modifying the principal axes to optimally decorrelate it. It is best explained through a simple example. Consider the correlation between any two adjacent pixels in a natural image and let us represent the amplitudes of these pixels by $\{x_0, x_1\}$. A scatter plot of such adjacent pixels from a typical correlated image might look like that shown in Figure 5.2.

In Figure 5.2 (left) we observe a strong correlation characterized by the relationship $\hat{x}_1 = x_0$, as would be expected from a natural image. We can exploit this correlation by introducing a new set of axes, c_0 and c_1, as shown in Figure 5.2 which are oriented such that c_0 aligns with the principal axes of the data. Using the annotations in Figure 5.3

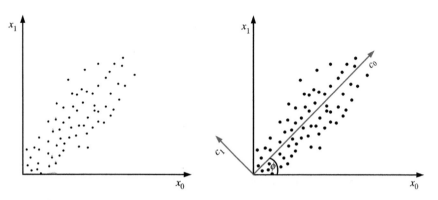

FIGURE 5.2

Plot of correlated adjacent pixel data (left) and decorrelation through rotation of principal axes (right).

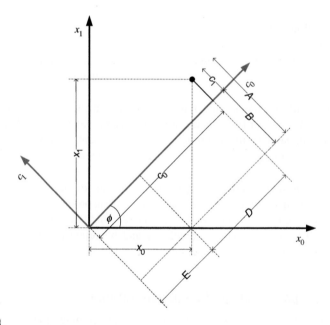

FIGURE 5.3

Relationship between original and transformed data.

we can deduce that:

$$A = x_1 \cos \phi$$
$$B = x_0 \sin \phi$$
$$D = x_1 \sin \phi$$
$$E = x_0 \cos \phi$$

Furthermore:

$$c_0 = D + E = x_0 \cos \phi + x_1 \sin \phi$$
$$c_1 = A - B = -x_0 \sin \phi + x_1 \cos \phi$$

Rearranging gives:

$$\begin{bmatrix} c_0 \\ c_1 \end{bmatrix} = \begin{bmatrix} \cos \phi & \sin \phi \\ -\sin \phi & \cos \phi \end{bmatrix} \begin{bmatrix} x_0 \\ x_1 \end{bmatrix} \tag{5.1}$$

or in matrix vector form:

$$\mathbf{c} = \mathbf{A}\mathbf{x}$$

For correlated pixels at $\phi = 45°$:

$$\mathbf{A} = \frac{1}{\sqrt{2}} \begin{bmatrix} 1 & 1 \\ -1 & 1 \end{bmatrix} \tag{5.2}$$

As we will see in Section 5.3.2, this bears a striking resemblance to the two-point Discrete Walsh Hadamard Transform. We will also see in Section 5.4.2 that this process is equivalent to extracting the eigenvectors for this data set, i.e. $(1/\sqrt{2}, 1/\sqrt{2})(-1/\sqrt{2}, 1/\sqrt{2})$.

From equation (5.2), we can see that the row vectors for this simple transform are: $\frac{1}{\sqrt{2}}[1 \quad 1]$ and $\frac{1}{\sqrt{2}}[-1 \quad 1]$. These are effectively sum and difference operations respectively, and can be interpreted as basic low- and high-pass filters. For example, in the high-pass case:

$$H(z) = \frac{1}{\sqrt{2}}\left(1 - z^{-1}\right) \tag{5.3}$$

which has a pole at $z = 0$ and a zero at $z = 1$. This alternative view of the transform will stand us in good stead when we consider wavelets and subband filters in Chapter 6.

Let us reflect on what we have achieved in this section. Firstly, we note that the operation of rotating the axes effectively decorrelates the data, with many of the c_1 values being very small. In practice, we may set small values to zero leaving only a sparse set of coefficients which embody and preserve the main attributes of the original data set. These can then be quantized to reduce their dynamic range and coded for transmission or storage. They can then be decoded, rescaled, and inverse transformed to yield an approximation to the original data which is optimum in the mean square error sense. This is the essence of transform-based data compression as depicted in Figure 5.1.

5.2 Unitary transforms

5.2.1 Basis functions and linear combinations

We know from our introduction to Fourier analysis in Chapter 3 that we can approximate a signal using a linear combination of basis functions, such as harmonically related sines and cosines or complex exponentials. Let us assume we have N such basis functions and indicate the contribution of each basis function $\mathbf{b}_k$ that is needed to approximate the block of data, by a value $c(k)$ where k represents the index of the basis function. Considering a one-dimensional $N \times 1$ block of data $\mathbf{x}$, we can write this relationship in matrix–vector form as follows:

$$
\begin{bmatrix} x[0] \\ x[1] \\ \vdots \\ x[N-1] \end{bmatrix} = \begin{bmatrix} b_{0,0} & b_{1,0} & \cdots & b_{N-1,0} \\ b_{0,1} & b_{1,1} & \cdots & b_{N-1,1} \\ \vdots & \vdots & \ddots & \vdots \\ b_{0,N-1} & b_{1,N-1} & \cdots & b_{N-1,N-1} \end{bmatrix} \begin{bmatrix} c(0) \\ c(1) \\ \vdots \\ c(N-1) \end{bmatrix} \tag{5.4}
$$

However, if we are given an appropriate set of basis functions and some input data, then what we usually want to compute are the coefficient values or weights $c(k)$ as it is these that we will be processing, storing, and transmitting in the context of image compression. So let us rewrite equation (5.4) as follows:

$$
\begin{bmatrix} c(0) \\ c(1) \\ \vdots \\ c(N-1) \end{bmatrix} = \begin{bmatrix} a_{0,0} & a_{0,1} & \cdots & a_{0,N-1} \\ a_{1,0} & a_{1,1} & \cdots & a_{1,N-1} \\ \vdots & \vdots & \ddots & \vdots \\ a_{N-1,0} & a_{N-1,1} & \cdots & a_{N-1,N-1} \end{bmatrix} \begin{bmatrix} x[0] \\ x[1] \\ \vdots \\ x[N-1] \end{bmatrix}
$$

$$
c(k) = \sum_{i=0}^{N-1} x[i] a_{k,i}; \quad k = 0 \cdots N - 1
$$

or

$$
\mathbf{c} = \mathbf{A}\mathbf{x} \tag{5.5}
$$

We have already defined the original signal in terms of our coefficients as:

$$
\mathbf{x} = \mathbf{B}\mathbf{c}
$$

But from equation (5.5) we can also see that the original signal can be reconstructed using the inverse transform:

$$
\mathbf{x} = \mathbf{A}^{-1}\mathbf{c} \tag{5.6}
$$

Hence $\mathbf{B} = \mathbf{A}^{-1}$. $\mathbf{A}$ is an $N \times N$ matrix referred to as the transform matrix and $\mathbf{c}$ is an $N \times 1$ vector of transform coefficients. The columns of the matrix $\mathbf{A}^{-1}$ are the basis functions of the transform. It can thus be observed that the signal $\mathbf{x}$ is represented by a linear combination of weighted basis functions where the weights are given by the coefficients in the column vector $\mathbf{c}$.

5.2.2 Orthogonality and normalization

Orthogonality ensures that the basis functions of a transform are independent and hence that they provide decorrelation of the input vector. Basis functions $\mathbf{a}_i$ and $\mathbf{a}_j$ are defined as orthogonal if their inner product is zero. That is:

$$
\mathbf{a}_i^{\mathrm{H}} \mathbf{a}_j = 0; \quad \forall i, j \ (i \neq j) \tag{5.7}
$$

Orthonormality imposes further constraints on the norm of the basis vectors in that they must have unit length, i.e. $\|\mathbf{a}_i\| = 1$. Orthonormality is useful when we wish to use the same operations for forward and inverse transformation as it preserves energy between domains, i.e.:

$$\sum_{k=0}^{N-1} c^2(k) = \sum_{i=0}^{N-1} x^2[i] \tag{5.8}$$

Furthermore, as shown above, it ensures that the forward and inverse basis function matrices are related through transposition:

$$\mathbf{a}_i^H \mathbf{a}_j = \begin{cases} 1; & i = j \\ 0; & i \neq j \end{cases} \tag{5.9}$$

If the transform is orthonormal (or unitary) then it has a unique and computable inverse given by:

$$\mathbf{A}^{-1} = \mathbf{A}^H \tag{5.10}$$

and hence $\mathbf{A}^H \mathbf{A} = \mathbf{I}$. Clearly if the $\mathbf{A}$ matrix is real valued then:

$$\mathbf{A}^{-1} = \mathbf{A}^T \tag{5.11}$$

and the inverse transform then becomes:

$$\mathbf{x} = \mathbf{A}^T \mathbf{c} \tag{5.12}$$

In this special case, the rows of the transform matrix are the basis functions. Unitary transforms have some useful properties:

1. The autocorrelation (or covariance for non-zero mean) matrices in the signal and transform domains are related by:

$$\mathbf{R}_c = \mathbf{A} \mathbf{R}_x \mathbf{A}^H \tag{5.13}$$

2. The total energy of the original and transformed vectors are equal (see Section 5.4.1).
3. If we select a subset of k transform coefficients, then the approximation error for a particular vector is minimized if we choose the k largest coefficients.

5.2.3 Extension to 2-D

In the case of a two-dimensional signal, let us assume that the transform is applied to an $N \times N$ block of real data $\mathbf{X}$. Assuming a separable transform, as discussed in Chapter 3 and Section 5.5.3:

$$\mathbf{C} = \mathbf{A} \mathbf{X} \mathbf{A}^T \tag{5.14}$$

and

$$\mathbf{X} = \mathbf{A}^T \mathbf{C} \mathbf{A} \tag{5.15}$$

Note here that equation (5.14) requires two matrix multiplications of size $N \times N$ instead of (in the non-separable case) one multiplication with a matrix of size $N^2 \times N^2$. As we will see later, the existence and exploitation of separability is important in reducing the complexity of forward and inverse transforms.

It follows that the basis functions of a 2-D transform are themselves 2-D functions, and the 2-D transform can be interpreted as expansions in terms of matrices that are obtained from the outer product of the individual 1-D basis functions, $\mathbf{a}_j$. So if the 2-D basis functions are denoted as $\boldsymbol{\alpha}_{i,j}$, and the 1-D basis functions are real-valued column vectors, then the 2-D functions are formed from the outer product of $\mathbf{a}_i$ and $\mathbf{a}_j$:

$$\boldsymbol{\alpha}_{i,j} = \mathbf{a}_i \mathbf{a}_j^{\mathsf{T}} \tag{5.16}$$

$$\boldsymbol{\alpha}_{i,j} = \begin{bmatrix} a_{i,0}a_{j,0} & a_{i,0}a_{j,1} & \cdots & a_{i,0}a_{j,N-1} \\ a_{i,1}a_{j,0} & a_{i,1}a_{j,1} & \cdots & a_{i,1}a_{j,N-1} \\ \vdots & \vdots & \ddots & \cdots \\ a_{i,N-1}a_{j,0} & a_{i,N-1}a_{j,1} & \cdots & a_{i,N-1}a_{j,N-1} \end{bmatrix} \tag{5.17}$$

Example 5.1 (Orthonormality of basis functions)
Consider the simple transform matrix from Section 5.1.2:

$$\mathbf{A} = \frac{1}{\sqrt{2}} \begin{bmatrix} 1 & 1 \\ -1 & 1 \end{bmatrix}$$

Show that this is a unitary transform.

Solution. The basis functions are orthogonal because:

$$\frac{1}{2} \begin{bmatrix} 1 & 1 \end{bmatrix} \begin{bmatrix} -1 \\ 1 \end{bmatrix} = 0$$

and all basis functions have unit length since:

$$\|\mathbf{a}_i\| = \frac{1}{\sqrt{2}} \sqrt{(1)^2 + (1)^2} = 1$$

5.3 Basic transforms

5.3.1 The Haar transform

Consider the simple Haar transform:

$$\mathbf{H}_4 = \frac{1}{2} \begin{bmatrix} 1 & 1 & 1 & 1 \\ 1 & 1 & -1 & -1 \\ \sqrt{2} & -\sqrt{2} & 0 & 0 \\ 0 & 0 & \sqrt{2} & -\sqrt{2} \end{bmatrix} \tag{5.18}$$

The first basis function creates a running sum of the input data, the second creates a difference between the first two and the second two data samples, the third creates a difference between the first two data points and similarly the basis function in the bottom row does the same for the last two data points.

Example 5.2 (Energy compaction and the Haar transform)
Consider the input vector:

$$\mathbf{x} = \begin{bmatrix} 1.0 & 0.5 & -0.5 & -1.0 \end{bmatrix}^{\mathrm{T}}$$

Show that the application of the Haar transform provides energy compaction.

Solution. A reduced number of non-zero coefficients results from the Haar transform, together with a modest amount of energy compaction, thus:

$$\mathbf{c} = \frac{1}{2} \begin{bmatrix} 0 & 3 & \frac{\sqrt{2}}{2} & \frac{\sqrt{2}}{2} \end{bmatrix}^{\mathrm{T}}$$

As we will see later, other transforms exist that can provide much greater energy compaction and decorrelation.

5.3.2 The Walsh–Hadamard transform

The Walsh and Walsh–Hadamard transforms are simple but effective ways of compressing data. They have the significant advantage that the basic transform requires no multiplications, only sums and differences. While their coding gain is lower than transforms such as the DCT which we will consider shortly, they do find application in modern video compression systems, such as H.264/AVC, where they are used in intra-frame coding.

The discrete Walsh–Hadamard transform (DWHT) is obtained from a simple rearrangement of the discrete Hadamard matrix. The Hadamard matrix is an $N \times N$ matrix with the property $\mathbf{HH}^{\mathrm{T}} = N\mathbf{I}$. Higher order matrices can be found by iteratively applying the following operation:

$$\mathbf{H}_{2N} = \begin{bmatrix} \mathbf{H}_N & \mathbf{H}_N \\ \mathbf{H}_N & -\mathbf{H}_N \end{bmatrix} \tag{5.19}$$

For example:

$$\mathbf{H}_1 = 1; \quad \mathbf{H}_2 = \begin{bmatrix} 1 & 1 \\ 1 & -1 \end{bmatrix}; \quad \mathbf{H}_4 = \begin{bmatrix} 1 & 1 & 1 & 1 \\ 1 & -1 & 1 & -1 \\ 1 & 1 & -1 & -1 \\ 1 & -1 & -1 & 1 \end{bmatrix} \tag{5.20}$$

The DWHT is simply obtained from the corresponding Hadamard matrix by normalization and rearranging the rows in sequence order (i.e. in terms of the number of

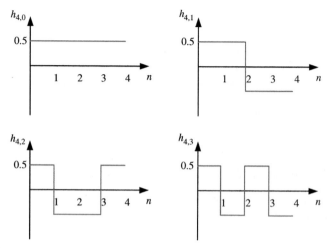

FIGURE 5.4

DWHT basis functions for $N = 4$.

sign changes). Therefore the four-point DWHT is given by:

$$\mathbf{H}_4 = \frac{1}{2} \begin{bmatrix} 1 & 1 & 1 & 1 \\ 1 & 1 & -1 & -1 \\ 1 & -1 & -1 & 1 \\ 1 & -1 & 1 & -1 \end{bmatrix} \tag{5.21}$$

The basis functions for the 1-D DWHT are depicted in Figure 5.4.

In the case of the 2-D transform, the basis functions are again formed from the outer product of the individual 1-D basis functions, $\mathbf{h}_j$. The basis functions for the 2-D 4×4 DWHT are shown in Figure 5.5.

Example 5.3 (1-D DWHT)

Compute the 1-D DWHT for the data vector $\mathbf{x} = \begin{bmatrix} 5 & 6 & 4 & 8 \end{bmatrix}^{\mathrm{T}}$:

$$\begin{bmatrix} c(0) \\ c(1) \\ c(2) \\ c(3) \end{bmatrix} = \frac{1}{2} \begin{bmatrix} 1 & 1 & 1 & 1 \\ 1 & 1 & -1 & -1 \\ 1 & -1 & -1 & 1 \\ 1 & -1 & 1 & -1 \end{bmatrix} \begin{bmatrix} 5 \\ 6 \\ 4 \\ 8 \end{bmatrix}$$

Solution.

$$\mathbf{c} = \begin{bmatrix} 11.5 & -0.5 & 1.5 & -2.5 \end{bmatrix}^{\mathrm{T}}$$

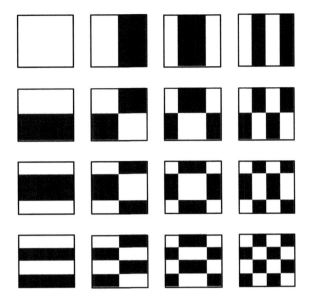

FIGURE 5.5

Basis functions for the 2-D DWHT.

Example 5.4 (2-D DWHT)

Compute the 2-D DWHT of the input image block **S** below:

$$\mathbf{S} = \begin{bmatrix} 5 & 6 & 8 & 10 \\ 6 & 6 & 5 & 7 \\ 4 & 5 & 3 & 6 \\ 8 & 7 & 5 & 5 \end{bmatrix}$$

Solution. Since the 2-D DWHT is separable, we can use equation (5.14). Letting $\mathbf{C} = \mathbf{C}'\mathbf{H}^{\mathrm{T}}$ where $\mathbf{C}' = \mathbf{H}\mathbf{S}$:

$$\mathbf{C}' = \frac{1}{2}\begin{bmatrix} 1 & 1 & 1 & 1 \\ 1 & 1 & -1 & -1 \\ 1 & -1 & -1 & 1 \\ 1 & -1 & 1 & -1 \end{bmatrix}\begin{bmatrix} 5 & 6 & 8 & 10 \\ 6 & 6 & 5 & 7 \\ 4 & 5 & 3 & 6 \\ 8 & 7 & 5 & 5 \end{bmatrix} = \begin{bmatrix} 11.5 & 12 & 10.5 & 14 \\ -0.5 & 0 & 2.5 & 3 \\ 1.5 & 1 & 2.5 & 1 \\ -2.5 & -1 & 0.5 & 2 \end{bmatrix}$$

Then

$$\mathbf{C} = \begin{bmatrix} 24 & -0.5 & 1.5 & 2 \\ 2.5 & 3 & 0 & -0.5 \\ 3 & -0.5 & -0.5 & 1 \\ -0.5 & -3 & 0 & -1.5 \end{bmatrix}$$

The energy compaction and decorrelation properties of the DWHT can clearly be seen in this simple example. Recall that, in the case of the original image block, the energy was distributed fairly uniformly across the block. After transformation, the data has been decorrelated horizontally and vertically and one dominant coefficient (top left) now contains some 93% of the energy.

Example 5.5 (2-D DWHT compression)

Taking the result of Example 5.4, set all small-valued coefficients (when $|c_{i,j}| \leq 2$) to zero and perform inverse transformation. What is the PSNR of the reconstruction if the original signal was represented with a 6 bit word length?

Solution. Now

$$
\tilde{\mathbf{C}} = \begin{bmatrix} 24 & 0 & 0 & 0 \\ 2.5 & 3 & 0 & 0 \\ 3 & 0 & 0 & 0 \\ 0 & -3 & 0 & 0 \end{bmatrix}
$$

and

$$
\tilde{\mathbf{S}} = \mathbf{H}^T \tilde{\mathbf{C}} \mathbf{H}
$$

so

$$
\tilde{\mathbf{S}} = \begin{bmatrix} 7 & 7 & 7 & 7 \\ 7 & 7 & 4 & 4 \\ 3 & 3 & 6 & 6 \\ 6 & 6 & 6 & 6 \end{bmatrix}
$$

Assuming a 6 bit wordlength, the MSE = 3 and the PSNR of the reconstruction is:

$$
\text{PSNR} = 10 \log_{10} \left[\frac{16(2^6 - 1)^2}{\sum_{\forall(i,j)} (s[i,j] - \tilde{s}[i,j])^2} \right] = 31.2 \text{ dB}
$$

5.3.3 So why not use the discrete Fourier transform?

As a starting point for selecting a transform that decorrelates data we could do worse than consider the familiar DFT. After all, we know that the DFT is very good at representing sinusoidal data with very few coefficients. However, things are not quite that simple.

With the DFT, a windowed finite length data sequence is naturally extended by periodic extension prior to transformation. Applying the Discrete Time Fourier Series (DTFS) to this new periodic function produces the DFT coefficients. The first problem with this is that it produces a set of complex coefficients. Although this is quite useful for capturing the magnitude and phase of the underlying signal in the frequency domain, it is not particularly useful for compression since (for the case of a real input signal) we are doubling the amount of information!

Putting the complex nature of the output to one side for the moment, periodic extension can be acceptable for long stationary sequences or for shorter sequences where the extension process does not introduce discontinuities. However, for shorter

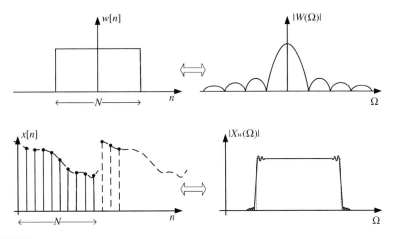

FIGURE 5.6

Spectral leakage due to windowing and periodic extension with the DFT: the rectangular window function and its magnitude spectrum (top); the input signal showing periodic extension and its spectrum (bottom).

sequences like those typical in compression, discontinuities are common and can have a major impact on performance as they produce ringing or spectral leakage in the frequency domain. This is exactly what we don't want in a compression system as it introduces more energy in the coefficient space, requiring more bits to code.

The effect of the rectangular window on the sampling and transformation processes is illustrated in Figure 5.6. Recall that multiplication of the input signal by the window function in the time (or spatial) domain is equivalent to the convolution of their spectra in the frequency domain, i.e.:

$$w[n]x[n] \Longleftrightarrow W(\Omega) * X(\Omega) \tag{5.22}$$

The top two subfigures show a length N window function, alongside its spectrum, while the lower two subfigures show a sampled and windowed input signal alongside its spectrum. Since the DFT is effectively the DTFS of the periodic signal obtained by concatenating an infinite number of windowed signal samples, discontinuities exist at the window boundaries as shown in Figure 5.6. These cause spectral leakage or ripples in the frequency domain. The effects of windowing on the resulting DFT coefficients can be clearly seen.

Ringing is normally addressed in spectral analysis applications by applying a non-rectangular window function (such as a Hamming window) to the input data prior to extension. Such windows exhibit reduced frequency domain sidelobes than the conventional rectangular window, but also have a wider central lobe. Hence they reduce spectral ringing but smear sharp frequency domain edges, distorting the characteristics of the underlying signal. Again, this is exactly what we don't want in a compression system!

5.4　Optimum transforms

5.4.1　Discarding coefficients

Let us assume that we want to define a transform based on a set of basis functions that pack the most energy into the fewest transform coefficients. If we then approximate our signal by a limited number (M) of coefficients, setting the remaining coefficients to constant value, k_i, firstly we ask—What should the value of k be? Following the approach of Clarke [1]:

$$\tilde{\mathbf{x}}_M = \sum_{i=0}^{M-1} c(i)\mathbf{a}_i + \sum_{i=M}^{N-1} k_i \mathbf{a}_i$$

and the error which results from this can be computed as follows:

$$\mathbf{e}_M = \mathbf{x}_M - \tilde{\mathbf{x}}_M = \sum_{i=M}^{N-1} (c(i) - k_i)\mathbf{a}_i$$

We now wish to compute the energy in this residual signal and attempt to minimize it:

$$E\left\{e_M^2\right\} = E\left\{(c(M) - k_M)^2 \mathbf{a}_M^T \mathbf{a}_M + (c(M) - k_M)(c(M+1) - k_{M+1})\mathbf{a}_M^T \mathbf{a}_{M+1} \right.$$
$$\left. + \cdots + (c(N-1) - k_{N-1})(c(N-1) - k_{N-1})\mathbf{a}_{N-1}^T \mathbf{a}_{N-1}\right\}$$

Assuming that the transform is orthonormal (and noting that the basis functions are column vectors here):

$$\mathbf{a}_i^T \mathbf{a}_j = \begin{cases} 1; & i = j \\ 0; & i \neq j \end{cases}$$

Hence the equation for error energy reduces to:

$$E\left[e_M^2\right] = E\left[\sum_{i=M}^{N-1} (c(i) - k_i)^2\right] \tag{5.23}$$

If we minimize this by partial differentiation with respect to k and set the result equal to 0, we can determine the optimal value of k:

$$\frac{\partial}{\partial k_i} E\left\{e_M^2\right\} = \frac{\partial}{\partial k_i} E\left\{\sum_{i=M}^{N-1} (c(i) - k_i)^2\right\} = E\left\{-2(c(i) - k_i)\right\} = 0 \text{ at minimum}$$

Hence at the minimum:

$$k_i = E\{c(i)\} = \mathbf{a}_i^T E\{\mathbf{x}\}$$

If we assume a zero mean sequence, the solution becomes:

$$k_i = 0 \ (i = M, \ldots, N-1)$$

and hence our best approximation is:

$$\tilde{\mathbf{x}}_M = \sum_{i=0}^{M-1} c(i)\mathbf{a}_i \tag{5.24}$$

Now we can compute the basis functions which minimize the error energy in equation (5.23). So:

$$E\left\{e_M^2\right\} = E\left\{\sum_{i=M}^{N-1}(c(i))^2\right\} \tag{5.25}$$

5.4.2 The Karhunen–Loeve transform (KLT)

Also known as the Hotelling or Eigenvector transform, the basis vectors of the KLT transformation matrix, **A**, are obtained from the eigenvectors of the signal autocorrelation matrix (note the similarity to Principal Component Analysis (PCA)). Assuming the signal is derived from a random process, the resulting transform coefficients will be uncorrelated. The KLT provides the best energy compaction of any transform and yields optimal performance in the statistical (mean square error) sense for a Gaussian source, by minimizing the geometric mean of the variance of the transform coefficients. For a signal **x** (biased to remove the mean for convenience), the basis functions $\mathbf{a}_k$ can be shown to be eigenvectors of the signal autocorrelation matrix, thus:

$$\mathbf{R}_x \mathbf{a}_k = \lambda_k \mathbf{a}_k \tag{5.26}$$

where λ_k are the corresponding eigenvalues. Consider a signal $\mathbf{x} = \{x[0], x[1], \ldots, x[N-1]\}$. Assuming that the signal is a stationary random zero-mean sequence, the autocorrelation matrix for this block of pixels is given by:

$$\mathbf{R}_x = \begin{bmatrix} r_{xx}(0) & r_{xx}(1) & \cdots & r_{xx}(N-1) \\ r_{xx}(1) & r_{xx}(0) & \cdots & r_{xx}(N-2) \\ \vdots & \vdots & \ddots & \vdots \\ r_{xx}(N-1) & r_{xx}(N-2) & \cdots & r_{xx}(0) \end{bmatrix} \tag{5.27}$$

where

$$r_{xx}(k) = E\{x[n+k]x[n]\} = r_{xx}(-k)$$

As stated above, the KLT is the unitary transform that uses the eigenvectors from equation (5.26) as its basis vectors. Replacing the basis function matrix **A** in equation (5.13) with the matrix of eigenvectors, Ø, we obtain:

$$\mathbf{R}_c = \text{Ø}^H \mathbf{R}_x \text{Ø} \tag{5.28}$$

$$= \begin{bmatrix} \boldsymbol{\phi}_1^H \\ \boldsymbol{\phi}_2^H \\ \vdots \\ \boldsymbol{\phi}_N^H \end{bmatrix} \mathbf{R}_x \begin{bmatrix} \boldsymbol{\phi}_1 & \boldsymbol{\phi}_2 & \cdots & \boldsymbol{\phi}_N \end{bmatrix}$$

but from equation (5.26) this can be rewritten as:

$$\mathbf{R}_c = \begin{bmatrix} \boldsymbol{\phi}_1^H \\ \boldsymbol{\phi}_2^H \\ \vdots \\ \boldsymbol{\phi}_N^H \end{bmatrix} \mathbf{R}_x \begin{bmatrix} \lambda_1\boldsymbol{\phi}_1 & \lambda_2\boldsymbol{\phi}_2 & \cdots & \lambda_N\boldsymbol{\phi}_N \end{bmatrix} = \begin{bmatrix} \lambda_1 & 0 & \cdots & 0 \\ 0 & \lambda_2 & 0 & 0 \\ \vdots & \vdots & \vdots & \vdots \\ 0 & 0 & 0 & \lambda_N \end{bmatrix} \quad (5.29)$$

Hence we can see that the autocorrelation matrix is diagonalized by the KLT. It can be shown, for any autocorrelation matrix with diagonal entries, $\sigma_{c,k}^2$, that the following inequality holds:

$$\det[\mathbf{R}_c] \le \prod_{\forall k} \sigma_{c,k}^2 \quad (5.30)$$

But, from equation (5.13) we can see that:

$$\det[\mathbf{R}_c] = \det[\mathbf{R}_x]$$

Therefore, for any unitary transform the following inequality holds:

$$\prod_{\forall k} \sigma_{c,k}^2 \ge \det[\mathbf{R}_x] \quad (5.31)$$

But, for the case of the KLT we can see that:

$$\prod_{\forall k} \sigma_{c,k}^2 = \det[\mathbf{R}_x] = \det[\mathbf{R}_c] \quad (5.32)$$

Hence the KLT minimizes the geometric mean and, as a consequence (see equation (5.43)), maximizes the coding gain for a Gaussian source. It also provides the minimum approximation error, according to equation (5.25), if a subset of coefficients is use to represent the signal.

5.4.3 The KLT in practice

The main drawback of the KLT is the dependence of the transform on the signal and thus the need to compute and transmit the transformation matrix to the decoder prior to signal reconstruction. This can be a significant overhead in the case of small block sizes and signal non-stationarity. Furthermore, for correlated data (such as an image), other transforms such as the *discrete cosine transform* (DCT) are a close approximation to the KLT. For these reasons, the DCT is the transform of choice for most image and video coding applications.

For further details on the KLT, the reader is referred to the excellent text of Clarke [1].

5.5 Discrete cosine transform (DCT)

5.5.1 Derivation of the DCT

The discrete cosine transform was first introduced by Ahmed et al. [2] and is the most widely used unitary transform for image and video coding applications. Like the discrete Fourier transform, the DCT provides information about a signal in the frequency domain. However, unlike the DFT, the DCT of a real-valued signal is itself real valued and importantly it also does not introduce artifacts due to periodic extension of the input data.

With the DFT, a finite length data sequence is naturally extended by periodic extension. Discontinuities in the time (or spatial) domain therefore produce ringing or spectral leakage in the frequency domain. This can be avoided if the data sequence is symmetrically (rather than periodically) extended prior to application of the DFT. This produces an even sequence which has the added benefit of yielding real-valued coefficients. The DCT is not as useful as the DFT for frequency domain signal analysis due to its deficiencies when representing pure sinusoidal waveforms. However, in its primary role of signal compression, it performs exceptionally well.

As we will see, the DCT has good energy compaction properties and its performance approaches that of the KLT for correlated image data. Furthermore, unlike the KLT, its basis functions are independent of the signal. The 1-D DCT, in its most popular form, is given by:

$$c(k) = \sqrt{\frac{2}{N}} \varepsilon_k \sum_{m=0}^{N-1} x[m] \cos\left(\frac{\pi k}{N}\left(m + \frac{1}{2}\right)\right) \tag{5.33}$$

and in 2-D by:

$$c(k, l) = 2\frac{\varepsilon_k \varepsilon_l}{\sqrt{NM}} \sum_{m=0}^{M-1}\sum_{n=0}^{N-1} x[m, n] \cos\left(\frac{\pi k}{N}\left(m + \frac{1}{2}\right)\right) \cos\left(\frac{\pi l}{N}\left(n + \frac{1}{2}\right)\right) \tag{5.34}$$

$$\varepsilon_k = \begin{cases} \frac{1}{\sqrt{2}} & k = 0 \\ 1 & \text{otherwise} \end{cases}$$

DCT derivation

Consider the 1-D signal shown in Figure 5.7. The lower subfigure illustrates how a 1-D signal can be symmetrically extended (or mirrored) to form a sequence of length $2N$. This has the clear advantage that no discontinuities are introduced and so, when we apply the DFT to create a frequency domain representation, ringing artifacts are absent.

All N elements of the original signal, $x[n]$, are duplicated to give a new sequence $x_1[n]$:

$$x_1[n] = \begin{cases} x[n]; & 0 \leq n \leq N - 1 \\ x[2N - 1 - n]; & N \leq n \leq 2N - 1 \end{cases}$$

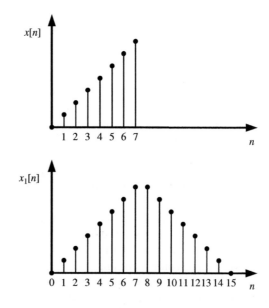

FIGURE 5.7

Symmetrical signal extension for the DCT.

The DFT of $x_1[n]$ is then given by:

$$\mathcal{X}_1(k) = \sum_{n=0}^{N-1} x[n] W_{2N}^{nk} + \sum_{n=N}^{2N-1} x[2N - 1 - n] W_{2N}^{nk}$$

$$= \sum_{n=0}^{N-1} x[n] \left(W_{2N}^{nk} + W_{2N}^{-(n+1)k} \right); \quad 0 \le k \le 2N - 1$$

Multiplying the numerator and denominator by $W_{2N}^{0.5k}$:

$$\mathcal{X}_1(k) = W_{2N}^{-k/2} \sum_{n=0}^{N-1} x[n] \left(W_{2N}^{(n+0.5)k} + W_{2N}^{-(n+0.5)k} \right)$$

$$= 2W_{2N}^{-k/2} \sum_{n=0}^{N-1} x[n] \cos \left(\frac{\pi (n + 0.5) k}{N} \right)$$

This is starting to look attractive as a basis for compression, apart from the rotation factor which makes the result complex and non-symmetric. Since this term contributes nothing in terms of compression performance, the DCT is normally defined without it as follows:

$$c(k) = W_{2N}^{k/2} \mathcal{X}_1(k)$$

and then normalized to give:

$$c(k) = \sqrt{\frac{2}{N}} \varepsilon_k \sum_{n=0}^{N-1} x[n] \cos\left(\frac{\pi k}{N}(n+0.5)\right) \tag{5.35}$$

where

$$\varepsilon_k = \begin{cases} \frac{1}{\sqrt{2}} & k=0 \\ 1 & \text{otherwise} \end{cases}$$

Similarly, the inverse DCT is defined by:

$$x[n] = \sqrt{\frac{2}{N}} \sum_{k=0}^{N-1} \varepsilon_k c(k) \cos\left(\frac{\pi k}{N}(n+0.5)\right) \tag{5.36}$$

5.5.2   DCT basis functions

The basis functions are the column vectors of the inverse transform matrix. It is these that are weighted by the transform coefficients and linearly combined to form the signal approximation. However, in the case of a real-valued orthonormal transform, we have already seen that these are identical to the row vectors in the forward transform matrix.

Using equation (5.35) we can see that the DCT basis functions are given by:

$$a(k, n) = \sqrt{\frac{2}{N}} \varepsilon_k \cos\left(\frac{\pi k}{N}(n+0.5)\right); \quad 0 \le k, n \le N-1 \tag{5.37}$$

Example 5.6 (Four-point DCT basis functions)
Compute the basis functions and transform matrix for the four-point DCT.

Solution.   Applying equation (5.37) we have:

$$\mathbf{A} = \sqrt{\frac{2}{N}} \varepsilon_k \begin{bmatrix} \cos(0) & \cos(0) & \cos(0) & \cos(0) \\ \cos\left(\frac{\pi}{8}\right) & \cos\left(\frac{3\pi}{8}\right) & \cos\left(\frac{5\pi}{8}\right) & \cos\left(\frac{7\pi}{8}\right) \\ \cos\left(\frac{\pi}{4}\right) & \cos\left(\frac{3\pi}{4}\right) & \cos\left(\frac{5\pi}{4}\right) & \cos\left(\frac{7\pi}{4}\right) \\ \cos\left(\frac{3\pi}{8}\right) & \cos\left(\frac{9\pi}{8}\right) & \cos\left(\frac{15\pi}{8}\right) & \cos\left(\frac{21\pi}{8}\right) \end{bmatrix}$$

$$= \sqrt{\frac{1}{2}} \varepsilon_k \begin{bmatrix} \cos(0) & \cos(0) & \cos(0) & \cos(0) \\ \cos\left(\frac{\pi}{8}\right) & \cos\left(\frac{3\pi}{8}\right) & -\cos\left(\frac{3\pi}{8}\right) & -\cos\left(\frac{\pi}{8}\right) \\ \cos\left(\frac{\pi}{4}\right) & -\cos\left(\frac{\pi}{4}\right) & -\cos\left(\frac{\pi}{4}\right) & \cos\left(\frac{\pi}{4}\right) \\ \cos\left(\frac{3\pi}{8}\right) & -\cos\left(\frac{\pi}{8}\right) & \cos\left(\frac{\pi}{8}\right) & -\cos\left(\frac{3\pi}{8}\right) \end{bmatrix} \tag{5.38}$$

Returning to the four-point DCT, let us convince ourselves that this transform is orthonormal and confirm the value of ε_k. Consider, for example, the first basis function. We can rewrite this as the vector $\mathbf{a}_0$, as follows:

$$\mathbf{a}_0 = k_0 \begin{bmatrix} 1 & 1 & 1 & 1 \end{bmatrix}$$

Now we know that for orthonormality the basis functions must have unity norm, thus:

$$\|\mathbf{a}_0\| = k_0\sqrt{1^2 + 1^2 + 1^2 + 1^2} = 2k_0 = 1$$

Hence:

$$k_0 = \frac{1}{2}$$

Similarly for the second basis function:

$$\|\mathbf{a}_1\| = k_1\sqrt{\cos^2\left(\frac{\pi}{8}\right) + \cos^2\left(\frac{3\pi}{8}\right) + \cos^2\left(\frac{3\pi}{8}\right) + \cos^2\left(\frac{\pi}{8}\right)} = \sqrt{2}k_1 = 1$$

Hence:

$$k_1 = \frac{1}{\sqrt{2}}$$

And it can also be shown that:

$$k_2 = k_3 = \frac{1}{\sqrt{2}}$$

So in general we have, as expected:

$$k_i = \frac{1}{\sqrt{2}}\varepsilon_i \quad \varepsilon_i = \begin{cases} 1/\sqrt{2} & i = 0 \\ 1 & i \neq 0 \end{cases}$$

Since we now know that the DCT is orthonormal, we can simplify computation of the inverse transform since it is straightforward to show from equation (5.11) that:

$$\mathbf{A}^{-1} = \mathbf{A}^{\mathrm{T}} = \sqrt{\frac{1}{2}}\begin{bmatrix} \frac{1}{\sqrt{2}} & \cos\left(\frac{\pi}{8}\right) & \cos\left(\frac{\pi}{4}\right) & \cos\left(\frac{3\pi}{8}\right) \\ \frac{1}{\sqrt{2}} & \cos\left(\frac{3\pi}{8}\right) & -\cos\left(\frac{\pi}{4}\right) & -\cos\left(\frac{\pi}{8}\right) \\ \frac{1}{\sqrt{2}} & -\cos\left(\frac{3\pi}{8}\right) & -\cos\left(\frac{\pi}{4}\right) & \cos\left(\frac{\pi}{8}\right) \\ \frac{1}{\sqrt{2}} & -\cos\left(\frac{\pi}{8}\right) & \cos\left(\frac{\pi}{4}\right) & -\cos\left(\frac{3\pi}{8}\right) \end{bmatrix}$$

Example 5.7 (Eight-point DCT basis functions)
We can easily compute the basis functions for other length transforms. For example, compute the eight-point DCT basis functions.

Solution. The eight-point DCT basis functions can be computed using equation (5.37). These are shown in Figure 5.8.

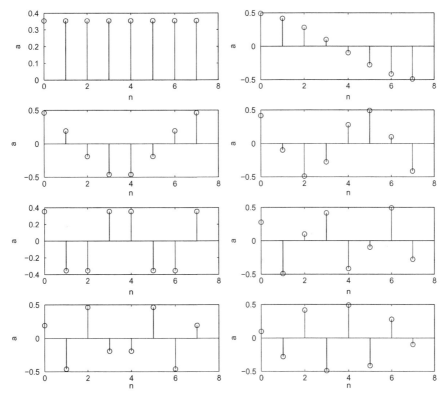

FIGURE 5.8

DCT basis functions for $N = 8$.

5.5.3 Extension to 2-D: separability

The 2-D DCT basis functions are, as described earlier, obtained from the outer product of the 1-D basis functions. These are shown for the case of the 8×8 DCT in Figure 5.9.

Let us now consider the issue of separability in the context of the 2-D DCT. Using the 8×8 transform as an example:

$$c_2(m, n) = \frac{\varepsilon_m \varepsilon_n}{4} \sum_{i=0}^{7} \sum_{j=0}^{7} s[i, j] \cos \left(\frac{(2i + 1)m\pi}{16} \right) \cos \left(\frac{(2j + 1)n\pi}{16} \right) \quad (5.39)$$

Now the 1-D DCT is given by:

$$c_1(m) = \frac{\varepsilon_m}{2} \sum_{i=0}^{7} s[i, j] \cos \left(\frac{(2i + 1)\pi m}{16} \right)$$

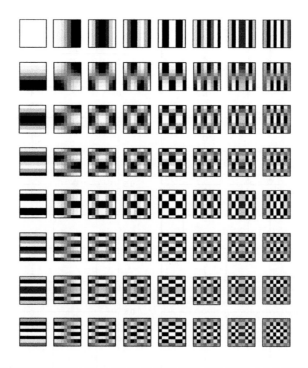

FIGURE 5.9

2-D DCT basis functions for $N = 8$.

Equation (5.39) can thus be rearranged as follows:

$$c_2(m, n) = \frac{\varepsilon_n}{2} \sum_{i=0}^{7} c_1(m) \cos\left(\frac{(2j + 1)\pi n}{16}\right)$$

In other words, the 2-D DCT can be formed in two passes, firstly in the m-direction and then in the n-direction, thus:

$$c_2(m, n) = \text{DCT}_n^{1\text{-D}}\left(\text{DCT}_m^{1\text{-D}}\right) \qquad (5.40)$$

Separable computation of the 2-D DCT (or indeed any separable transform) can be visualized as shown in Figure 5.10.

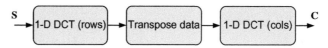

FIGURE 5.10

Separable computation of the forward DCT.

Separability has two main advantages:

1. The number of calculations is reduced—for the 8×8 case, each pass requires 8×64 multiply accumulate operations (MACs), giving a total of $2 \times 8 \times 64 = 1024$ MACs.
2. The 1-D DCT can be further decomposed to yield even greater savings (see Section 5.8.2).

Example 5.8 (2-D DCT calculation)

Calculate the DCT of the following 2-D image block:

$$\mathbf{S} = \begin{bmatrix} 21 & 19 \\ 15 & 20 \end{bmatrix}$$

Solution. The 2-D DCT transform matrix can be computed as:

$$\mathbf{A} = \frac{1}{\sqrt{2}} \begin{bmatrix} 1 & 1 \\ 1 & -1 \end{bmatrix}$$

Recalling that:

$$\mathbf{C}_2 = \mathbf{A}\mathbf{S}\mathbf{A}^{\mathrm{T}}$$

we have

$$\mathbf{C}_2 = \frac{1}{2} \begin{bmatrix} 1 & 1 \\ 1 & -1 \end{bmatrix} \begin{bmatrix} 21 & 19 \\ 15 & 20 \end{bmatrix} \begin{bmatrix} 1 & 1 \\ 1 & -1 \end{bmatrix}$$
$$= \begin{bmatrix} 37.5 & -1.5 \\ 2.5 & 3.5 \end{bmatrix}$$

Now if we perform the inverse DCT on this result, we should recover the original signal:

$$\tilde{\mathbf{S}} = \frac{1}{2} \begin{bmatrix} 1 & 1 \\ 1 & -1 \end{bmatrix} \begin{bmatrix} 37.5 & -1.5 \\ 2.5 & 3.5 \end{bmatrix} \begin{bmatrix} 1 & 1 \\ 1 & -1 \end{bmatrix} = \begin{bmatrix} 21 & 19 \\ 15 & 20 \end{bmatrix} = \mathbf{S}$$

5.6 Quantization of DCT coefficients

As we have discussed previously, the process of energy compaction in itself does not provide any data compression. For example, if we have an 8×8 block of spatial domain pixel values with a dynamic range of 8 bits and we transform these to an 8×8 block of transform domain coefficients, also requiring 8 bits each, then we have gained little. What we do achieve, however, is a concentration of energy into a small number of high-valued coefficients. These larger coefficients are likely to have a higher psychovisual significance than their low-valued counterparts and can be coded accordingly.

Quantization is an important step in lossy compression; however, it is irreversible and causes signal degradation. As we saw in Chapter 3, the quantizer comprises a set of decision levels and a set of reconstruction levels. One of the challenges is to perform quantization in such a way as to minimize its psychovisual impact.

The primary advantage of transform coding is that it provides both energy compaction and decorrelation. This means that the transform coefficients can be approximated as originating from a memoryless source (e.g. i.i.d. Laplacian) for which simple quantizers exist. Intra-frame transform coefficients are therefore normally quantized using a uniform quantizer, with the coefficients pre-weighted to reflect the frequency-dependent sensitivity of the HVS. A general expression which captures this is given in equation (5.41), where Q is the quantizer step-size, k is a constant and $\mathbf{W}$ is a coefficient-dependent weighting matrix obtained from psychovisual experiments:

$$c_Q(i, j) = \left\lfloor \frac{kc(i, j)}{Qw_{i,j}} \right\rfloor \tag{5.41}$$

After transmission or storage, we must rescale the quantized transform coefficients prior to inverse transformation, thus:

$$\tilde{c}(i, j) = \frac{c_Q(i, j)Qw_{i,j}}{k} \tag{5.42}$$

First let us consider a simple example (Example 5.9), before progressing to the more realistic 8×8 case (Example 5.10).

Example 5.9 (Recalculating Example 5.8 with quantization)
Quantize the DCT result from Example 5.8 using the following quantization matrix (assume $k = 1$ and $Q = 1$):

$$\mathbf{W} = \begin{bmatrix} 4 & 8 \\ 8 & 8 \end{bmatrix}$$

Solution. Now

$$\mathbf{C}_Q = \lfloor \mathbf{C}_2 / W \rceil$$

So

$$\mathbf{C}_Q = \begin{bmatrix} 9 & 0 \\ 0 & 0 \end{bmatrix}$$

Now compute the inverse 2-D DCT after rescaling:

$$\tilde{\mathbf{S}} = \frac{1}{2} \begin{bmatrix} 1 & 1 \\ 1 & -1 \end{bmatrix} \begin{bmatrix} 36 & 0 \\ 0 & 0 \end{bmatrix} \begin{bmatrix} 1 & 1 \\ 1 & -1 \end{bmatrix} = \begin{bmatrix} 18 & 18 \\ 18 & 18 \end{bmatrix} \neq \mathbf{S}$$

This leaves an error signal of:

$$\mathbf{E} = \begin{bmatrix} 3 & 1 \\ -3 & 2 \end{bmatrix}$$

Assuming 6 bit signals—what is the PSNR?

Example 5.10 (8 × 8 2-D DCT calculation with quantization)
Consider the following 8 × 8 image block, **S**:

$$
\mathbf{S} =
\begin{bmatrix}
168 & 163 & 161 & 150 & 154 & 168 & 164 & 154 \\
171 & 154 & 161 & 150 & 157 & 171 & 150 & 164 \\
171 & 168 & 147 & 164 & 164 & 161 & 143 & 154 \\
164 & 171 & 154 & 161 & 157 & 157 & 147 & 132 \\
161 & 161 & 157 & 154 & 143 & 161 & 154 & 132 \\
164 & 161 & 161 & 154 & 150 & 157 & 154 & 140 \\
161 & 168 & 157 & 154 & 161 & 140 & 140 & 132 \\
154 & 161 & 157 & 150 & 140 & 132 & 136 & 128
\end{bmatrix}
$$

First of all calculate the 8 × 8 block of transform coefficients, **C**, then quantize the resulting DCT coefficients using the quantization matrix, **W** (this is actually the default JPEG quantization matrix). Assuming $k = 1$ and $Q = 1$:

$$
\mathbf{W} =
\begin{bmatrix}
16 & 11 & 10 & 16 & 24 & 40 & 51 & 61 \\
12 & 12 & 14 & 19 & 26 & 58 & 60 & 55 \\
14 & 13 & 16 & 24 & 40 & 57 & 69 & 56 \\
14 & 17 & 22 & 29 & 51 & 87 & 80 & 62 \\
18 & 22 & 37 & 56 & 68 & 109 & 103 & 77 \\
24 & 35 & 55 & 64 & 81 & 104 & 113 & 92 \\
49 & 64 & 78 & 87 & 103 & 121 & 120 & 101 \\
72 & 92 & 95 & 98 & 112 & 100 & 103 & 99
\end{bmatrix}
$$

Finally compute the rescaled DCT coefficient matrix and perform the inverse DCT to obtain the reconstructed image block, $\tilde{\mathbf{S}}$.

Solution.

$$
\mathbf{C} =
\begin{bmatrix}
1239 & 50 & -3 & 20 & -10 & -1 & 0 & -6 \\
35 & -24 & 11 & 13 & 4 & -3 & 14 & -6 \\
-6 & -3 & 8 & -9 & 2 & -3 & 5 & 10 \\
9 & -10 & 4 & 4 & -15 & 10 & 5 & 6 \\
-12 & 5 & -1 & -2 & -15 & 9 & -6 & -2 \\
5 & 10 & -7 & 3 & 4 & -7 & -14 & 2 \\
2 & -2 & 3 & -1 & 1 & 3 & -3 & -4 \\
-1 & 1 & 0 & 2 & 3 & -2 & -4 & -2
\end{bmatrix}
$$

Quantize the DCT coefficients using the quantization matrix, $\mathbf{W}$. Assuming $k = 1$ and $Q = 1$, we have:

$$
\mathbf{C}_Q = \begin{bmatrix}
77 & 5 & 0 & 1 & 0 & 0 & 0 & 0 \\
3 & -2 & 1 & 1 & 0 & 0 & 0 & 0 \\
0 & 0 & 1 & 0 & 0 & 0 & 0 & 0 \\
1 & -1 & 0 & 0 & 0 & 0 & 0 & 0 \\
-1 & 0 & 0 & 0 & 0 & 0 & 0 & 0 \\
0 & 0 & 0 & 0 & 0 & 0 & 0 & 0 \\
0 & 0 & 0 & 0 & 0 & 0 & 0 & 0 \\
0 & 0 & 0 & 0 & 0 & 0 & 0 & 0
\end{bmatrix}
$$

Rescaling the DCT coefficient matrix:

$$
\tilde{\mathbf{C}} = \begin{bmatrix}
1232 & 55 & 0 & 16 & 0 & 0 & 0 & 0 \\
36 & -24 & 14 & 19 & 0 & 0 & 0 & 0 \\
0 & 0 & 16 & 0 & 0 & 0 & 0 & 0 \\
14 & -17 & 0 & 0 & 0 & 0 & 0 & 0 \\
-18 & 0 & 0 & 0 & 0 & 0 & 0 & 0 \\
0 & 0 & 0 & 0 & 0 & 0 & 0 & 0 \\
0 & 0 & 0 & 0 & 0 & 0 & 0 & 0 \\
0 & 0 & 0 & 0 & 0 & 0 & 0 & 0
\end{bmatrix}
$$

and inverse transforming gives the reconstructed image block $\tilde{\mathbf{S}}$:

$$
\tilde{\mathbf{S}} = \begin{bmatrix}
173 & 162 & 150 & 149 & 158 & 164 & 164 & 160 \\
176 & 166 & 156 & 154 & 160 & 163 & 160 & 154 \\
173 & 165 & 158 & 156 & 160 & 157 & 150 & 143 \\
163 & 159 & 155 & 154 & 154 & 150 & 142 & 135 \\
158 & 157 & 156 & 157 & 155 & 151 & 143 & 138 \\
161 & 161 & 161 & 160 & 157 & 152 & 146 & 143 \\
163 & 163 & 161 & 156 & 149 & 143 & 139 & 137 \\
161 & 160 & 156 & 148 & 138 & 131 & 127 & 126
\end{bmatrix}
$$

As we can see, the resulting matrix after quantization is close in value to the original signal. The key thing to notice is that the matrix $\tilde{\mathbf{C}}$ is now sparse in non-zero values and, as expected (because the input data is highly correlated), the energy in the DCT coefficients is compacted in the top left corner of the matrix.

As an additional exercise, calculate the PSNR for this approximation—assuming a signal wordlength of 8 bits.

An example of the effects of coefficient quantization on reconstruction quality for a range of block types is shown in Figure 5.11. It can be observed, as expected, that more textured blocks require larger numbers of coefficients in order to create a good approximation to the original content. The best reconstruction is achieved with fewer coefficients for the case of untextured blocks, as shown for the left-hand block in the figure.

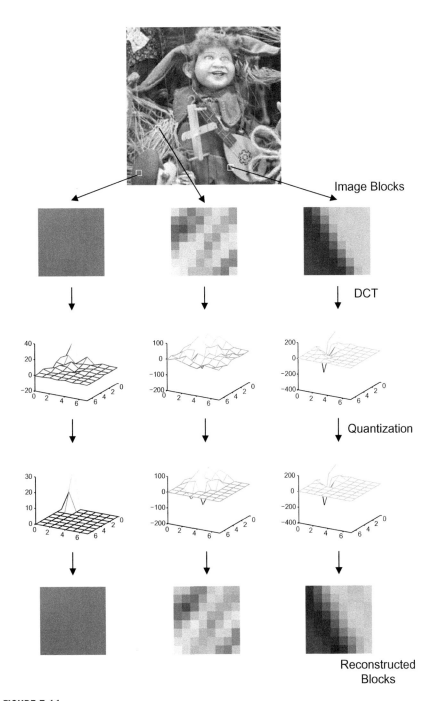

Effects of coefficient quantization on various types of data block.

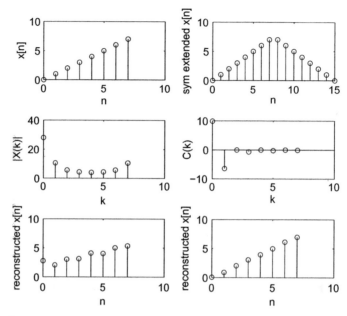

FIGURE 5.12

Comparison of DFT and DCT for compression. Top: input signal $x[n]$ and the symmetrically extended sequence $x_1[n]$. Middle: eight-point DFT$\{x[n]\}$ (magnitude only) and the eight-point DCT$\{x[n]\}$. Bottom: IDFT$\{X(k)\}$ (magnitude only) and IDCT$\{C(k)\}$.

5.7 Performance comparisons

5.7.1 DCT vs DFT revisited

Earlier we explained why the DFT does not perform well as a basis for compression. We are now in a position to compare the performance of the DCT and the DFT. Figure 5.12 shows an input signal $x[n]$ which is compressed with both an eight-point DFT and an eight-point DCT. In both cases compression is achieved by truncating the transform coefficient vector to leave only the first four coefficients, the remainder being set to zero. It can be clearly seen that the energy compaction of the DCT is superior to the DFT as is the reconstructed signal quality for a given compression strategy. Recall also that the DFT coefficients are complex and thus require approximately double the bandwidth of the DCT coefficients.

5.7.2 Comparison of transforms

Figure 5.13 shows the effects of transformation on a natural image (256×256) using the DFT, DWHT, and DCT. As expected, the DFT magnitude spectrum exhibits multiple energy concentrations, whereas the performance of the DCT provides significant energy compaction without any of the disadvantages of the DFT. The DWHT also provides good performance on this image despite its very simple structure.

Acer original image	DFT (magnitude)
DWHT	DCT

FIGURE 5.13

Comparison of KLT, DFT, DWHT, DST, and DCT on a natural image.

Computation of the KLT for a 256×256 image is impractical, so Figure 5.14 compares its performance with the DCT based on the use of 8×8 transform blocks. This figure shows two cases—the first where only nine of the 64 coefficients in each block are retained for reconstruction, and the second where 25 are used. As can be seen, the DCT comes close to the KLT in terms of quality, but importantly without the need for the transmission of basis functions. This provides additional justification of why the DCT remains so popular in image and video compression applications.

5.7.3 Rate–distortion performance of the DCT

The benefit of block-transform-based coding with the DCT is that, for correlated data, the transform compacts most of the energy in the original signal into a small number of significant coefficients. The coding gain for the block transform is defined in terms of the ratio of the arithmetic and geometric means of the transformed block variances as follows:

$$G_{TC} = 10 \log_{10} \left[\frac{\frac{1}{N} \sum_{i=0}^{N-1} \sigma_i^2}{\left(\prod_{i=0}^{N-1} \sigma_i^2 \right)^{1/N}} \right] \tag{5.43}$$

DCT first 9 coeffs: PSNR=27.7dB   KLT first 9 coeffs: PSNR=28.4dB

DCT first 25 coeffs: PSNR=33.5dB   KLT first 25 coeffs: PSNR=34.5dB

FIGURE 5.14

Comparison of KLT and DCT performance.

FIGURE 5.15

DCT performance. Left: Lena original 256 × 256. Right: compressed at 0.3 bpp.

Figure 5.15 shows an example of the performance of DCT compression (8 × 8 blocks) for a 256 × 256 resolution image at 0.3 bpp, whereas Figure 5.16 shows a comparison between the same 256 × 256 image and an equivalent 512 × 512 image. It can clearly be seen that, for the same compression ratio, the larger image is much less distorted. There are two fairly obvious reasons for this: (i) the 512 × 512 image,

FIGURE 5.16

DCT performance. Left: Lena 256 × 256 0.3 bpp. Right: Lena 512 × 512 0.3 bpp.

FIGURE 5.17

DCT rate–distortion performance for 256 × 256 Lena image using Huffman encoding.

despite being compressed at the same ratio, has more bits and (ii) the 512×512 image will exhibit higher correlation over an 8×8 block.

Figure 5.17 shows an example of the rate–distortion performance of the DCT coder after entropy coding (see Chapter 7) based on the Lena image.

5.8 DCT implementation

5.8.1 Choice of transform block size

What is the optimum block size for transform coding? This question is best answered by considering the data being processed and the complexity of the transform itself. We normally seek the best compromise between decorrelation performance and complexity. As we will see below, the typical size is 8×8 pixels, although sizes down to 4×4 are used in H.264/AVC and block sizes up to 64×64 are available in H.265/HEVC. Variable block sizes are now becoming commonplace and the size is selected as part of the rate–distortion optimization (RDO) process (see Chapter 10).

It is also worth noting that the choice of transform can be data dependent. For example, the *discrete sine transform* (DST) can outperform the DCT for cases of low correlation.

DCT complexity

Using the standard 2-D DCT, the computation of each coefficient in an $N \times N$ block requires N^2 multiplications and N^2 additions. This is indicated by a complexity of order $O(N^4)$. This can be reduced to $2N$ multiplications per coefficient for the case when separability is exploited (i.e. $O(N^3)$). Hence for the case of a 4×4 block we require 8 multiplications per coefficient, for an 8×8 block we require 16 multiplications per coefficient and for a 16×16 block we require 32 multiplications per coefficient.

Taking account of both complexity and autocorrelation surface characteristics, a choice between 4×4 and 16×16 is normally used in practice.

5.8.2 DCT complexity reduction

Due to its ubiquity, many fast algorithms exist for computing the DCT. Early fast algorithms operated indirectly, exploiting the Fast Fourier Transform (FFT), with $O(N \log_2 N)$ complexity. It can be shown [1,2] that the 1-D DCT can be expressed as:

$$c(k) = \frac{2}{N} \mathbb{R} \left[e^{\frac{-jk\pi}{2N}} \sum_{n=0}^{2N-1} x[n] W^{kn} \right] \tag{5.44}$$

This type of approach is not so popular now due to the requirement for complex multiplications and additions. Other approaches take into account the unitary property of the DCT which allows it to be factorized into products of relatively sparse matrices [4]. Recursive algorithms have also been proposed [3]. The complexity of these approaches is typically $O(N^2 \log_2 N)$.

Later algorithms took direct advantage of the properties of the transform matrix, either through factorization, matrix partitioning, or by exploiting its symmetries in a divide and conquer methodology. One such approach is described below.

McGovern algorithm

In McGovern's algorithm [5], the matrix rows are decomposed, exploiting odd and even symmetries using a divide and conquer approach. The aim of this method is to reduce the number of multiplications per coefficient:

$$
\begin{bmatrix} c(0) \\ c(1) \\ c(2) \\ c(3) \\ c(4) \\ c(5) \\ c(6) \\ c(7) \end{bmatrix} =
\begin{bmatrix}
a_4 & a_4 & a_4 & a_4 & a_4 & a_4 & a_4 & a_4 \\
a_1 & a_3 & a_5 & a_7 & -a_7 & -a_5 & -a_3 & -a_1 \\
a_2 & a_6 & -a_6 & -a_2 & -a_2 & -a_6 & a_2 & a_2 \\
a_3 & -a_7 & -a_1 & -a_5 & a_5 & a_1 & a_7 & -a_3 \\
a_4 & -a_4 & -a_4 & a_4 & a_4 & -a_4 & -a_4 & a_4 \\
a_5 & -a_1 & a_7 & a_3 & -a_3 & -a_7 & a_1 & -a_5 \\
a_6 & -a_2 & a_2 & -a_6 & -a_6 & a_2 & -a_2 & a_6 \\
a_7 & -a_5 & a_3 & -a_1 & a_1 & -a_3 & a_5 & -a_7
\end{bmatrix}
\begin{bmatrix} x[0] \\ x[1] \\ x[2] \\ x[3] \\ x[4] \\ x[5] \\ x[6] \\ x[7] \end{bmatrix}
\tag{5.45}
$$

where:

$$
a_i = \cos\left(i\frac{\pi}{16}\right)
$$

$$
\begin{bmatrix} c(0) \\ c(2) \\ c(4) \\ c(6) \end{bmatrix} =
\begin{bmatrix}
a_4 & a_4 & a_4 & a_4 \\
a_2 & a_6 & -a_6 & -a_2 \\
a_4 & -a_4 & -a_4 & a_4 \\
a_6 & -a_2 & a_2 & -a_6
\end{bmatrix}
\begin{bmatrix} x[0]+x[7] \\ x[1]+x[6] \\ x[2]+x[5] \\ x[3]+x[4] \end{bmatrix}
\tag{5.46}
$$

$$
\begin{bmatrix} c(1) \\ c(3) \\ c(5) \\ c(7) \end{bmatrix} =
\begin{bmatrix}
a_1 & a_3 & a_5 & a_7 \\
a_3 & -a_7 & -a_1 & -a_5 \\
a_5 & -a_1 & a_7 & a_3 \\
a_7 & -a_5 & a_3 & -a_1
\end{bmatrix}
\begin{bmatrix} x[0]-x[7] \\ x[1]-x[6] \\ x[2]-x[5] \\ x[3]-x[4] \end{bmatrix}
\tag{5.47}
$$

The decomposition in equation (5.46) reduces the number of multiplications required from $64 \ (=N^2)$ in equation (5.45) to only $32 \ (= 2(N/2)^2)$. Note that the matrix required to compute the even components of $\mathbf{x}$ is the same as that required to compute the four-point DCT. The resultant matrices can be further decomposed into a sparse form as follows:

$$
\begin{bmatrix} c(0) \\ c(4) \\ c(2) \\ c(6) \\ c(7) \\ c(5) \\ c(1) \\ c(3) \end{bmatrix} =
\begin{bmatrix}
a_4 & 0 & 0 & 0 & 0 & 0 & 0 & 0 \\
0 & a_4 & 0 & 0 & 0 & 0 & 0 & 0 \\
0 & 0 & a_6 & a_2 & 0 & 0 & 0 & 0 \\
0 & 0 & -a_2 & a_6 & 0 & 0 & 0 & 0 \\
0 & 0 & 0 & 0 & a_7 & a_5 & a_1 & a_3 \\
0 & 0 & 0 & 0 & a_5 & -a_1 & a_3 & a_7 \\
0 & 0 & 0 & 0 & a_1 & a_3 & -a_7 & a_5 \\
0 & 0 & 0 & 0 & a_3 & a_7 & -a_5 & -a_1
\end{bmatrix}
\begin{bmatrix} g_0 \\ g_1 \\ g_2 \\ g_3 \\ g_4 \\ g_5 \\ g_6 \\ g_7 \end{bmatrix}
\tag{5.48}
$$

where:

$$
g_0 = x[0] + x[1] + x[2] + x[3] + x[4] + x[5] + x[6] + x[7]
$$
$$
g_1 = x[0] + x[7] + x[3] + x[4] - x[1] - x[6] - x[2] - x[5]
$$
$$
g_2 = x[1] + x[6] - x[2] - x[5]
$$
$$
g_3 = x[0] + x[7] - x[3] - x[4]
$$

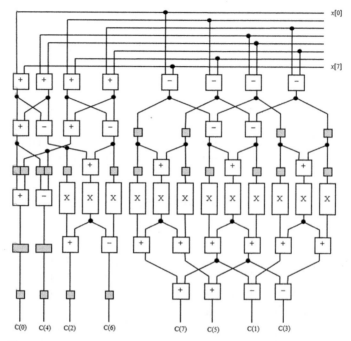

FIGURE 5.18

Efficient DCT implementation: McGovern's algorithm.

$$g_4 = x[0] - x[7]$$
$$g_5 = x[6] - x[1]$$
$$g_6 = x[3] - x[4]$$
$$g_7 = x[2] - x[5]$$

This reduces the number of multiplications required from 64 to 22. This number can be further reduced to 14 using *standard rotator products*:

$$\begin{bmatrix} x & w \\ w & -y \end{bmatrix} \begin{bmatrix} r_0 \\ r_1 \end{bmatrix} = \begin{bmatrix} 1 & 1 & 0 \\ 0 & 1 & -1 \end{bmatrix} \begin{bmatrix} (x-w)r_0 \\ w(r_0+r_1) \\ (w+y)r_1 \end{bmatrix}$$

and still further reduced to 12 multiplications if we divide through by a_4. A bit-serial architecture for this algorithm is shown in Figure 5.18 [5] where the gray boxes represent synchronizing delay operators.

5.8.3 Field vs frame encoding for interlaced sequences

In the case of interlaced image sequences, if motion occurs between the field scans, large valued spurious DCT coefficients can occur at high spatial frequencies. These will be coarsely quantized during compression, reducing the quality of the encoded sequence. Standards since MPEG-2 have supported the option of performing a DCT either on each field independently or on the combined frame. In the case of field pictures, all the blocks in every macroblock come from one field, whereas with an (interlaced) frame picture, frame or field DCT coding decisions can be made adaptively on a macroblock-by-macroblock basis.

When an interlaced macroblock from an interlaced frame picture is frame DCT coded, each of its four blocks has pixels from both fields. However, if the interlaced macroblock from an interlaced frame picture is field coded, each block consists of pixels from only one of the two fields.

Martin and Bull [6] produced an efficient algorithm using the primitive operator methodology of Bull and Horrocks [7]. This supports coding either one 8×8 DCT or two independent 4×4 DCT blocks. The result was compared with alternative implementations using an area–time metric and significant complexity savings were observed.

5.8.4 Integer transforms

While the DCT is in principle lossless, it can suffer from numerical mismatches between the encoder and decoder, leading to drift. This can be overcome through the use of integer transforms where integer arithmetic is employed to ensure drift-free realizations. This approach is employed in the basic H.264/AVC 4×4 integer transform which is itself derived from the 4×4 DCT. This transform is given in equation (5.49) and is discussed further in Chapter 9:

$$\mathbf{A} = \begin{bmatrix} 1 & 1 & 1 & 1 \\ 2 & 1 & -1 & -2 \\ 1 & -1 & -1 & 1 \\ 1 & -2 & 2 & -1 \end{bmatrix} \tag{5.49}$$

5.8.5 DCT demo

Figure 5.19 shows the graphical user interface for an informative DCT demonstration. This simple Matlab® software offers users the opportunities to visualize the effect of selecting and deselecting DCT coefficients, showing their influence on the quantized DCT matrix and on the reconstructed image. Readers can download this as part of an integrated demonstration package from http://www.bristol.ac.uk/vi-lab/demos.

5.9 JPEG

The JPEG standard, created originally in 1992 [8], is still ubiquitous. Despite the emergence of JPEG2000 as a serious contender with superior performance in 2000, JPEG has maintained its position as the dominant codec for internet and consumer use. JPEG has four primary modes:

- Baseline (sequential DCT + Huffman coding).
- Sequential lossless (predictive coding plus Huffman or arithmetic coding).

DCT demo GUI.

- Progressive DCT (bit plane progressive).
- Hierarchical.

Baseline JPEG is based on a simple block-based DCT with scalar quantization and Huffman coding. It is applicable to gray-scale and color images with up to 8 bits per sample and is described in Figure 5.20. We will examine the complete JPEG coding process in more detail in Chapter 7, after describing entropy coding.

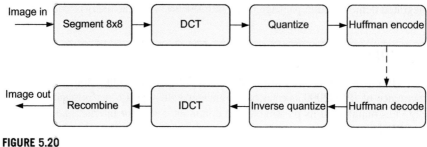

FIGURE 5.20
JPEG baseline system diagram.

5.10 **Summary**

In this chapter we have introduced the reader to transform coding, explaining its basic concepts and mathematical underpinnings. We have shown that the optimum transform, while providing the best energy compaction for a given data set, suffers when the data statistics are non-stationary, because of the need to continually update the transform matrix and transmit this to the decoder. Alternative transforms, such as the DCT, provide near optimum performance for correlated images with fixed bases and are the preferred choice for practical image and video compression. We have derived the DCT and demonstrated its excellent performance and its low complexity implementations.

In the next chapter we will explore another data transformation based on wavelet filterbanks and in Chapter 7 we will describe lossless coding and show how decorrelating transforms can be combined with entropy (symbol) encoding to achieve impressive compression ratios while retaining excellent perceptual image quality.

References

[1] R. Clarke, Transform Coding of Images, Academic Press, 1985.

[2] N. Ahmed, T. Natarajn, K. Rao, Discrete cosine transform, IEEE Transactions on Computers 23 (1974) 90–93.

[3] B. Lee, A new algorithm to compute the discrete cosine transform, IEEE Transactions on Acoustics, Speech, and Signal Processing 32 (1984) 1243–1245.

[4] W.-H. Chen, C. Smith, S. Fralick, A fast computational algorithm for discrete cosine transform, IEEE Transactions on Communications 25 (1977) 1004–1009.

[5] F. McGovern, R. Woods, M. Yan, Novel VLSI implementation of (8 × 8) point 2-D DCT, Electronics Letters 30 (8) (1994) 624–626.

[6] F. Martin, D. Bull, An improved architecture for the adaptive discrete cosine transform, in: IEEE International Symposium on Circuits and Systems, 1996, pp. 742–745.

[7] D. Bull, D. Horrocks, Primitive operator digital filters, IEE Proceedings G (Circuits, Devices and Systems) 138 (3) (1991) 401–412.

[8] ISO/IEC International Standard 10918-1, Information Technology—Digital and Coding of Continuous-Tone Still Images Requirements and Guidelines, 1992.

Filter Banks and Wavelet Compression

CHAPTER OUTLINE

6.1 Introduction to multiscale processing **172**
    6.1.1 The short-time Fourier transform and the Gabor transform 173
    6.1.2 So what is a wavelet? 174
    6.1.3 Wavelet and filter bank properties 177
6.2 Perfect reconstruction filter banks **178**
    6.2.1 Filter and decomposition requirements 178
    6.2.2 The 1-D filter bank structure 178
6.3 Multirate filtering **182**
    6.3.1 Upsampling 183
    6.3.2 Downsampling 183
    6.3.3 System transfer function 186
    6.3.4 Perfect reconstruction 189
    6.3.5 Spectral effects of the two-channel decomposition 191
6.4 Useful filters and filter banks **191**
    6.4.1 Quadrature mirror filters 192
    6.4.2 Wavelet filters 194
    6.4.3 Multistage (multiscale) decompositions 197
    6.4.4 Separability and extension to 2-D 200
    6.4.5 Finite length sequences, edge artifacts, and boundary extension . . 202
    6.4.6 Wavelet compression performance 203
6.5 Coefficient quantization and bit allocation **204**
    6.5.1 Bit allocation and zonal coding 204
    6.5.2 Hierarchical coding 207
6.6 JPEG2000 . . . . **207**
    6.6.1 Overview 207
    6.6.2 Architecture—bit planes and scalable coding 207
    6.6.3 Coding performance 208
    6.6.4 Region of interest coding 208
    6.6.5 Benefits and status 210

Communicating Pictures. http://dx.doi.org/10.1016/B978-0-12-405906-1.00006-4

6.7 Summary **210**
References **210**

Wavelets are mathematical functions that can be used as an alternative to transforms such as the DCT to split a signal into different frequency bands (subbands) prior to processing and quantization. Wavelets are usually implemented as banks of filters that enable each subband to be analyzed at a resolution matched to its scale. This, as we will see later, is normally achieved by combining filtering operations with sub-sampling, which also has the benefit of maintaining a constant overall sample rate at all stages in the decomposition. This approach has significant advantages over Fourier-based methods, especially when signals are not harmonically related, are of short duration with discontinuities or are non-stationary. It also has benefits in terms of subband coefficient quantization since the artifacts produced tend to be perceptually less annoying than the blocking artifacts associated with the DCT.

This chapter firstly examines the basic two-channel filter bank structure and then investigates the filter properties that enable perfect reconstruction in the absence of quantization. We then go on to consider the more general case of multirate filtering, showing how, through the appropriate combination of filters and up- and downsampling operations, we can achieve critical sampling. We next examine specific cases of useful subband and wavelet filters and then go on to extend the basic architecture to the case of a multistage (multiscale) decomposition and to two dimensions as required for image compression applications. Finally, we examine bit allocation strategies and scalability in the context of the JPEG2000 still image coding standard.

6.1 Introduction to multiscale processing

As the name suggests, wavelets are small waves with compact support. They can be designed to have specific and useful properties and, in the same way as Fourier or DCT bases, can be linearly combined to synthesize a signal. The difference with wavelets is that the basis functions at different scales are related through translation and dilation of the basic building block functions. Wavelets can thus be localized in both spatial (or temporal) and frequency domains and, as such, offer important multiresolution properties. They provide much more flexibility in tiling the time–frequency or spatial frequency plane than conventional transforms, offering a means of analyzing signals at different scales and with different resolutions. For example, the tiles can vary in size to better match the characteristics of the human visual system for image coding (or the auditory system in the case of speech coding), thus creating a basis for improved quantization strategies and, hence, improved compression performance.

Historically, development of the theory of wavelets (although they were not originally called wavelets) preceded that of filter banks due to the early work of Haar over 100 years ago [1]. The area was then dormant for many years until the emergence of subband coding methods for speech, audio, and image processing in the 1980s. One

For colour versions of Figures 6.24, 6.27 and 6.31 please refer to the electronic version or the website.

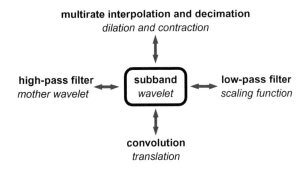

FIGURE 6.1

Equivalences between subband filtering and wavelet nomenclature.

of the major challenges identified at that time was the design of appropriate channel splitting filters, and the earliest contributions in this respect were by Crosier et al., who introduced the concept of the quadrature mirror filter (QMF) [2]. This then led to the development of perfect reconstruction filters with significant contributions from Smith and Barnwell [3,4], Vetterli [5], Woods and O'Neill [6], and Vaidyanathan [7].

Work on wavelets developed in parallel, perhaps starting with the contribution in 1984 from Grossman and Morlet [8], who were the first to use the term "wavelet". This initiated a major body of work on the topic, with major contributions from Daubechies [9] and several others. Wavelet theories and their associated design methods were developed into a comprehensive framework by Mallat [10,11] in 1989.

Although we will primarily adopt a subband filtering rather than a wavelet oriented nomenclature here, it is worth comparing the terms commonly used in each domain. This is illustrated in Figure 6.1. Many excellent texts exist on the topics of wavelets and subband filtering [12–14] and the reader is referred to these for further information.

6.1.1 The short-time Fourier transform and the Gabor transform

One of the main problems with conventional transforms (such as the Fourier transform) is that they are constructed around a set of infinite length basis functions such as sinusoids or complex exponentials. This means that any temporally (or spatially) localized changes or features are spread over the whole frequency domain. This is a particular problem for non-stationary signals such as those typical of images and videos.

One approach to overcome this has been to divide the input signal into shorter time duration sections. While this solves the issue of non-stationarity, it introduces additional problems due to boundary effects and edge distortions. As we saw in Chapter 5, these can distort the underlying spectrum and, in the context of compression, make perfect reconstruction of the original signal impossible. To reduce the influence of these boundary effects it has been conventional in spectrum analysis to window each block of data. The window function reduces to zero at the edges of the block, hence eliminating temporal discontinuities and any associated spectral ringing effects. If we consider the case of the short-time Fourier transform (STFT), this is defined by equation (6.1) where, if the window function $w(t)$ is a Gaussian function, then this is referred to as the Gabor transform.

$$X(\omega, \tau) = \int_{-\infty}^{\infty} x(t)w(t - \tau)e^{-j\omega t}dt \qquad (6.1)$$

A primary issue with the STFT is that the window size is fixed, so any variation in the temporal extent of any non-stationarities cannot be adequately represented. Also, no good basis functions exist for the STFT. Another problem is that the window function distorts the underlying signal which means that perfect reconstruction is not possible unless overlapping windows are used.

Clearly, because of the uncertainty principle, we cannot have both arbitrarily high temporal resolution and frequency resolution. We can however trade off temporal and frequency resolution by altering the window duration used and this is considered next.

6.1.2 So what is a wavelet?

Ideally we would like a set of basis functions that can isolate localized high frequency components in a signal while, at the same time, capturing lower frequency components of longer duration. Consider the example in Figure 6.2, where two signals—a sinusoid and an impulse—are analyzed in the frequency and temporal domains. The time

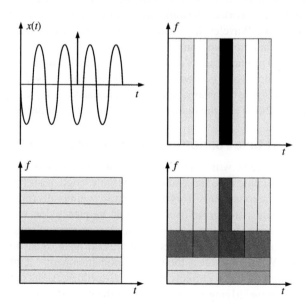

FIGURE 6.2

Comparison of time domain (top right), frequency domain (bottom left), and time–frequency analysis (bottom right) applied to the combination of a sinusoidal signal and an impulse function (top left). (Example adapted from Vetterli and Kovacevic [12].)

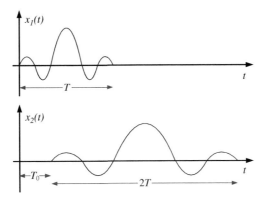

FIGURE 6.3

Wavelet translation and dilation.

domain analysis clearly picks out the impulse, whereas the frequency domain analysis localizes the sine wave. In contrast, frequency domain analysis cannot localize the impulse and time domain analysis cannot localize the sine wave. If we were able to tile the time–frequency plane more flexibly, as shown in the bottom right subfigure, then it would be possible to pick out both short duration high frequency signal components as well as longer duration low frequency components.

In practice, we can do this by combining short high frequency basis functions with longer lower frequency ones. Wavelets achieve this in a very elegant way, through the dilation and translation of a single prototype function. Consider Figure 6.3, where a basic function of duration T is dilated by a factor of 2 and translated by an amount T_0. Clearly the relationship between $x_1(t)$ and $x_2(t)$ in the figure is:

$$x_2(t) = x_1 \left(\frac{t - T_0}{2} \right)$$

We can imagine a linear combination of weighted versions of these translated and dilated functions being used to approximate a signal of interest. However, unlike conventional basis functions constructed from cosine terms or complex exponentials, these exhibit compact support as well as oscillatory behavior. They thus offer the potential to capture the local features as well as the large scale characteristics of a signal.

The continuous wavelet transform (CWT)

We can define a single general prototype function (equation (6.2)) that can be stretched and shifted to form new functions which can be linearly combined to synthesize a signal of interest. This is the basic wavelet transform as given in equation (6.3).

$$h_{a,b}(t) = \frac{1}{\sqrt{a}} h \left(\frac{t - b}{a} \right) \tag{6.2}$$

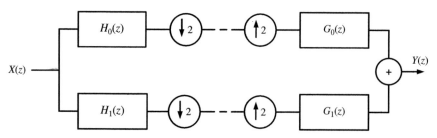

FIGURE 6.4

Two-channel perfect reconstruction filter bank.

$$X(a, b) = \frac{1}{\sqrt{a}} \int_{-\infty}^{\infty} h\left(\frac{t - b}{a}\right) x(t) \mathrm{d}t \tag{6.3}$$

Here a is the scale parameter, b is the translation parameter, and $h(t)$ is called the mother wavelet (i.e. the prototype function from which all the other bases are formed).

This ability to deal with local variations as well as big image features in the same framework is clearly attractive for image compression. However, in order to apply it, we need to be able to evaluate it in discrete form. The continuous time wavelet transform is highly redundant, so the scale space can be sampled dyadically using a wavelet series approach. This however is not a true discrete transform, but fortunately one does exist and we will consider that next.

The discrete wavelet transform (DWT)

Although the CWT can be computed discretely as described above, this is still only a sampled version of the CWT and not a proper discrete transform. The discrete wavelet transform (DWT) is instead based on a set of discrete functions or filters that are convolved with the input signal to effect translation and that are effectively dilated by sub-sampling the signal as it passes through each scale. It is normal to use octave band decompositions with dyadic sampling. This means that the wavelet filters form a set of band-pass responses that provide constant-Q filtering. As the width of the octave bands reduce, more localized high frequency details are captured by the representation. To achieve this we therefore, in equation (6.3), constrain the scale and translation terms to $a = 2^j$ and $b = k2^j$ respectively, where j and k are integers.

The wavelet decomposition is usually realized in the form of a filter bank as shown (for the case of a simple two-band split) in Figure 6.4. The input signal is spectrally decomposed into distinct bands in an analysis section which uses carefully designed filters in conjunction with downsampling[1] to split the signal without increasing the effective sample rate. After analysis, the information in each band can be processed (e.g. quantized) independently according to the characteristics of the source and/or the application. The signal is then passed though a matched synthesis section that

---

[1]Downsampling is sometimes referred to as decimation.

combines filtering with upsampling to reconstruct the original signal or (in the case of lossy compression) an approximation to it.

The wavelet transform and its filter bank realization do not simply divide a signal into multiple frequency bands, but it does this iteratively, normally only iterating the low-pass output at each scale. It thus produces a series of band-pass outputs which are, in fact, the wavelet transform coefficients. The role of the wavelet is played by the high-pass filter and the cascade of low-pass filters followed by a high-pass filter. This will be discussed in more detail later in Section 6.4.3 and is illustrated in Figure 6.19.

As indicated above, this framework has been formalized by Mallat and others, and the reader is referred to Refs. [10, 12] for further details.

6.1.3 Wavelet and filter bank properties

As we have discussed above, the DWT is normally implemented as a filter bank and, in this respect, it is identical to the conventional subband filtering structure. Wavelet filter banks can be considered as a subclass of subband filterbanks where the filters used have some specific properties associated with regularity. This means that, as the filters are iterated, their characteristic shape converges to a regular function, i.e. one that is continuous or continuously differentiable. A necessary condition in this respect is that the filter should have sufficient zeros at the half sampling frequency (i.e. at $z = -1$). This ensures that the spectral images introduced during each filtering stage are attenuated. An excellent review of the requirements for convergence is provided by Rioul and Vetterli [15].

Before we look in detail at the construction of wavelet filter banks, let us summarize some of their advantages. Wavelet coding offers a number of potential benefits over other transforms. These include:

- **Constant relative bandwidth processing:** As illustrated in Figure 6.2, this facilitates more flexible and perceptually accurate tiling of the spatial frequency (or time–frequency) plane.
- **Improved perceptual quality:** As discussed previously, the main disadvantage of transform coding is the subjective degradation at lower bit rates where the viewer is able to perceive the outline of the transform blocks. Subband approaches are however (generally) free of block distortions. By decomposing the signal into a number of individual (possibly overlapping) frequency bands, each band may be quantized differently according to subjective HVS or energy distribution criteria.
- **Scalability:** As well as improved perceptual performance, wavelet methods also provide a basis for scalability,[2] both for images—using for example Set Partitioning (SPIHT) [16] or Embedded Block Truncation Coding (EBCOT) [17], and for video using 3-D wavelets and motion compensated temporal filtering (MCTF) [18].

---

[2]Scalability refers to the property where, after encoding, a compressed bitstream can be manipulated to reduce its bit rate without decoding. This is useful in order to adapt to prevailing channel conditions without the need for explicit transcoding.

- **Error resilience:** A single error during transmission can be modeled as an impulse that excites the synthesis filters to produce a weighted impulse response, which is superimposed on the correct response. With smooth short filters, this produces a localized ringing effect which is perceptually preferable to the blocking errors common with the DCT. Of course error propagation due to prediction loops and variable length coding is still possible. Techniques such as wavelet-based Pyramid Vector Quantization (PVQ) [19] overcome this rather successfully by using fixed length codes (see Chapter 11).

Although not selected for incorporation in standards such as JPEG, MPEG-2/4, H.263/4 or HEVC, subband (wavelet) decomposition techniques are widely acknowledged to offer improved performance and flexibility (with similar implementation complexity) compared with their block-transform counterparts, such as the DCT. This is particularly true for image coding, especially at lower bit rates. These benefits have been acknowledged through the adoption of wavelet coding methods in the most recent still image coding standard, JPEG2000 [20,21].

6.2 Perfect reconstruction filter banks

6.2.1 Filter and decomposition requirements

A simple two-channel filter bank is shown in Figure 6.4. Here, as we will shortly see, $H_0(z)$ and $H_1(z)$ are the low- and high-pass analysis filters and $G_0(z)$ and $G_1(z)$ are the corresponding synthesis filters. The circles in Figure 6.4 represent the downsampling (decimation) and upsampling (interpolation) operators. These are non-linear operators that:

1. **Ensure critical sampling:** The effective sampling rate throughout the filter bank is kept constant. This is important in compression applications as otherwise there is an expansion in the amount of data being processed.
2. **Are responsible for introducing aliasing and imaging:** This, as we will see later, must be managed through the careful design of the filters used.

In designing filters for filter bank use, we must therefore consider two important issues:

1. The filters used in practice do not have perfect transition characteristics and so there will be spectral cross-talk across the bands.
2. The sample rate changing operations will introduce aliasing and spectral imaging that must be eliminated during reconstruction if perfect reconstruction (a property essential for lossless compression) is to be achieved.

6.2.2 The 1-D filter bank structure

The fundamental building block of subband or wavelet filtering is the two-channel filter bank of Figure 6.4. We will examine this structure and its properties in detail later. In the meantime, let us consider how it comes about using a very simple approach to filtering and sampling.

Intuitive development of the two-channel filter bank

The basic idea of the two-band split is to divide a signal into high-pass and low-pass components such that they can be analyzed or quantized independently in a manner related to signal scale. Consider a simple signal that comprises a steadily changing function with some superimposed random fluctuations, for example that given below:

$$x[n] = \{2, 4, 8, 8, 10, 14, 12, 6, 8, 4\}$$

We could estimate the underlying function through low-pass filtering and isolate the noise through high-pass filtering. Perhaps the simplest low-pass filter is the two-tap moving average as given by equation (6.4). The residual signal, $r[n]$, remaining after applying equation (6.4) is the high-pass component and can be characterized using equation (6.5).

$$l[n] = \frac{x[n] + x[n-1]}{2} \tag{6.4}$$

$$r[n] = x[n] - l[n] = \frac{x[n] - x[n-1]}{2} \tag{6.5}$$

Let us examine the frequency domain characteristics of these two simple filters in more detail, firstly the moving average or low-pass filter. From equation (6.4) we can see that:

$$L(z) = 0.5 \left(X(z) + X(z)z^{-1} \right)$$

and thus:

$$H_l(z) = 0.5 \left(\frac{z+1}{z} \right) \tag{6.6}$$

This, as expected, provides a simple low-pass filter response due to its pole at the origin in the z plane and its zero at $\Omega = \pi$. Similarly, the difference operation or high-pass filter is given by equation (6.7) which also has a pole at the origin and a zero at $\Omega = 0$.

$$R(z) = 0.5 \left(X(z) - X(z)z^{-1} \right)$$

$$H_r(z) = 0.5 \left(1 - z^{-1} \right) = 0.5 \left(\frac{z-1}{z} \right) \tag{6.7}$$

The frequency responses of this complementary pair are shown in Figure 6.5. These responses together form an all-pass function—which is not entirely surprising since their time domain equivalents are also complementary, adding to exactly reconstruct the original signal.

So we now begin to see a relationship between perfect reconstruction and a pair of complementary filters that cross at $\Omega = \pi/2$. It can be seen that the filter responses

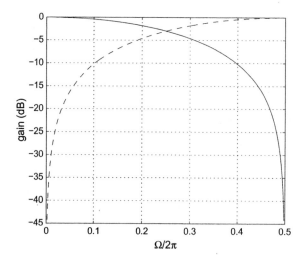

FIGURE 6.5

Complementary half-band filters from equations (6.4) (solid) and (6.5) (dotted).

are symmetrical, crossing with -3 dB attenuation at $\Omega = \pi/2$. Filters with this characteristic are often referred to as half-band filters or, more specifically, quadrature mirror filters (QMF). We will explore this relationship further when we examine filter design and performance later.

We can apply these simple filters to our example data sequence, $x[n]$, to produce the following outputs (assuming that $x[-1] = 0$):

$$l[n] = \{1, 3, 6, 8, 9, 12, 13, 9, 7, 6\}$$
$$r[n] = \{1, 1, 2, 0, 1, 2, -1, -3, 1, -2\}$$

The signals $l[n]$ and $r[n]$ together completely characterize the signal $x[n]$ and could be used as a representation for $x[n]$ that emphasizes its low-pass and high-pass (trend and difference) components. However, one obvious problem arises: $l[n]$ and $r[n]$ together contain twice as many samples as $x[n]$ alone. Doubling the size of the input signal is clearly not a great starting point for compression!

So what can we do about this? Firstly, there is obviously redundancy in the $\{l[n], r[n]\}$ combination—this must be the case as we have doubled the size without introducing any new information. We can observe this redundancy by splitting $l[n]$ and $r[n]$ into even and odd sequences thus: $\{l[2n-1], l[2n]\}$ and $\{r[2n-1], r[2n]\}$. Considering, for example, only the even sequences, we can reconstruct the original sequence $x[n]$ as follows:

$$l[2n] = \frac{x[2n] + x[2n-1]}{2}$$
$$r[2n] = \frac{x[2n] - x[2n-1]}{2}$$

Thus:

$$x[2n] = l[2n] + r[2n] \tag{6.8}$$

$$x[2n - 1] = l[2n] - r[2n] \tag{6.9}$$

Hence, we can perfectly reconstruct the original sequence $x[n]$, using only half of the data produced by the filters, by combining the downsampled sequence $x[2n - 1]$ with the one-sample delayed version, $x[2n]$. The system diagram for this process is given in Figure 6.6. This shows, on the left of the diagram, the analysis stage which produces the filtered outputs from equations (6.4) and (6.5) and downsamples these by a factor of 2 in order to maintain a constant overall bit rate. On the right side of the diagram is the synthesis section. This recombines the filtered and downsampled samples, through upsampling and reconstruction based on equations (6.8) and (6.9).

A refined system diagram, obtained through reconfiguration of the synthesis section to yield a more recognizable filter bank structure, is given in Figure 6.7. Let us now validate the operation of this system through an example (Example 6.1).

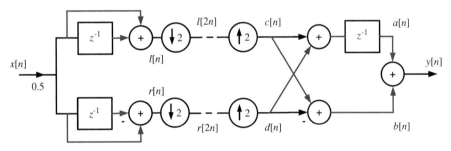

FIGURE 6.6

Basic system diagram for the two-channel filter bank.

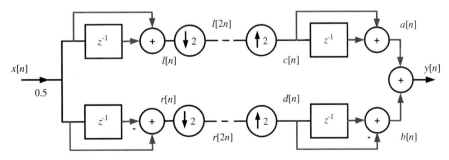

FIGURE 6.7

The two-channel filter bank.

Example 6.1 (Operation of the simple two-channel perfect reconstruction filter bank)

Verify the operation of the filter bank in Figure 6.7 using the example data sequence: $x[n] = \{2, 4, 8, 8, 10, 14, 12, 6, 8, 4\}$.

Solution. The operation of the system for this data is captured in the table below. This illustrates that the filter bank does indeed provide perfect reconstruction, with the output delayed by one sample with respect to the input.

n	0	1	2	3	4	5	6	7	8	9
$x[n]$	2	4	8	8	10	14	12	6	8	4
$l[n]$	1	3	6	8	9	12	13	9	7	6
$r[n]$	1	1	2	0	1	2	−1	−3	1	−2
$l[2n]$	1	6	9	13	7	0	0	0	0	0
$r[2n]$	1	2	1	−1	1	0	0	0	0	0
$c[n]$	1	0	6	0	9	0	13	0	7	0
$d[n]$	1	0	2	0	1	0	−1	0	1	0
$a[n]$	1	1	6	6	9	9	13	13	7	7
$b[n]$	−1	1	−2	2	−1	1	1	−1	−1	1
$y[n]$	0	2	4	8	8	10	14	12	6	8

6.3 Multirate filtering

Subband decomposition techniques are based on multirate processing. Multirate theory enables a signal to be represented in the form of multiple band-pass components whilst maintaining a constant overall bit rate. Consider, for example, the two-band split. Since the convolution of a signal with a filter impulse response does not affect bit rate,[3] for an input bit rate of k bps, the overall output bit rate without sub-sampling will be 2 kbps. Multirate theory however enables us to sub-sample the filter outputs in order to maintain a constant bit rate while preserving overall signal content (this is known as critical decimation or critical sampling).

The two primary operations used in multirate systems are interpolation and decimation. Decimation is normally employed in the encoder (or analysis stage) to reduce the sampling rate by a factor of M at each node relative to the input rate, whereas interpolation is normally used in the decoder (or synthesis stage) to increase it during reconstruction. These operations facilitate the removal of redundant information and hence help to manage the processing requirements.

---

[3] Assuming an infinite length sequence.

6.3.1 Upsampling

Increasing the sample rate of a signal is often referred to as interpolation and is shown diagrammatically in Figure 6.8. The operation of the upsampler is characterized according to its upsampling ratio, L, and is defined by equation (6.10):

$$x_u[n] = \begin{cases} x\left[\frac{n}{L}\right] & n = mL \\ 0 & \text{otherwise} \end{cases} \tag{6.10}$$

So, for example with $L = 2$, we insert $L - 1 = 1$ zeros between each of the original samples. Thus if $x[n] = \{1, 2, 3, 4, 5, 6\}$ then $x_u[n] = \{1, 0, 2, 0, 3, 0, 4, 0, 5, 0, 6\}$. The z transform of the upsampled signal is then given by:

$$X_u(z) = \sum_{n=-\infty}^{\infty} x_u[n]z^{-n} = \sum_{n=-\infty}^{\infty} x\left[\frac{n}{L}\right]z^{-n}; \quad n/L \text{ integer} \tag{6.11}$$

Substituting $m = \frac{n}{L}$ in equation (6.11):

$$= \sum_{m=-\infty}^{\infty} x_u[m]z^{-mL} = X\left(z^L\right) \tag{6.12}$$

Let us consider what the implications of this are in the frequency domain. In the case of $L = 2$, the mapping $X\left(z^2\right)$ provides a factor of 2 "compression" of the spectrum in the frequency domain. This can be verified by evaluating equation (6.12) on the unit circle in the z plane, i.e.:

$$X\left(z^2\right)\Big|_{z=e^{j\Omega}} = X\left(e^{j2\Omega}\right)$$

and plotting the result, as illustrated in Figure 6.9. This figure shows that an additional spectral image is introduced at $\Omega = \pi$ as a result of upsampling. This must be removed using a subsequent interpolation filter, as shown in Figure 6.10. An example signal showing the spectral effect of upsampling is shown in Figure 6.11.

6.3.2 Downsampling

Decreasing the sample rate of a signal is often referred to as decimation (although strictly this should refer only to downsampling by a factor of 10). The downsampling operation is shown diagrammatically in Figure 6.12. The operation of the

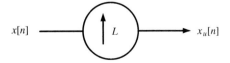

FIGURE 6.8

Upsampling by a factor L.

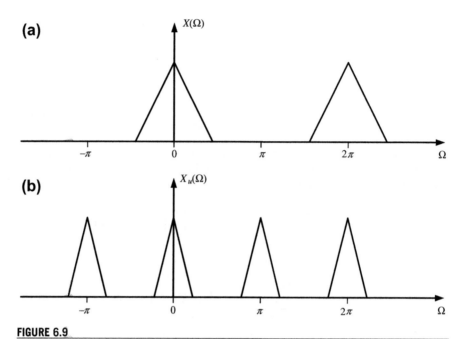

FIGURE 6.9

Spectral effect of upsampling: (a) input; (b) output.

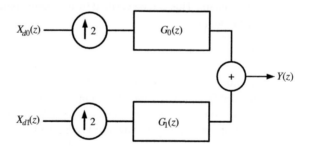

FIGURE 6.10

Two-channel synthesis filter bank.

downsampler is characterized according to its downsampling ratio, M, and is defined by equation (6.13):

$$x_d[n] = x[nM] \tag{6.13}$$

So, for example with $M = 2$, we delete every Mth sample in the original sequence. Thus if $x[n] = \{1, 2, 3, 4, 5, 6\}$ then $x_d[n] = \{1, 3, 5\}$. The z transform of the down-sampled signal is then given by:

$$X_d(z) = \sum_{n=-\infty}^{\infty} x_d[n]z^{-n} = \sum_{n=-\infty}^{\infty} x[nM]z^{-n} \tag{6.14}$$

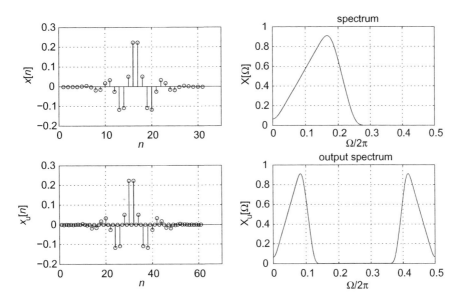

FIGURE 6.11

Spectral effects of upsampling where $L = 2$.

FIGURE 6.12

Downsampling by a factor of M.

Substituting $m = nM$ in equation (6.14):

$$X_d(z) = \sum_{m=-\infty}^{\infty} x'[m]z^{-m/M} = X'\left(z^{1/M}\right) \tag{6.15}$$

Note that $x'[n]$ is not the same as $x[n]$. They can however be related by $x'[n] = c[n]x[n]$, where:

$$c[n] = \begin{cases} 1 & n = 0, \pm M, \pm 2M \ldots \\ 0 & \text{otherwise} \end{cases}$$

where $c[n]$ can be represented, using the DFT, as:

$$c[n] = \frac{1}{M} \sum_{k=0}^{M-1} W_M^{nk} \tag{6.16}$$

So, for example, in the case of $M = 2$, we have:

$$c[n] = \frac{1}{2} \left(1 + (-1)^n\right)$$

Now the z-transform of $x'[n]$ is given by:

$$X'(z) = \frac{1}{M} \sum_{n=-\infty}^{\infty} \left(\sum_{k=0}^{M-1} W_M^{nk}\right) x[n] z^{-n} = \frac{1}{M} \sum_{k=0}^{M-1} \left(\sum_{n=-\infty}^{\infty} x[n] W_M^{nk} z^{-n}\right) \quad (6.17)$$

$$= \frac{1}{M} \sum_{k=0}^{M-1} X\left(z W_M^{-k}\right) \quad (6.18)$$

Equation (6.18) can now be substituted into equation (6.15) to give:

$$X_d(z) = \frac{1}{M} \sum_{k=0}^{M-1} X\left(z^{1/M} W_M^{-k}\right) \quad (6.19)$$

Consider the case where $M = 2$:

$$X_d(z) = \frac{1}{2} \left(X\left(z^{1/2}\right) + X\left(-z^{1/2}\right)\right) \quad (6.20)$$

Evaluating this on the unit circle in the z-plane gives the frequency response:

$$X_d\left(e^{j\Omega}\right) = \frac{1}{2} \left(X\left(e^{j\Omega/2}\right) + X\left(-e^{j\Omega/2}\right)\right) \quad (6.21)$$

or equivalently:

$$X_d\left(e^{j\Omega}\right) = \frac{1}{2} \left(X\left(e^{j\Omega/2}\right) + X\left(e^{j(\Omega/2+\pi)}\right)\right) \quad (6.22)$$

A plot of equation (6.22) is shown in Figure 6.13. This shows that the original spectrum is expanded by a factor of 2 in the frequency domain, but also that an additional spectral image is introduced at $\Omega = 2\pi$ as a result of the second term in equation (6.22). It can be seen that, in order to avoid aliasing which would occur if the original spectrum extended beyond $\Omega = \pi/2$, the input signal must be anti-alias filtered prior to downsampling. This is achieved using the analysis filter in Figure 6.14. An example of downsampling with an actual signal is shown in Figure 6.15.

6.3.3 System transfer function

We are now in a position to derive a transfer function for the two-channel filter bank that was introduced in Figure 6.4. This combines the analysis stage of Figure 6.14 and the synthesis stage of Figure 6.10. Considering the top path in the analysis stage only at first, we have:

$$X_0(z) = X(z)H_0(z) \quad (6.23)$$

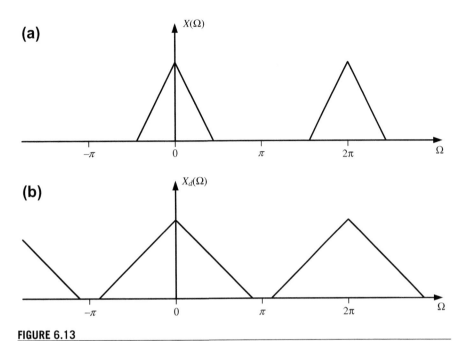

FIGURE 6.13

Spectral effects of downsampling: (a) input; (b) output.

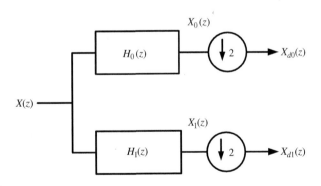

FIGURE 6.14

Two-channel analysis filter bank.

and

$$X_{d0}(z) = \frac{1}{2} \left(X_0 \left(z^{1/2} \right) + X_0 \left(-z^{1/2} \right) \right) \tag{6.24}$$

Substituting equation (6.23) in equation (6.24) gives:

$$X_{d0}(z) = \frac{1}{2} \left(X \left(z^{1/2} \right) H_0 \left(z^{1/2} \right) + X \left(-z^{1/2} \right) H_0 \left(-z^{1/2} \right) \right) \tag{6.25}$$

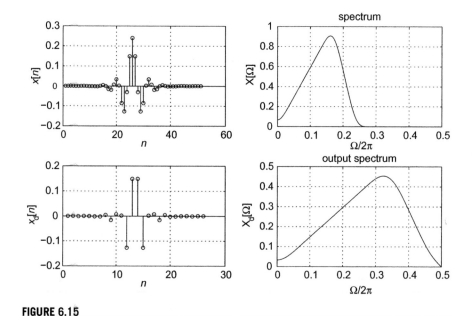

FIGURE 6.15

Spectral effects of downsampling with $M = 2$.

Now considering the synthesis stage:

$$X_{ud0}(z) = X_{d0}\left(z^2\right)$$

$$X_{ud0}(z) = \frac{1}{2}\left(X(z)H_0(z) + X(-z)H_0(-z)\right)$$

Defining the filtered output on the upper path as $Y_0(z)$:

$$Y_0(z) = X_{ud0}(z)G_0(z) \tag{6.26}$$

$$Y_0(z) = \frac{1}{2}\left(X(z)H_0(z)G_0(z) + X(-z)H_0(-z)G_0(z)\right) \tag{6.27}$$

Similarly for the filtered output on the lower path, we have:

$$Y_1(z) = \frac{1}{2}\left(X(z)H_1(z)G_1(z) + X(-z)H_1(-z)G_1(z)\right) \tag{6.28}$$

Finally, combining equations (6.27) and (6.28) at the output of the filter bank gives:

$$Y(z) = \frac{1}{2}X(z)\left[H_0(z)G_0(z) + H_1(z)G_1(z)\right]$$

$$+ \frac{1}{2}X(-z)\left[H_0(-z)G_0(z) + H_1(-z)G_1(z)\right] \tag{6.29}$$

So we now have the ability to analyze the end-to-end operation of the two-channel filter bank. Let us consider this further in Example 6.2.

Example 6.2 (A simple filter bank transfer function)
Consider the two-band split in Figure 6.4, where the filters are given as follows:

$$H_0(z) = 1 + z^{-1} \quad H_1(z) = 1 - z^{-1}$$
$$G_0(z) = 1 + z^{-1} \quad G_1(z) = -1 + z^{-1}$$

These filters are identical to those defined in our earlier example, except that the multiplicative factor of 0.5 in the analysis stage has been omitted for numerical convenience. Using equation (6.29), compute the output of the filter bank for an input $x[n] = \{1, 2, 3, 4\}$.

Solution. The effects of analysis filtering, downsampling, upsampling, and synthesis filtering can be seen in the diagram below, yielding an output that is the same as the input, but scaled by a factor of 2 and delayed by one sample.

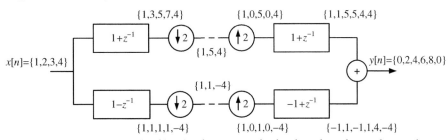

Rather than compute the filter bank response in the time domain, as shown above, we can instead verify its operation in the z-domain using equation (6.29). Thus we have:

$$Y(z) = \frac{1}{2}\left(1 + 2z^{-1} + 3z^{-2} + 4z^{-3}\right)\left[\left(1 + 2z^{-1} + z^{-2}\right) + \left(-1 + 2z^{-1} - z^{-2}\right)\right]$$

$$+ \cdots + \frac{1}{2}\left(1 - 2z^{-1} + 3z^{-2} - 4z^{-3}\right)\left[\left(1 - z^{-2}\right) + \left(-1 + z^{-2}\right)\right]$$

$$= \frac{1}{2}\left(1 + 2z^{-1} + 3z^{-2} + 4z^{-3}\right)\left[\left(4z^{-1}\right)\right] + \frac{1}{2}\left(1 - 2z^{-1} + 3z^{-2} - 4z^{-3}\right)[(0)]$$

$$= \left(2z^{-1} + 4z^{-2} + 6z^{-3} + 8z^{-4}\right)$$

Note that the leading and trailing zeros shown in the filter bank output are redundant and due to the expansive nature of the convolution operation.

6.3.4 Perfect reconstruction

Equation (6.29) can be rewritten as:

$$Y(z) = F_0(z)X(z) + F_1(z)X(-z) \tag{6.30}$$

where

$$F_0(z) = \frac{1}{2}\left[H_0(z)G_0(z) + H_1(z)G_1(z)\right]$$

and

$$F_1(z) = \frac{1}{2}\left[H_0(-z)G_0(z) + H_1(-z)G_1(z)\right]$$

$F_1(z)$ represents the alias components produced by the overlapping filter frequency responses. If $F_1(z) = 0$ then the filter bank is alias-free. $F_0(z)$ represents the quality

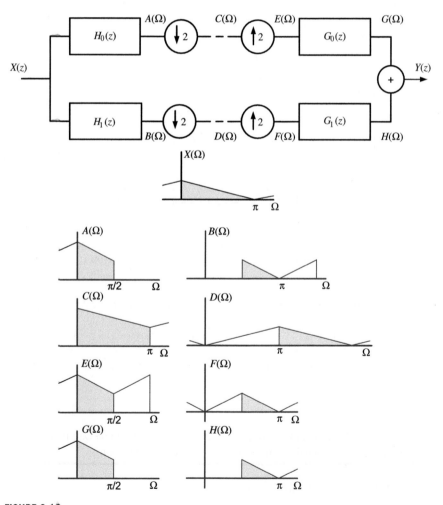

FIGURE 6.16

Spectral relationships within the two-channel perfect reconstruction filter bank. The spectral characteristics at signal points A–H are shown.

of the reconstruction. If $\left| F_0\left(e^{j\Omega}\right)\right| = 1$ for all frequencies, then the filter bank is free of amplitude distortion. If $F_0\left(e^{j\Omega}\right)$ is linear phase then the filter bank is free of phase distortion. In general we say that, if $F_1(z) = 0$ and $F_0(z) = cz^{-k}$ (where c and k are constants), then the filter bank provides perfect reconstruction (PR).

6.3.5 Spectral effects of the two-channel decomposition

A perhaps more intuitive visualization of the operation of the two-channel PR filter bank is provided by Figure 6.16. This shows the spectral characteristics of a signal, $x[n]$, which is bandlimited to the Nyquist frequency, as it passes through the filter bank. For simplicity in the diagram, we assume that the filters have brickwall responses with no leakage between bands. However, as we have already seen, with appropriate filter design, this is not a requirement for perfect reconstruction.

Working across the system from left to right we can see firstly that $H_0(z)$ and $H_1(z)$ select the low-pass and high-pass bands respectively. In order to preserve critical sampling, these signals are then downsampled by a factor of 2. The results of this operation, $C(\Omega)$ and $D(\Omega)$, present spectra that are expanded by a factor of 2 and that also include aliases as expected. The signals would then typically be quantized for transmission, but we ignore this step at the moment and move straight to the decoder.

At the decoder, the signals are upsampled in preparation for recombination. The effect of this is to "compress" the spectra by a factor of 2 along the frequency axis as shown by $E(\Omega)$ and $F(\Omega)$. Finally these signals are passed through the interpolation filters $G_0(z)$ and $G_1(z)$, which remove the aliases by low-pass and high-pass filtering respectively, to yield $G(\Omega)$ and $H(\Omega)$. As can be observed, $G(\Omega)$ and $H(\Omega)$ sum at the output to provide a perfect reconstruction of the input signal.

6.4 Useful filters and filter banks

For compression applications it is desirable for the filters to exhibit certain characteristics. These are:

1. **Perfect reconstruction:** To enable distortion-free reconstruction of the original signal at the output of the filter bank, in the absence of quantization.
2. **Compact support:** To ensure that distortions due to quantization and transmission loss are localized.
3. **Smoothness:** To better match the characteristics of natural images.
4. **Regularity:** To ensure convergence and maintain smoothness when iterated.
5. **Orthogonality and orthonormality:** To enable matching of analysis and synthesis sections and to remove transform redundancy.[4]

Some commonly used filters are examined below.

---

[4]*Note:* The imposition of orthogonality conditions places severe constraints on the filters (or basis functions) used in the filter bank. An alternative condition, known as *biorthogonality*, can however be satisfied more generally, where the analysis filters are independent but do not form an orthogonal pair. This means that the inverse or synthesis transform is no longer given by the transpose of the analysis transform. In the case of biorthogonality, a dual basis can be found where projection on to an orthogonal set is possible.

6.4.1 Quadrature mirror filters

Quadrature mirror filters (QMF) are an important class of alias-cancelation filters that were used in early subband processing applications [2].

Aliasing elimination

QMF filters start by eliminating the aliasing term in equation (6.30) by using the following relationships:

$$G_0(z) = H_1(-z); \quad G_1(z) = -H_0(-z)$$

QMF filters are based on a pair of analysis filters that are symmetrical about $\Omega = \pi/2$. This can be achieved if $H_0(z) = H(z)$ and $H_1(z) = H(-z)$. We can therefore base our QMF filter design on the following relationships:

$$H_0(z) = H(z); \quad H_1(z) = H(-z)$$
$$G_0(z) = H(z); \quad G_1(z) = -H(-z)$$

If we substitute these relationships into equation (6.29), we can see that the aliasing term is zero, thus:

$$F_1(z) = \frac{1}{2}\left[H(-z)H(z) - H(z)H(-z)\right] = 0 \tag{6.31}$$

Amplitude distortion

The distortion term for the QMF filter bank is given by:

$$F_0(z) = \frac{1}{2}\left[H(z)H(z) - H(-z)H(-z)\right] \tag{6.32}$$

Furthermore, if the prototype filter is linear phase: $H(z) = A(z)z^{-(N-1)/2}$, where $A\left(e^{j\Omega}\right)$ is a real zero-phase frequency response, then:

$$F_0(z) = \left[A^2(z) + A^2(-z)\right]z^{-(N-1)/2}$$

The QMF filter bank will therefore exhibit amplitude distortion unless the magnitude of $F_0\left(e^{j\Omega}\right)$ is constant for all Ω. As we will see in Example 6.3, QMF filters can be designed that exhibit perfect reconstruction.

Example 6.3 (Simple QMF filters)
Consider the following example prototype filter:

$$H(z) = \frac{1}{\sqrt{2}}\left(1 + z^{-1}\right)$$

Generate the corresponding set of QMF filters and show that this filter bank exhibits perfect reconstruction.

Solution.   This prototype yields the following set of filters, which hopefully the reader will recognize as the half-band filters from our previous example:

$$H_0(z) = \frac{1}{\sqrt{2}}\left(1 + z^{-1}\right); \quad H_1(z) = \frac{1}{\sqrt{2}}\left(1 - z^{-1}\right)$$

$$G_0(z) = \frac{1}{\sqrt{2}}\left(1 + z^{-1}\right); \quad G_1(z) = \frac{1}{\sqrt{2}}\left(-1 + z^{-1}\right)$$

We can see from equation (6.31) that this filter bank is alias-free. We can further show, from equation (6.32) that the system provides a constant delay with unity magnitude since:

$$F_0(z) = \frac{1}{4}\left[\left(1 + z^{-1}\right)\left(1 + z^{-1}\right) - \left(1 - z^{-1}\right)\left(1 - z^{-1}\right)\right] = z^{-1}$$

In fact these are the only causal FIR QMF analysis filters yielding exact perfect reconstruction.

Practical QMF filters

In practice, approximate QMF filters can be usefully designed where the characteristic does not exactly provide perfect reconstruction. In such cases, the amplitude distortion can be minimized using an optimization procedure to iteratively adjust the filter coefficients. We ideally want the magnitude response of the filter characteristic to have a sharp transition so that subbands are spectrally distinct while providing an overall system response that is approximately all-pass. This means that the filters have to be designed so that each branch delivers an attenuation of 6 dB (0.5) at the point where the low- and high-pass responses intersect—at one quarter of the sampling frequency. Each of the four filters must therefore have an attenuation of 3 dB at the half Nyquist frequency. Such filters have been designed by Johnson [22] and others. Johnson's method minimizes the residual amplitude distortion using an objective measure, based upon stop-band attenuation and the requirement to maintain the power symmetry property of the filters. Low amplitude distortions, typically less than 0.02 dB, result. Example Johnson filters are:

$$h_0[0] = -0.006443977 = h_0[11]; \quad h_0[1] = 0.02745539 = h_0[10];$$
$$h_0[2] = -0.00758164 = h_0[9]; \quad h_0[3] = -0.0013825 = h_0[8];$$
$$h_0[4] = 0.09808522 = h_0[7]; \quad h_0[5] = 0.4807962 = h_0[6]$$

The mirror responses of these filters are shown in Figure 6.17. It can be seen that the overall distortion characteristic, although small, is not zero. Hence these filters do not provide perfect reconstruction. In practice, for compression applications where significant quantization error is introduced prior to transmission of the subband coefficients, this small distortion may not be a major issue. However, we will see shortly how useful filters can be designed that do offer perfect reconstruction.

FIGURE 6.17

QMF filter pair frequency response.

6.4.2 Wavelet filters

The wavelet filter bank can form a perfect reconstruction system that is completely defined by the low-pass filters. The high-pass filters are obtained by shifting and modulating the low-pass filters, thus:

$$H_1(z) = z^{-l}G_0(-z) \text{ and } G_1(z) = z^l H_0(-z) \tag{6.33}$$

and provided that l is odd, then:

$$F_1(z) = \frac{1}{2}\left[H_0(-z)G_0(z) + (-z)^{-l}G_0(z)z^l H_0(-z)\right] = 0$$

i.e. the system is alias-free.

We can now address the distortion term $F_0(z)$. If $F_0(z) = \frac{1}{2}[H_0(z)G_0(z) + H_1(z)G_1(z)] = cz^{-k}$, then the filter bank will exhibit perfect reconstruction. If we let $P(z) = H_0(z)G_0(z)$, and we assume zero-phase response filters, then we can rewrite $F_0(z)$ in the form (absorbing the $\frac{1}{2}$ term into the constant, c):

$$P(z) + P(-z) = c \tag{6.34}$$

Since all terms in odd powers of z will cancel, for equation (6.34) to hold true, with the exception of the z^0 term, all the terms in even powers of z must sum to zero.

To design filters we can now factorize the polynomial $P(z)$ into the two low-pass filters, $H_0(z)$ and $G_0(z)$, resulting in a completely defined filter bank. The question is, how do we factorize it, and how do we ensure that we have good filter characteristics? We have already discussed the desirable property of regularity and the associated requirement to have multiple roots at $z = -1$. This gives us a good starting point. So if we have:

$$P(z) = \left(1 + 2z^{-1} + z^{-2}\right)$$

then this can be factorized into our familiar simple PR filter bank where both low-pass filters are identical, thus:

$$P(z) = \left(1 + z^{-1}\right)\left(1 + z^{-1}\right)$$

There are several filter combinations which (in the absence of quantization) offer perfect reconstruction. A performance comparison of these is provided by Villasenor [23] and a good overview of design methods is provided by Kingsbury [24]. We will consider two example sets of filters here—both of which have been selected for use in JPEG2000. These are the LeGall 5/3 wavelet and the Daubechies 9/7 wavelet. They are both compact, have symmetric responses (this helps to avoid boundary artifacts) and are highly regular (they have a maximum number of vanishing moments).

LeGall 5/3 filters

One example of a simple, yet effective, set of wavelet filters is the biorthogonal LeGall (5/3) pair [25]. These can be designed using the above approach and are defined in terms of their low-pass filter responses as follows:[5]

$$H_0(z) = \frac{1}{8}z\left(1 + z^{-1}\right)^2\left(-z - z^{-1} + 4\right)$$

$$G_0(z) = \frac{1}{2}z\left(1 + z^{-1}\right)^2 \tag{6.35}$$

The filter coefficients are thus:

$$h_0[n] = \{-1, 2, 6, 2, -1\}/8; \quad g_0[n] = \{1, 2, 1\}/2;$$
$$h_1[n] = \{1, -2, 1\}/2; \quad g_1[n] = \{1, 2, -6, 2, 1\}/8$$

Recalling from equation (6.29) that:

$$Y(z) = \frac{1}{2}X(z)\left[G_0(z)H_0(z) + G_1(z)H_1(z)\right]$$
$$+ \frac{1}{2}X(-z)\left[G_0(z)H_0(-z) + G_1(z)H_1(-z)\right]$$
$$= F_0(z)X(z) + F_1(z)X(-z) \tag{6.36}$$

for perfect reconstruction we require $F_0(z) = constant$ and $F_1(z) = 0$. So for the distortion term we have:

$$G_0(z)H_0(z) = \left(-1 + 0z^{-1} + 9z^{-2} + 16z^{-3} + 9z^{-4} + 0z^{-5} - z^{-6}\right)/16$$

$$G_1(z)H_1(z) = \left(+1 + 0z^{-1} - 9z^{-2} + 16z^{-3} - 9z^{-4} + 0z^{-5} + z^{-6}\right)/16$$

Thus the distortion is constant for all frequencies:

$$F_0(z) = z^{-3}$$

---

[5]*Note:* The LeGall filters are sometimes defined where both are pre-multiplied by a factor 1/4 rather than 1/8 and 1/2.

In the case of the alias term we have:

$$G_0(z)H_0(-z) = \left(-1 - 4z + z^{-2} + 8z^{-3} + z^{-4} - 4z^{-5} - z^{-6}\right)\Big/16$$

$$G_1(z)H_1(-z) = \left(1 + 4z - z^{-2} - 8z^{-3} - z^{-4} + 4z^{-5} + z^{-6}\right)\Big/16$$

Thus the alias term is given by:

$$F_1(z) = 0$$

Hence the filter bank is also alias free and satisfies the conditions required for perfect reconstruction. The frequency responses of the four LeGall (5/3) biorthogonal filters are shown in Figure 6.18.

The LeGall wavelet has rational coefficients and is recommended for reversible or lossless compression in JPEG2000.

Daubechies 9/7 filters

The Daubechies 9/7 filters [26] are recommended for lossy compression in JPEG2000. They are defined as:

$$H_0(z) = \frac{2^{-5}}{\frac{64}{5\rho}-6+\rho}z^2\left(1+z^{-1}\right)^4\left(z^2+z^{-2} - (8-\rho)\left(z+z^{-1}\right) + \frac{128}{5\rho}+2\right)$$

$$G_0(z) = \frac{2^{-3}}{\rho-2}z^2\left(1+z^{-1}\right)^4\left(-z-z^{-1}+\rho\right) \tag{6.37}$$

where $\rho \approx 3.3695$. In comparisons of short length filters by Villasenor [23], the Daubechies 9/7 filters were ranked as the best performing for lossy image compression.

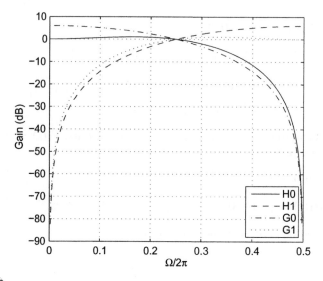

FIGURE 6.18

LeGall filter frequency responses.

6.4.3 **Multistage (multiscale) decompositions**

One important class of subband decomposition, commonly (although, as described previously, not always accurately) referred to as a discrete wavelet transform (DWT), is achieved by cascading a series of two-band systems, recursively applied to the low band. Using this method, the high frequency bands have good spatial resolution and poorer frequency resolution, while the low frequency bands have good frequency localization but with inferior spatial localization.

A 1-D three-stage decomposition is shown in Figure 6.19 and the associated time–frequency tilings at all stages of decomposition are given in Figure 6.20. This figure

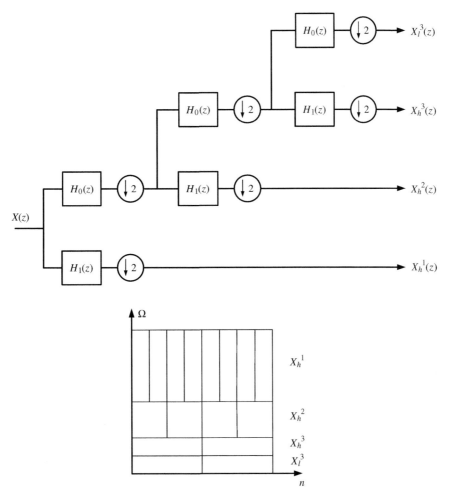

FIGURE 6.19

1-D three stage wavelet decomposition. Top: block diagram. Bottom: associated time–frequency tiling.

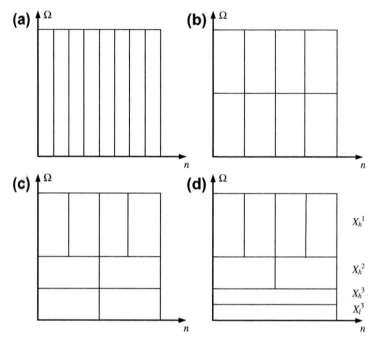

FIGURE 6.20

Time–frequency tiling for progressive stages of decomposition in a 1-D wavelet filter bank: (a) input of eight samples; (b) after the first stage; (c) after the second stage; (d) after the third stage.

shows an input of eight consecutive time domain samples, subsequently filtered and downsampled to yield four samples in each of two bands (b). The lower frequency band is then further filtered and downsampled to yield two more bands, each with two samples (c). Finally in (d), the low-pass band is once again filtered and downsampled to provide the tiling shown. This clearly shows the property of critical sampling as the number of samples (eight in this case) is identical at each decomposition stage.

In general, for an input sequence of length N, the relationship between the bandwidth and the number of samples in each subband is given in Table 6.1. This is shown for a five-stage decomposition, although the pattern for more or fewer stages should be obvious.

An alternative view of multistage decomposition

We can see from Figure 6.19 that each output from the analysis section is a series of filtering and downsampling operations. This can be restructured as shown in Figure 6.21. This is straightforward to show, based on the two alternative configurations at the bottom of Figure 6.21. For example, in the case of the downsampler followed by a filter $H(z)$, we have:

$$Y(z) = \frac{1}{2}H(z)\left(X\left(z^{\frac{1}{2}}\right) + X\left(-z^{\frac{1}{2}}\right)\right)$$

Table 6.1 Relationship between subband frequency range and number of samples for each subband in a five-stage decomposition.

Subband	Bandwidth	No. samples
X_h^1	$\pi/2 \cdots \pi$	$N/2$
X_h^2	$\pi/4 \cdots \pi/2$	$N/4$
X_h^3	$\pi/8 \cdots \pi/4$	$N/8$
X_h^4	$\pi/16 \cdots \pi/8$	$N/16$
X_h^5	$\pi/32 \cdots \pi/16$	$N/32$
X_l^5	$0 \cdots \pi/32$	$N/32$

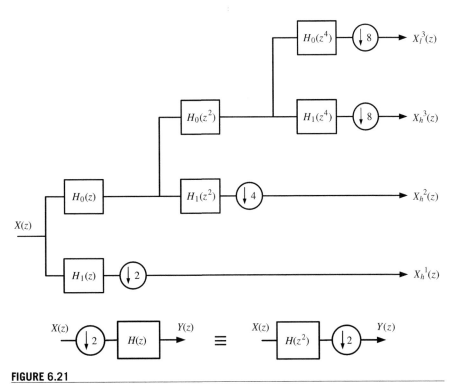

FIGURE 6.21

1-D three-stage wavelet decomposition—modified downsampling structure.

and, in the case of the modified filter followed by the downsampler, we have:

$$Y(z) = \frac{1}{2}\left(X\left(z^{\frac{1}{2}}\right)H(z) + X\left(-z^{\frac{1}{2}}\right)H(z)\right)$$

A similar relationship exists for an upsampler–filter combination.

6.4.4 Separability and extension to 2-D

A two-level subband decomposition applied in two dimensions (separably) is shown in Figure 6.22 and the corresponding tiling of the spatial frequency plane is shown in Figure 6.23. The upper left subband in Figure 6.23 contains the lowest frequency

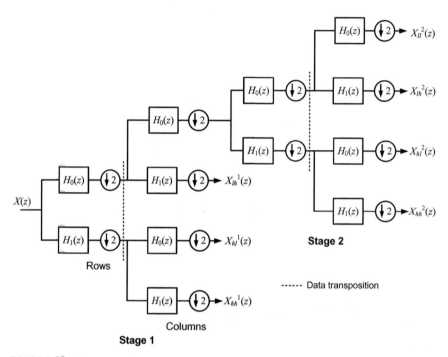

FIGURE 6.22

2D 2 channel 2 stage wavelet decomposition. Data transposition (as shown) is required between row and column processing.

FIGURE 6.23

2-D wavelet—frequency plane tiling.

FIGURE 6.24

Frame from Akiyo sequence (left) wavelet transformed (right) using a three-stage decomposition with LeGall 5/3 filters.

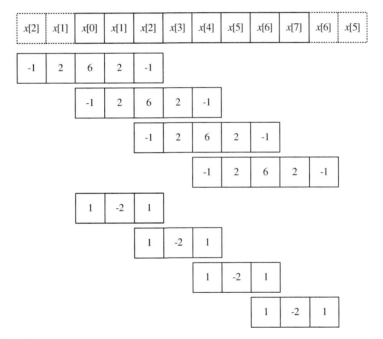

FIGURE 6.25

Illustration of boundary extension.

information and this would be recognizable as a small sub-sampled version of the original image. Intuitively therefore, this subband is important from an HVS point of view and must be coded accurately (usually using DPCM). The other subbands correspond to mid- and high-range spatial frequencies and usually contain relatively little energy. Thus the encoder can allocate fewer (or no) bits to coding this information. We will look at bit allocation strategies and entropy coding of these bands a little later.

An example showing a three-level 2-D wavelet decomposition of one frame in the Akiyo sequence is shown in Figure 6.24. This decomposition employs the LeGall 5/3 filters described earlier.

6.4.5 Finite length sequences, edge artifacts, and boundary extension

When we discussed critical sampling previously, we have assumed the context of infinitely long sequences. It is clear in the case of finite length sequences, such as those typical in image coding, that the number of subband samples will be greater than the number of input samples. If we maintain strict critical sampling, then distortion would be introduced by the filter bank, due to discontinuities at the signal boundaries, and perfect reconstruction would not be achieved. This can be reduced by extending the boundary samples either periodically or symmetrically or by zero padding. Symmetrical extension is normally preferred as this does not create discontinuities and has the effect of introducing lower boundary distortions.

We can easily maintain critical sampling with boundary extension and an example for an eight-sample input with 5/3 filters is shown in Figure 6.25. This illustrates the effective convolution shifts for each filter taking account of sub-sampling.

Example 6.4 (Boundary effects and critical sampling)
Consider a 1-D two-channel single stage filter bank based on the LeGall 5/3 filters. Compute the signal values at all intermediate nodes for the case of an input signal: $x[n] = \{1, 2, 3, 4, 5, 6, 7, 8\}$.

Solution. We can compute the convolution of the filters with the input sequence at all intermediate nodes, as defined in the following figure:

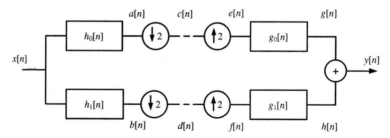

Let us consider first the case where critical sampling is not strictly observed. The signal values at each intermediate node are given in the following table:

$x[n]$	1 2 3 4 5 6 7 8
$a[n]$	−1 0 7 16 24 32 40 48 65 56 9 −8
$b[n]$	1 0 0 0 0 0 0 0 −9 8
$c[n]$	−1 7 24 40 65 9
$d[n]$	1 0 0 0 −9

$e[n]$	$-1\ 0\ 7\ 0\ 24\ 0\ 40\ 0\ 65\ 0\ 9$
$f[n]$	$1\ 0\ 0\ 0\ 0\ 0\ 0\ 0\ -9$
$g[n]$	$-1\ -2\ 6\ 14\ 31\ 48\ 64\ 80\ 105\ 130\ 74\ 18\ 9\ 0\ 0$
$h[n]$	$1\ 2\ -6\ 2\ 1\ 0\ 0\ 0\ -9\ -18\ 54\ -18\ -9$
$y[n]$	**$0\ 0\ 0\ 16\ 32\ 48\ 64\ 80\ 96\ 112\ 128\ 0\ 0$**

Note that the filter gain terms have been ignored here for convenience. However, it is clear that the answer is correct when we divide the output by 16. The problem with this is that we have 11 samples at the output of the analysis stage (c and d) rather than 8. If we maintain strict critical sampling, then distortion would be introduced by the filter bank due to discontinuities at the signal boundaries and perfect reconstruction would not be achieved.

Let us now consider what happens if we keep only eight samples for processing by the synthesis section. Let us discard the first two samples of the a and b signals as these are associated with the filter transients, thus:

$c[n]$	$7\ 24\ 40\ 65$
$d[n]$	$0\ 0\ 0\ -9$
$e[n]$	$7\ 0\ 24\ 0\ 40\ 0\ 65$
$f[n]$	$0\ 0\ 0\ 0\ 0\ 0\ -9$
$g[n]$	$7\ 14\ 31\ 48\ 64\ 80\ 105\ 130\ 65\ 0\ 0$
$h[n]$	$0\ 0\ 0\ 0\ 0\ 0\ -9\ -18\ 54\ -18\ -9$
$y[n]$	**$7\ 14\ 31\ 48\ 64\ 80\ 96\ 112\ 119\ -18\ -9$**

We can now see that the filter bank output is corrupted at its boundaries. We can improve this by boundary extension using, for example, symmetric extension.

6.4.6 Wavelet compression performance

An excellent comparison between the performances of various types of wavelet filter is provided by Villasenor [23]. Unser and Blu [27] also provide a rigorous comparison between the LeGall 5/3 and the Daubechies 9/7 filters in the context of JPEG2000. Both papers show that the 9/7 filters provide excellent performance with compact support for lossy coding. Considering their size, the 5/3 filters also perform exceptionally well in an image coding context, with the added advantage that lossless reconstruction can always be guaranteed because of their use of integer coefficients.

A comparison between wavelet and DCT compression methods (based on PSNR) is provided for the case of the 256×256 Lena image in Figure 6.26. Although this shows clear benefits of wavelet methods over their DCT counterparts, we must be careful of reading too much into this graph because of the significant differences in the coding methods used and in the artifacts produced by them. In subjective testing, however, it has been widely reported that wavelet coding (for example in JPEG2000) significantly outperforms DCT coding (as in JPEG). Subjective performance comparisons are provided in Section 6.6 (see Figure 6.27).

FIGURE 6.26

Compression performance, DCT vs wavelet, for the 256×256 luma Lena image.

FIGURE 6.27

Comparisons of wavelet-coded images, 256×256 Lena. Left: LeGall 5/3 wavelets at 0.3 bpp. Right: LeGall 5/3 wavelets at 0.5 bpp.

6.5 Coefficient quantization and bit allocation

6.5.1 Bit allocation and zonal coding

In order to quantize subband coefficients we can, in a similar manner to our approach with the DCT, apply different quantizer weights to different subbands according to their perceptual importance. This is shown in Figure 6.28, where different weights, W_0 to W_6, are used to differentiate both subband orientation and scale.

FIGURE 6.28

Subband weighting.

FIGURE 6.29

Scanning of wavelet subbands.

Similar to the way that we applied zig-zag scanning to a DCT block in Chapter 5, we can also scan subband coefficients from high to low subbands and in a manner that is matched to the orientation of energy in the subband. This is shown in Figure 6.29

where, for each of the three orientations, a data structure is used that supports efficient quantization. For example, subbands 2, 5, and 8 are likely to be highly correlated since subband 2 is a coarse approximation of subband 5 and subband 5 is a coarse approximation of subband 8. Similarly for subbands 4, 7, and 10 and subbands 3, 6, and 9. We can observe that the scanning patterns have been devised to exploit these relationships.

In the case of subband 1 (DC), coefficients will typically be encoded differentially and then {size, value} symbols entropy coded. For the AC coefficients, after scanning, a {run length/size, value} format, similar to that used with JPEG, would normally be employed to yield symbols for entropy coding.

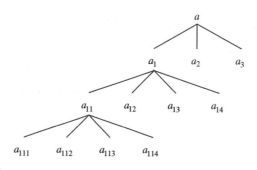

FIGURE 6.30

Tree structures for embedded wavelet coding.

6.5.2 **Hierarchical coding**

Alternative scanning methods have been proposed in the context of wavelet compression, that provide the basis for scalable encoding using embedded bitstreams. In particular the reader is referred to the work of Shapiro [28], who introduced the concept of scalable coding using embedded zero-tree coding (EZW), and that of Said and Pearlman [16], who extended this idea into an approach called set partitioning in hierarchical trees (SPIHT). We will not consider these methods in detail, but provide a brief introduction below.

Embedded coding methods generally follow the approach illustrated in Figure 6.30, where multiscale dependencies (self-similarities) are exploited across the wavelet subbands when coding the coefficients. The tree structure in the lower subfigure captures these dependencies. In EZW, Shapiro showed how these dependencies could be exploited using a succession of coding passes through the tree, in each pass coding significant coefficients that exceed a threshold. As the threshold decreases, more bits are included and a natural bit-plane data structure results. In EZW, arithmetic coding is essential in order to compress the data resulting from the significance passes. Said and Pearlman extended EZW with SPIHT which provides more efficient subset partitioning and which can perform exceptionally well, even without arithmetic coding.

6.6 **JPEG2000**

6.6.1 **Overview**

JPEG has been a big success with around 80% of all images still stored in this format. However, in the late 1990s, JPEG's limited coding efficiency, the presence of annoying visual blocking artifacts at high compression ratios, limited color gamut, and limited resolution were all reasons why work on an improved standard began. Furthermore, many applications were emerging that required functionalities not supported by JPEG, for example: spatial scalability, SNR scalability, and region of interest (ROI) coding.

The aim of JPEG2000 was to achieve a 30% bit rate saving for the same quality compared to JPEG, supporting 2–16 million colors and both lossless and lossy compression within the same architecture at bit depths greater than 8 bits. Additional features to support region of interest coding, error resilience, and data security were also included and a high emphasis was placed on scalability and progressive transmission. JPEG2000 Part 1 became an International Standard (ISO/IEC 15444-1) in December 2000 [21,29,30].

6.6.2 **Architecture—bit planes and scalable coding**

In JPEG2000, each subband is split into a number of code blocks. The encoder uses Embedded Block Coding with Optimal Truncation (EBCOT) [17,21] to code the quantized coefficients. Each bit plane of each code block is encoded during three passes (Significance Propagation, Magnitude Refinement and Cleanup), the first encoding bits (and signs) of insignificant coefficients with significant neighbors, the

Table 6.2   Technical overview of JPEG2000.

Property	Attribute
Wavelet transform	Dyadic decomposition
Filters	5/3 Integer filter (enables strictly reversible compression) [25]
	9/7 Floating point filter [26]
Block structure	Each subband is divided into codeblocks
Quantization	Scalar quantization
Entropy coding	Block-based (EBCOT) and context-based arithmetic coding
Regions of Interest	Maxshift with multiple regions supported—ROI shape not encoded
Error resilience	JPWL for wireless transmission
Security tools	JPSEC
Streaming tools	JPIP

second refining bits of significant coefficients and finally coding coefficients without significant neighbors. In the JPEG2000 lossless mode, all bit planes must be encoded, whereas in lossy mode some of the bit planes can be dropped. The bits selected during EBCOT are passed to a binary MQ encoder (a context-driven binary arithmetic coder), where the context of each coefficient is related to the state of its nine neighbors in the code block.

An overview of the main features of JPEG2000 is given in Table 6.2.

6.6.3 Coding performance

Examples of JPEG2000 coding are provided in Figure 6.31. This shows the Puppet test image (512×512), encoded and decoded at compression ratios of 64:1 and 128:1, compared to JPEG. It can be seen that the JPEG2000 codec significantly outperforms JPEG at high compression ratios.

In terms of compression performance, JPEG2000 achieves similar results to the intra mode of H.264/AVC and is generally inferior to HEVC. The latter two methods benefit from variable block sizes and sophisticated intra-prediction techniques. This perhaps also provides an explanation of why wavelet methods have not been widely adopted for video coding, where advanced prediction modes and adaptive coding structures, used in combination with block-based transforms, can often provide equivalent or superior results.

6.6.4 Region of interest coding

ROI processing in JPEG2000 enables the coding of important portions of an image at higher qualities than the surrounding content (or background). Before entropy coding, the bit planes of the coefficients belonging to the ROI mask are shifted up. This scaling corresponds to a local increase of dynamic range, which is then prioritized in the rate allocation process. At the decoder, the quantized wavelet coefficients within each ROI are scaled back to their original values. This process is illustrated in Figure 6.32.

FIGURE 6.31

JPEG2000 encoding of *Puppet* test image (512 × 512). Top left to bottom right: JPEG at 64:1; JPEG2000 at 64:1; JPEG at 128:1; JPEG2000 at 128:1.

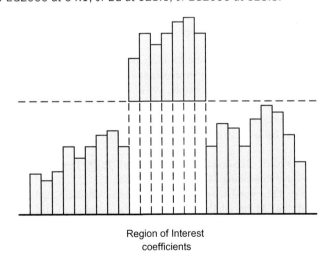

Region of Interest coefficients

FIGURE 6.32

JPEG2000 ROI coefficient coding using Maxshift.

6.6.5 Benefits and status

Although the core image coding part of JPEG2000 is intended by ITU-T to be royalty-free, it is not patent-free. It has been widely reported that the reason why the standard has not gained the success it perhaps deserves is because of fears of patent infringement. JPEG2000 has however gained favor in many military and surveillance applications, where its improved performance and its ability to offer scalable delivery have proved attractive.

6.7 Summary

This chapter has explored an approach to image decorrelation and source coding based on multiscale analysis and synthesis. Through the combination of appropriate filters with up- and downsampling, we have seen how critical constant sample rates can be maintained throughout the filter bank. We have also demonstrated how practical filters can be designed that, in the absence of quantization, achieve perfect reconstruction. We finally considered specific cases of wavelet filters, particularly in the context of the JPEG2000 still image coding standard.

When used in a lossy image compression system, wavelet-based coding generally produces less annoying visual artifacts than block-based transforms and it can also cope better with localized non-stationarities.

References

[1] A. Haar, Zur theorie der orthogonalen funktionensysteme, Mathematical Analysis 69 (1910) 331–371.

[2] A. Crosier, D. Esteban, C. Galand, Perfect channel splitting by use of interpolation/decimation techniques, in: Proceedings of the IEEE Conference on Information Science and Systems, 1976.

[3] M. Smith, T. Barnwell, A procedure for designing exact reconstruction filterbanks for tree-structured sub-band coders, in: Proceedings of the IEEE International Conference on Acoustics, Speech and Signal Processing, 1984.

[4] M. Smith, T. Barnwell, Exact reconstruction techniques for tree structured subband coders, IEEE Transactions on Acoustics, Speech and Signal Processing 34 (1986) 434–441.

[5] M. Vetterli, Filter banks allowing perfect reconstruction, Signal Processing 10 (3) (1986) 219–244.

[6] J. Woods, S. O'Neill, Sub-band coding of images, IEEE Transactions on Acoustics, Speech and Signal Processing 34 (1986) 1278–1288.

[7] P. Vaidyanathan, Theory and design of M-channel maximally decimated quadrature mirror filters with arbitrary M, having the perfect reconstruction property, IEEE Transaction on Acoustics, Speech and Signal Processing 35 (4) (1987) 476–492.

[8] A. Grossman, J. Morlet, Decomposition of hardy functions into square-integrable wavelets of constant shape, SIAM Journal on Mathematical Analysis 15 (4) (1984) 723–736.

[9] I. Daubechies, Orthonormal bases of compactly supported wavelets, Communications on Pure and Applied Mathematics 41 (1988) 909–996.

[10] S. Mallat, A theory for multiresolution signal decomposition: the wavelet representation, IEEE Transactions on Pattern Analysis and Machine Intelligence 11 (7) (1989) 674–693.

[11] S. Mallat, Multifrequency channel decomposition of images and wavelet models, IEEE Transactions on Acoustics, Speech and Signal Processing 37 (12) (1989) 2091–2110.

[12] M. Vetterli, J. Kovacevic, Wavelets and Subband Coding, Prentice Hall, 1995 (reissued 2007).

[13] R. Rao, A. Bopardikar, Wavelet Transforms – Introduction to Theory and Applications, Addison Wesley, 1998.

[14] G. Strang, T. Nguyen, Wavelets and Filter Banks, Wellesley-Cambridge Press, 1996.

[15] O. Rioul, M. Vetterli, Wavelets and signal processing, IEEE Signal Processing Magazine 8 (4) (1991) 14–38.

[16] A. Said, W. Pearlman, A new, fast, and efficient image codec based on set partitioning in hierarchical trees, IEEE Transactions on Circuits and Systems for Video Technology 6 (3) (1996) 243–250.

[17] D. Taubman, High performance scalable image compression with EBCOT, IEEE Transactions on Image Processing 9 (7) (2000) 1151–1170.

[18] D. Taubman, A. Secker, Highly scalable video compression with scalable motion coding, IEEE Transactions on Image Processing 13 (8) (2004) 1029–1041.

[19] S. Bokhari, A. Nix, D. Bull, Rate-distortion-optimized video transmission using pyramid vector quantization, IEEE Transactions on Image Processing 21 (8) (2012) 3560–3572.

[20] P. Schelkens, A. Skodras, T. Ebrahimi, The JPEG 2000 Suite, Wiley-IS&T Series in Imaging Science and Technology, Wiley, 2009.

[21] D. Taubman, M. Marcellin, JPEG 2000: Image Compression Fundamentals, Standards and Practice, Kluwer International Series in Engineering and Computer Science, 2001.

[22] J. Johnson, A filter family designed for use in quadrature mirror filter banks, in: Proceedings of the IEEE International Conference on Acoustics, Speech and Signal Processing, 1980, pp. 291–294.

[23] J. Villasenor, B. Belzer, J. Liao, Wavelet filter evaluation for image compression, IEEE Transactions on Image Processing 4 (8) (1995) 1053–1060.

[24] N. Kingsbury, Good Filters/ Wavelets, Connexions. <http://cnx.org/content/m11139/2.3/>, June 8, 2005.

[25] D. LeGall, A. Tabatabai, Sub-band coding of digital images using symmetric short kernel filters and arithmetic coding techniques, in: Proceedings of the IEEE International Conference on Acoustics, Speech and Signal Processing, 1988, pp. 761–765.

[26] M. Antonini, M. Barlaud, P. Mathieu, I. Daubechies, Image coding using wavelet transform, IEEE Transactions on Image Processing 1 (2) (1992) 205–220.

[27] M. Unser, T. Blu, Mathematical properties of the JPEG2000 wavelet filters, IEEE Transactions on Image Processing 12 (9) (2003) 1080–1090.

[28] J. Shapiro, Embedded image coding using zerotrees of wavelet coefficients, IEEE Transactions on Signal Processing 41 (12) (1993) 3445–3462.

[29] JPEG2000 Core Coding System, ISO/IEC Standard ISO/IEC 15444-1, 2000.

[30] <http://www.itu.int/rec/T-REC-T.800/en>.

Lossless Compression Methods

CHAPTER OUTLINE

7.1 **Motivation for lossless image compression** **214**
    7.1.1 Applications 215
    7.1.2 Approaches 216
    7.1.3 Dictionary methods 216
7.2 **Symbol encoding** **217**
    7.2.1 A generic model for lossless compression 217
    7.2.2 Entropy, efficiency, and redundancy 217
    7.2.3 Prefix codes and unique decodability 218
7.3 **Huffman coding** **219**
    7.3.1 The basic algorithm 219
    7.3.2 Minimum variance Huffman coding 221
    7.3.3 Huffman decoding 223
    7.3.4 Modified Huffman coding 224
    7.3.5 Properties of Huffman codes 224
    7.3.6 Adaptive methods for Huffman encoding 225
7.4 **Symbol formation and encoding** **229**
    7.4.1 Dealing with sparse matrices 229
    7.4.2 Symbol encoding in JPEG 230
    7.4.3 JPEG performance examples 235
    7.4.4 Lossless JPEG mode 235
    7.4.5 Context-Adaptive Variable Length Coding (CAVLC)
         in H.264/AVC 236
7.5 **Golomb coding** **239**
    7.5.1 Unary codes 239
    7.5.2 Golomb and Golomb–Rice codes 239
    7.5.3 Exponential Golomb codes 240
7.6 **Arithmetic coding** **241**
    7.6.1 The basic arithmetic encoding algorithm 241
    7.6.2 The basic arithmetic decoding algorithm 244
    7.6.3 Advantages of arithmetic coding 246
    7.6.4 Binary arithmetic coding 247
    7.6.5 Tag generation with scaling 249
    7.6.6 Context-adaptive arithmetic coding 250

Communicating Pictures. http://dx.doi.org/10.1016/B978-0-12-405906-1.00007-6
© 2014 Elsevier Ltd. All rights reserved.

7.7 **Performance comparisons—Huffman vs arithmetic coding**   251
7.8 **Summary**   252
References   252

This chapter describes methods for coding images without loss; this means that, after compression, the input signal can be exactly reconstructed. These have two primary uses. Firstly, they are useful when it is a requirement of the application that loss of any kind is unacceptable. Examples include some medical applications, where compression artifacts could be considered to impact upon diagnoses, or perhaps in legal cases where documents are used as evidence. Secondly, and more commonly, they are used as a component in a lossy compression technique, in combination with transformation and quantization, to provide entropy-efficient representations of the resulting coefficients.

We first review the motivations and requirements for lossless compression and then focus attention on the two primary methods, Huffman coding and arithmetic coding. Huffman coding is used widely, for example in most JPEG codecs, and we show it can be combined with the DCT to yield a compressed bitstream with minimum average codeword length. We explore the properties and limitations of Huffman coding in Section 7.3 and then introduce an alternative method—arithmetic coding—in Section 7.6. This is an elegant and flexible, yet simple, method for encoding strings of symbols rather than single symbols as in Huffman coding. Finally, Section 7.7 presents some performance comparisons between Huffman and arithmetic coding.

7.1 Motivation for lossless image compression

For certain image compression applications, it is important that the decoded signal is an exact replica of the original. Such methods are known as "lossless" since there is no loss of information between the encoder and the decoder. They do not exploit psychovisual redundancy, but they do exploit statistical redundancy to achieve a bit rate close to the entropy of the source.

The arguments for using lossless compression can be summarized as follows:

- With some applications there can be uncertainty as to what image features are important and how these would be affected by lossy image coding. For example, in critical medical applications, errors in a diagnosis or interpretation could be legally attributed to artifacts introduced by the image compression process.
- In some applications it is essential that data is preserved intact—for example, text or numerical information in a computer file. In such cases, if the data is corrupted, it may lose value or malfunction.
- Lossless compression methods can provide an efficient means of representing the sparse matrices that result from quantization of transform coefficients in a lossy compression system. Sparsity is introduced through the use of decorrelation and

For colour and higher quality versions of Figures 7.9 and 7.10 please refer to the electronic version or the website.

quantization to exploit visual redundancy and lossless methods can further exploit the statistics of symbols, to achieve a bit rate close to the source entropy.

7.1.1 Applications

A good example of early lossless compression is the Morse code. In order to reduce the amount of data sent over the telegraph system, Samuel Morse, in the 19th century, developed a technique where shorter codewords were used to represent more common letters and numbers. His binary (digraph) system was based on dots and dashes with a space to delimit letters and an even longer space to indicate boundaries between words. For example, "a" and "e" are common letters while "q" and "j" occur less frequently, hence Morse used the mappings:

$$e \mapsto . \quad a \mapsto . - \quad q \mapsto - - - . - \quad j \mapsto . - - -$$

In practice, Morse signaling was effected by switching on and off a tone from an oscillator, with a longer tone representing a dash and a shorter duration for a dot. The full structure of Morse code for the letters of the English alphabet is given in Figure 7.1. Not shown are the other symbols represented such as numbers and special characters, including common pairs of letters such as "CH." Note the tree-structured representation used in the figure with left and right branches labeled as dots and dashes respectively. We will see a similar structure later when we examine Huffman codes. It should also be noted that the Morse mapping demands a space between codewords— because otherwise the reader could not differentiate between two "E"s and an "I." We will see later how Huffman encoding cleverly overcomes this limitation of Morse's variable length coding.

Example 7.1 (Morse code)
What is the Morse code for "MORSE CODE."

Solution. It can be seen from Figure 7.1 that the phrase MORSE CODE is represented by:

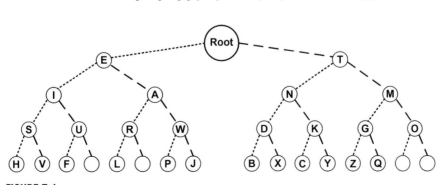

FIGURE 7.1

Symbol-code mappings for the 26 letters from the English alphabet in Morse code.

If we compare this with, let's say, a 5 bit fixed-length coding (actually there are 66 symbols in the Morse alphabet), which would require 45 bits (including a symbol for "space"), the Morse representation requires only 23 bits, a saving of approximately 50%.

Unknown to many, this encoding system is still widely used today—every radio navigation beacon used in aviation has a Morse identifier, as does every licensed airfield. Any pilot who selects such a beacon will always identify it prior to using it for navigation. This is to ensure that it is operational and that they have selected the correct beacon frequency!

7.1.2 Approaches

Several methods exist that can exploit data statistics to provide lossless compression. These include:

- **Huffman coding [1]:** This is a method for coding individual symbols at a rate close to the first order entropy, often used in conjunction with other techniques in a lossy codec.
- **Arithmetic coding [2–4]:** This is a more sophisticated method which is capable of achieving fractional bit rates for symbols, thereby providing greater compression efficiency for more common symbols.
- **Predictive coding:** This exploits data correlations, as opposed to symbol redundancies, and can be used without quantization to provide lossless image compression. As we saw in Chapter 3 and will examine further in this chapter, predictive coding is often used as a pre-processor for entropy coding.
- **Dictionary-based methods:** Algorithms such as LZ and LZW (Lempel–Ziv–Welch) [5,6] are well suited to applications where the source data statistics are unknown. They are frequently used in text and file compression.

Frequently, the above methods are used in combination. For example, DC DCT coefficients are often encoded using a combination of DPCM and either Huffman or arithmetic coding. Furthermore, as we will see in Chapter 8, motion vectors are similarly encoded using a form of predictive coding to condition the data prior to entropy coding.

Lossless encoding alone cannot normally provide the compression ratios required for most modern storage or transmission applications—we saw in Chapter 1 that requirements for compression ratios of 100:1 or even 200:1 are not uncommon. In contrast, lossless compression methods, if used in isolation, are typically restricted to ratios of around 4:1.[1]

7.1.3 Dictionary methods

Dictionary methods such as LZW (including LZ77, LZ78) are, in general, less suitable for image coding applications where the data statistics are known or can be estimated

---

[1]This is highly data- and application-dependent.

in advance. They do however find widespread use in more general data compression applications for text and file compression, in methods such as ZIP, UNIX compress, and GIF (Graphics Interchange Format). While these methods do not perform particularly well for natural images, methods such as PNG (Portable Network Graphics) improve upon them by pre-processing the image to exploit pixel correlations. PNG outperforms GIF and, for some images, can compete with arithmetic coding. We will not consider these techniques further here. The interested reader is referred to Sayood's book [7] for an excellent introduction.

7.2 Symbol encoding

7.2.1 A generic model for lossless compression

A generic model for lossless encoding is shown in Figure 7.2. This comprises a means of mapping the input symbols from the source into codewords. In order to provide compression, the symbol to codeword mapper must be conditioned by a representative probability model. A probability model can be derived directly from the input data being processed, from a priori assumptions or from a combination of both. It is important to note that compression is not guaranteed—if the probability model is inaccurate then no compression or even data expansion may result. As discussed in Chapter 3, the ideal symbol to code mapper will produce a codeword requiring $\log_2(1/P_i)$ bits.

7.2.2 Entropy, efficiency, and redundancy

As described in Chapter 3, H is known as the source entropy, and represents the minimum number of bits per symbol required to code the signal. For a given encoder which codes a signal at an average rate (i.e. an average codeword length) $\bar{l}$, the coding efficiency can be defined as:

$$E = \frac{H}{\bar{l}} \qquad (7.1)$$

Thus, if we code a source in the most efficient manner and the code is unambiguous (uniquely decodable) then the rate of the code will equal its entropy.

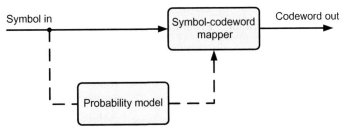

FIGURE 7.2

Generic model of lossless coding.

Table 7.1 Alternative codes for a given probability distribution.

Letter	Probability	C_1	C_2	C_3
a_1	0.5	0	0	0
a_2	0.25	0	1	10
a_3	0.125	1	00	110
a_4	0.125	10	11	111
Average length		1.125	1.25	1.75

A coding scheme can also be characterized by its redundancy, R:

$$R = \frac{\bar{l} - H}{H} \cdot 100\% \tag{7.2}$$

7.2.3 Prefix codes and unique decodability

We noted earlier that Morse code needs spaces to delineate between its variable length codewords. In modern digital compression systems we cannot afford to introduce a special codeword for "space" as it creates overhead. The way we overcome this is to use what are referred to as prefix codes. These have the property of ensuring that no codeword is the prefix of any other, and hence (in the absence of errors) that each codeword is uniquely decodable. Consider the alternative codes for the four-letter alphabet $\{a_1, a_2, a_3, a_4\}$ in Table 7.1.

Let us examine the three codes in this table. We can see that, in the case of C_1, there is ambiguity as $a_2 a_3 = a_1 a_3$. Similarly for C_2, $a_1 a_1 = a_3$. In fact, only C_3 is uniquely decodable and this is because it is a prefix code, i.e. no codeword is formed by a combination of other codewords. Because of this, it is self-synchronizing—a very useful property. The question is, how do we design self-synchronizing codes for larger alphabets? This is addressed in the next section.

Example 7.2 (Coding redundancy)
Consider a set of symbols with associated probabilities and codewords as given in the table below. Calculate the first order entropy of the alphabet and the redundancy of the encoding.

Symbol	Probability	Codeword
s_0	0.06	0110
s_1	0.23	10
s_2	0.3	00
s_3	0.15	010
s_4	0.08	111
s_5	0.06	0111
s_6	0.06	1100
s_7	0.06	1101

Solution.

1. **Average codeword length**
   The table is repeated below including the average length of each codeword. The average length $\bar{l}$ in this case is 2.71 bits/symbol.

Symbol	Probability	Code	Average length
s_0	0.06	0110	0.24
s_1	0.23	10	0.46
s_2	0.3	00	0.6
s_3	0.15	010	0.45
s_4	0.08	111	0.24
s_5	0.06	0111	0.24
s_6	0.06	1100	0.24
s_7	0.06	1101	0.24
Overall average length			**2.71**

2. **First order entropy**
   The first order entropy for this alphabet is given by:

   $$H = -\sum P \log_2 P$$
   $$= -(0.06 \times \log_2 0.06 + 0.023 \times \log_2 0.23 + 0.3$$
   $$\times \log_2 0.3 + \cdots + 0.06 \times \log_2 0.06)$$
   $$= 2.6849 \text{ bits/symbol}$$

3. **Redundancy**
   The redundancy of this encoding is:

   $$R = \frac{\bar{l} - H}{H} \times 100 = \frac{2.71 - 2.6849}{2.6849} \approx 1\%$$

7.3 Huffman coding

7.3.1 The basic algorithm

Huffman coding represents each symbol s_i with a variable length binary codeword c_i. The length of c_i is determined by rounding H_i up to the nearest integer. Huffman codes are prefix codes, i.e. no valid codeword is the prefix of any other. The decoder can thus operate without reference to previous or subsequent data and does not require explicit synchronization signals.

Huffman codes are generated using a tree-structured approach as will be described below. The decoder decodes codewords using a decoding tree of a similar form—this means that the exact tree structure must be transmitted prior to decoding or must be

---

Algorithm 7.1 Huffman tree formation.

1. Create ranked list of symbols $s_0 \cdots s_N$ in decreasing probability order;
2. REPEAT
3.   Combine pair of symbols with lowest probabilities. i.e. those at bottom of list (indicated as s_i and s_j here);
4.   Update probability of new node: $P(s_i') = P(s_i) + P(s_j)$;
5.   Sort intermediate nodes: order s_i' in the ranked list;
6. UNTIL a single root node is formed with $P = 1$;
7. Label the upper path of each branch node with a binary 0 and the lower path with a 1;
8. Scan tree from root to leaves to extract Huffman codes for each symbol;
9. END.

---

known a priori. For each symbol, the decoder starts at the tree root and uses each bit to select branches until a leaf node is reached which defines the transmitted symbol.

Huffman coding provides an efficient method for source coding provided that the symbol probabilities are all reasonably small. For coding of subband data (e.g. within JPEG) typical efficiencies between 90% and 95% are achievable. If any one of the symbol probabilities is high, then H will be small, and the efficiency of the codec is likely to be lower. Huffman coding requires that both the encoder and decoder have the same table of code words. In some systems (e.g. JPEG) this table is transmitted prior to the data (see Figure 7.7).

The procedure for creating a basic Huffman tree is as described by Algorithm 7.1. It should be clear that this procedure ensures that the result is a prefix code because, unlike Morse code, each symbol is represented by a unique path through the graph from root to leaf.

Example 7.3 (Basic Huffman encoding)
Consider a source which generates five symbols $\{s_1, s_2, s_3, s_4, s_5\}$ where $P(s_1) = 0.2$, $P(s_2) = 0.4$, $P(s_3) = 0.2$, $P(s_4) = 0.15$, $P(s_5) = 0.05$. Compute the Huffman tree for this alphabet.

Solution. The solution is described in stages below. In the first stage the symbols are ordered in terms of their probability of occurrence and the bottom two symbols on the list are combined to form a new node $s_4'(0.2)$ where the value in parentheses is the sum of the probabilities of the two constituent nodes. We then ensure that the newly formed list of nodes is also in descending probability order as shown below:

$$s_2(0.4) \longrightarrow s_2(0.4)$$

$$s_1(0.2) \longrightarrow s_1(0.2)$$

$$s_3(0.2) \longrightarrow s_3(0.2)$$

$$s_4(0.15) \rightharpoondown s_4'(0.2)$$

$$s_5(0.05) \longrightarrow$$

In stage 2, we repeat this process on the newly formed list of nodes, combining the bottom two nodes to form a new node $s_3'(0.4)$. This new node is placed in the output list in a position to ensure descending probability values. This is shown below:

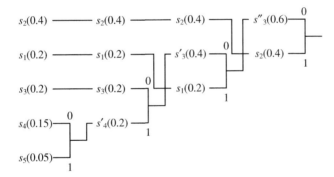

Finally in the third stage we form $s_3''(0.6)$ and combine this with the only other remaining node $s_2(0.4)$ to form the single root node with, as expected, a combined probability of 1:

By tracing each path through the graph, we can see that the Huffman codewords for this alphabet are:

$$s_1 = 01, \quad s_2 = 1, \quad s_3 = 000, \quad s_4 = 0010, \quad s_5 = 0011$$

This gives an average codeword length of 2.2 bits/symbol.

7.3.2 Minimum variance Huffman coding

Minimum variance Huffman coding minimizes the length of the longest codeword, thus minimizing dynamic variations in transmitted bit rate. The procedure for achieving this is almost identical to the basic approach, except that when forming a new ranked list of nodes, the newly formed node is inserted in the list as high up as possible, without destroying the ranking order. This ensures that its reuse will be delayed as long as possible without destroying the ranking, thus minimizing the longest path through the graph. The revised algorithm is given in Algorithm 7.2.

Algorithm 7.2 Minimum variance Huffman tree formation.

1. Create ranked list of symbols $s_0 \cdots s_N$ in decreasing probability order;
2. REPEAT
3.   Combine pair of symbols with lowest probabilities, i.e. those at the bottom of the list (indicated as s_i and s_j here);
4.   Update probability of new node: $P(s'_i) = P(s_i) + P(s_j)$;
5.   Sort intermediate nodes: insert s_i as high up as possible in the ranked list;
6. UNTIL a single root node is formed with $p = 1$;
7. Label the upper path of each branch node with a binary 0 and the lower path with a 1;
8. Scan tree from root to leaves to extract Huffman codes for each symbol;
9. END.

Example 7.4 (Minimum variance Huffman encoding)

Consider the same problem as in Example 7.3, where a source generates five symbols $\{s_1, s_2, s_3, s_4, s_5\}$ with $P(s_1) = 0.2$, $P(s_2) = 0.4$, $P(s_3) = 0.2$, $P(s_4) = 0.15$, $P(s_5) = 0.05$. Compute the minimum variance set of Huffman codes for this alphabet.

Solution. In the first stage the symbols are ordered in terms of their probability of occurrence and the bottom two symbols on the list are combined to form a new node $s'_4(0.2)$ where the value in parentheses is the sum of the probabilities of the two constituent nodes. We ensure that the newly formed node is inserted in the ordered list as high up as possible as shown below:

$s_2(0.4)$ ——————— $s_2(0.4)$

$s_1(0.2)$ ———————┐ $s'_4(0.2)$

$s_3(0.2)$ ———————┤ $s_1(0.2)$

$s_4(0.15)$———┐ $s_3(0.2)$

$s_5(0.05)$———┘

This process continues in a similar manner in stage 2:

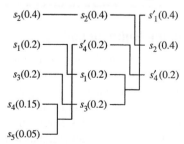

Until the complete tree is formed as shown below:

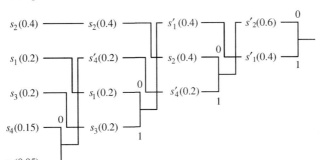

By tracing each path through the graph, we can see that the Huffman codewords for this alphabet are:

$$s_1 = 10, \quad s_2 = 00, \quad s_3 = 11, \quad s_4 = 010, \quad s_5 = 011$$

This again gives an average codeword length of 2.2 bits/symbol. However, this time the minimum variance approach has reduced the maximum codeword length to 3 bits rather than the 4 bits in the previous case.

7.3.3 Huffman decoding

Huffman decoding is a simple process which takes the input bitstream and, starting at the root node, uses each bit in sequence to switch between the upper and lower paths in the graph. This continues until a leaf node is reached, when the corresponding symbol is decoded. This is described for a bitstream of K bits representing a sequence of encoded symbols, by Algorithm 7.3. For ease of understanding, this algorithm

---

Algorithm 7.3 Huffman decoding process.

1. INPUT bitstream representing a sequence of symbols: $b[0], b[1] \cdots b[K-1]$; INPUT Huffman tree structure;
2. $i=0$;
3. REPEAT
4.   REPEAT
5.   Start at root node;
6.   IF $b[i]=0$ THEN select upper branch, ELSE select lower branch;
7. textit $i=i+1$;
8.   UNTIL leaf node reached;
9.   OUTPUT *symbol*;
10. UNTIL $i=K$;
11. END.

references the Huffman tree as the data structure during decoding. It should however be noted that this will normally be implemented using a state machine.

7.3.4 Modified Huffman coding

In practical situations, the alphabet for Huffman coding can be large. In such cases, construction of the codes can be laborious and the result is often not efficient, especially for biased probability distributions. For example, consider an 8 bit source with the following probabilities:

Value (x)	$P(x)$	$I(x)$	Code	Bits
0	0.2	2.232	10	2
+1	0.1	3.322	110	3
−1	0.1	3.322	1110	4
+2	0.05	4.322	11110	5
−2	0.05	4.322	11111	5
All other values	0.5	1	0 + value	9

This implies that we could generate short and efficient codewords for the first five symbols and assign an extra 1 bit flag as an escape code to indicate any of the symbols that is not one of the first five. These other values can then be fixed-length coded with (in this case) an 8 bit value. The average code length is then 2.05 bits.

An alternative approach, sometimes referred to as *extended Huffman coding*, can also be used to reduce the coding redundancy and code the alphabet at a rate closer to its entropy. Rather than coding each symbol independently, this method groups symbols together prior to code formation. As we saw in Chapter 3, this can provide significant advantages. However, there are more efficient methods for doing this such as *arithmetic coding* and we will look at these in Section 7.6.

7.3.5 Properties of Huffman codes

The Kraft inequality

An alphabet with N symbols $S = \{s_i\}$, if encoded using a binary prefix code and ranked according to their probability, P_i, will have $l_1 \le l_2 \le l_3 \le \cdots \le l_N$. The Kraft inequality states that these lengths must satisfy the following expression:

$$\sum_{i=1}^{N} 2^{-l_i} \le 1 \tag{7.3}$$

This follows from the information content of each symbol being:

$$l_i = -\log_2 P_i = \log_2 1/P_i$$

In the case of a dyadic source, where all codeword lengths are integers, then:

$$\bar{l} = \sum l_i P_i = H(S)$$

Length of Huffman codes

The average length of a codeword in a Huffman encoded alphabet will depend primarily on the distribution of probabilities in the alphabet and on the number of symbols in the alphabet. We have seen above that, for integer codeword lengths, the lower bound on the average codeword length is the source entropy, H. However, in cases where the codeword length is non-integer, for practical purposes they must be rounded up to the next largest integer value. Hence:

$$\bar{l} = \sum \lceil \log_2 1/P_i \rceil P_i < \sum \left(\log_2 \left(1/P_i\right) + 1\right) P_i < H(S) + 1$$

The Huffman code for an alphabet, S, therefore has an average length $\bar{l}$ bounded by:

$$H(S) \leq \bar{l} \leq H(S) + 1 \tag{7.4}$$

Optimality of Huffman codes

In order for the code to be optimum it must satisfy the following conditions [7,8]:

1. For any two symbols s_i and s_j in an alphabet S, if $P(s_i) > P(s_j)$ then $l_i \leq l_j$ where l represents the length of the codeword.
2. The two symbols with the lowest probabilities have equal and maximum lengths.
3. In a tree corresponding to the optimum code, each intermediate node must have two branches.
4. If we collapse several leaf nodes, connected to a single intermediate node, into a single composite leaf node then the tree will remain optimal.

It can clearly be seen that Huffman codes satisfy these conditions and represent an optimum representation for coding of individual symbols. We will however see later that greater efficiencies can be obtained by encoding sequences of symbols rather than individual symbols.

7.3.6 Adaptive methods for Huffman encoding

There are two basic limitations of Huffman coding:

1. It encodes one symbol at a time.
2. It depends on a pre-computed encoding and decoding structure.

As we will see later, both of these limitations are addressed by *arithmetic coding*. However, *adaptive Huffman coding* methods have also been devised which enable dynamic updating of the Huffman tree according to prevailing source statistics. Methods include those by Gallagher [9] and more recently by Vitter [10]. The essential element, as with any adaptive system, is that the encoder and decoder models initialize

FIGURE 7.3

Adaptive entropy coding.

and update sympathetically, using identical methods. The basis for this approach is shown in Figure 7.3.

In a simple case, both encoder and decoder would initialize their trees with a single node and both would know the symbol set in use. The update process assigns a weight W to each leaf node which equals the number of times that symbol has been encountered. Symbols are initially transmitted using a predetermined set of basic codes and, as frequencies of occurrence change, the tree is updated according to an agreed procedure.

An example of this procedure is illustrated in Algorithm 7.4. We first compute the number of nodes, M, required to represent all N symbols (i.e. $2N - 1$). We label each node with a number $1 \cdots M$ and append the current W value according to the frequency of occurrence of the corresponding symbol (intermediate nodes are assigned values equal to the sum of their siblings). We define an Escape node, initially as the root node with label N and weight $W = 0$. When the first symbol s_i is encountered, it is encoded using the pre-agreed initial encoding for the alphabet. The ESC node then creates a new root node (label N) with two siblings: that for s_i labeled $N - 1$ and a new ESC node labeled $N - 2$. The new codeword for s_i is now 1 and that for ESC is 0. So when a new symbol is encountered, it must be encoded as ESC (0 currently), followed by the pre-agreed code. This process continues, spawning new nodes as new symbols are encountered.

As the frequency count of a node increments, the tree will need to be restructured to account for this. This is achieved by swapping a node with weight W with the highest numbered node with weight $W - 1$ and then rearranging nodes to provide

---

Algorithm 7.4 Adaptive Huffman coding: tree updating.

1. INPUT symbol s_i;
2. If symbol s_i is new then create new node with two siblings: one for s_i and a new ESC node;
3. Update weight for node s_i: $W = W + 1$;
4. Swap node with weight W with the highest numbered node with weight $W - 1$ and then rearrange nodes to provide a minimum variance structure.

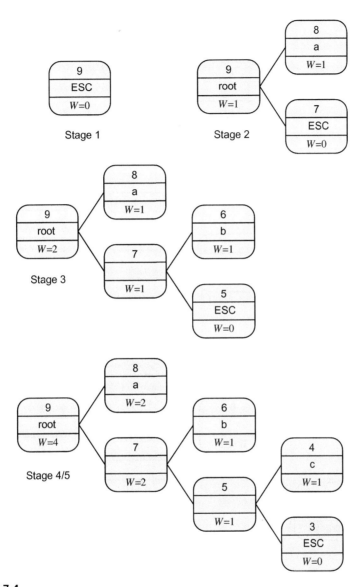

FIGURE 7.4

Example of adaptive Huffman encoding tree evolution.

a minimum variance structure. This process is capable of continually adapting to dynamic variations in symbol statistics, providing optimum codewords.

Figure 7.4 provides a simple example of the process of tree evolution for an alphabet of five symbols $\{a,b,c,d,e\}$, with the transmitted sequence $\{a,b,c,a\}$. Figure 7.5 illustrates the case of an established tree where node swapping is used to update the tree after the weight for symbol a is incremented by five counts.

FIGURE 7.5

Example of adaptive Huffman encoding showing node swapping (bottom) after weight updating symbol *a* by five counts.

7.4 Symbol formation and encoding

7.4.1 Dealing with sparse matrices

Many data sources contain long strings of one particular symbol. For example, the quantized subband data generated by a transform or wavelet-based compression system is typically sparse. We have seen in Chapter 5 that, after applying a forward transform and quantization, the resulting matrix contains a relatively small proportion of non-zero entries with most of its energy compacted toward the lower frequencies (i.e. the top left corner of the matrix). In such cases, *run-length coding* (RLC) can be used to efficiently represent long strings of identical values by grouping them into a single symbol which codes the value and the number of repetitions. This is a simple and effective method of reducing redundancies in a sequence. A slight variant on this for sparse matrices is to code runs of zeros in combination with the value of the next non-zero entry.

In order to perform run-length encoding we need to convert the 2-D coefficient matrix into a 1-D vector and furthermore we want to do this in such a way that maximizes the runs of zeros. Consider for example the 6×6 block of data and its transform coefficients in Figure 7.6. If we scan the matrix using a zig-zag pattern, as shown in the figure, then this is more energy efficient than scanning by rows or columns.

Example 7.5 (Run-length encoding of sparse energy-compact matrices)
Consider the 4×4 quantized DCT matrix below. Perform zig-zag scanning to produce a compact 1-D vector and run-length code this using $\{run/value\}$ symbols.

$$\mathbf{C}_Q = \begin{bmatrix} 8 & 5 & 2 & 0 \\ 3 & 0 & 0 & 0 \\ 0 & 0 & 0 & 0 \\ 1 & 0 & 0 & 0 \end{bmatrix}$$

Solution. Perform zig-zag scanning to produce a 1-D vector:

$$\mathbf{c}_Q = \begin{bmatrix} 8 & 5 & 3 & 0 & 0 & 2 & 0 & 0 & 0 & 1 & 0 & 0 & 0 & 0 & 0 \end{bmatrix}$$

Next represent this vector using $\{run/value\}$ symbols:

$$\mathbf{s}_Q = \begin{bmatrix} (0/8) & (0/5) & (0/3) & (2/2) & (3/1) & \text{EOB} \end{bmatrix}$$

FIGURE 7.6

Zig-zag scanning prior to variable length coding.

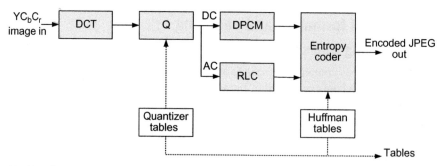

FIGURE 7.7

JPEG baseline image encoder architecture.

Note that, to avoid encoding the last run of zeros, it is normal to replace these with an End of Block (EOB) symbol.

7.4.2 Symbol encoding in JPEG

JPEG remains the universally adopted still-image coding standard. It was first introduced in 1992 and has changed little since then [11]. The operation of the baseline JPEG encoder is shown in Figure 7.7. First of all the input image, usually in 4:2:0 YC_bC_r format, is segmented into non-overlapping 8×8 blocks and these are decorrelated using an 8×8 DCT and quantized as discussed in Chapter 5. The entropy coding of these quantized coefficients is normally based on Huffman coding[2] and proceeds as follows:

- **DC coefficients** are pulse code modulated (DPCM) differentially with respect to the corresponding value from the previous block. The *size* category for the prediction residual (as defined in Table 7.4) is then Huffman encoded using a set of tables specifically designed for DC coefficients. The amplitude of the prediction residual is then appended to this codeword in ones complement form.
- **AC coefficients** are first of all zig-zag scanned as described above and then run-length encoded. The run-length symbols used for each non-zero AC coefficient are (*run/size*), where *run* is the number of zeros preceding the next non-zero coefficient and *size* relates to the value of that coefficient, as defined in Table 7.4. These symbols are then entropy encoded using a further set of Huffman tables designed for the AC coefficients and the coefficient value is again appended to the codeword in ones complement form.

These processes are described in more detail in Algorithms 7.5 and 7.6. The approach is justified by the graph in Figure 7.8, which shows the distribution of size values

---

[2]It should be noted that the JPEG standard also supports arithmetic coding but this is rarely used in practice.

Algorithm 7.5 JPEG DC coefficient symbol encoding.

1. Form DPCM residual for the DC coefficient in the current block DC_i, predicted relative to the DC coefficient from the previously encoded block, DC_{i-1} : $DC_{i(pred)} = DC_i - DC_{i-1}$;
2. Produce $symbol_{DC} = (size)$ for the DC coefficient prediction;
3. Huffman encode $symbol_{DC}$ using the Table for DC coefficient differences;
4. OUTPUT bits to JPEG file or bitstream;
5. OUTPUT *amplitude* value of the coefficient difference in ones complement format to the file or bitstream;
6. END.

Algorithm 7.6 JPEG AC coefficient symbol encoding.

1. Form a vector of zig-zag scanned AC coefficient values from the transformed data block as described in Figure 7.6;
2. $i=0$;
3. REPEAT
4. $i = i + 1$;
5.   Produce $symbol_{AC,i} = (run/size)$ for the next non-zero AC coefficient in the scan;
6.   Huffman encode $symbol_{AC,i}$ using Table for AC coefficients;
7.   OUTPUT codeword bits to the JPEG file or bitstream;
8.   OUTPUT *amplitude* value of the non-zero coefficient in ones complement form to the JPEG file or bitstream;
9. UNTIL $symbol_{AC,i-1} = (0/0)$;
10. END.

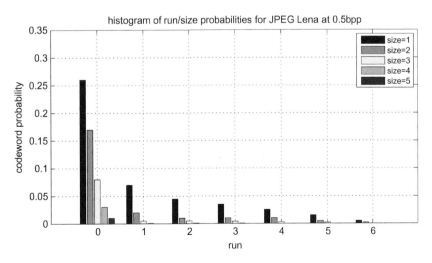

FIGURE 7.8

Distribution of *run/size* values in a typical JPEG encoding.

for various run lengths for a JPEG encoded Lena image. The figure shows statistics for 0.5 bpp but similar characteristics are obtained for other compression ratios. It can be seen that probabilities fall off rapidly as *run*-length increases and as *size* decreases. Hence longer codewords can be justified in such cases, with shorter codewords reserved for lower values of *run* and *size*.

JPEG actually restricts the number of bits allocated to run and size to 4 bits each, giving a range of 0–15 for both. If a run is greater than 15 zeros (which is unlikely) then it is divided into multiples of 15 plus the residual length. This is effective in constraining the size of the entropy code tables. Hence in JPEG, the *run/size* symbol is actually represented by: *RRRRSSSS*. Thus the only *run/size* symbols with a size of 0 are EOB (0/0) and ZRL (15/0).

Subsets of the default JPEG Huffman codes for DC and AC luminance coefficients are provided in Tables 7.2 and 7.3. A full set can be obtained from Ref. [11].

Example 7.6 (JPEG coefficient encoding)

Consider an AC luminance coefficient value of -3 preceded by five zeros. What is the JPEG symbol for this run and coefficient value? Using the default JPEG Huffman tables, compute the Huffman code for this coefficient and the resulting output bitstream.

Solution. From Table 7.4, it can be seen that the $size(-3) = 2$. The symbol is thus $s = (5/2)$.

From the default JPEG Huffman table for luminance AC coefficients, the Huffman code for (5/2) is 11111110111. The value of the non-zero coefficient is -3, which in ones complement format is 00.

The output bitstream produced for this sequence is thus:

$$1111111011100$$

Table 7.2 Table for luminance DC coefficient differences.

Category	Code length	Codeword
0	2	00
1	3	010
2	3	011
3	3	100
4	3	101
5	3	110
6	4	1110
7	5	11110
8	6	111110
9	7	1111110
10	8	11111110
11	9	111111110

Table 7.3 Table for luminance AC coefficients, showing *run/size* (*R/S*) values, Huffman codes, and their length (*L*).

R/S	L	Codeword
0/0	4	1010 (EOB)
0/1	2	00
0/2	2	01
0/3	3	100
0/4	4	1011
0/5	5	11010
0/6	7	1111000
0/7	8	11111000
0/8	10	1111110110
0/9	16	1111111110000010
0/A	16	1111111110000011
1/1	4	1100
1/2	5	11011
1/3	7	1111001
1/4	9	111110110
1/5	11	11111110110
1/6	16	1111111110000100
1/7	16	1111111110000101
1/8	16	1111111110000110
1/9	16	1111111110000111
1/A	16	1111111110001000
2/1	5	11100
2/2	8	11111001
2/3	10	1111110111
2/4	12	111111110100
2/5	16	1111111110001001
2/6	16	1111111110001010
2/7	16	1111111110001011
2/8	16	1111111110001100
2/9	16	1111111110001101
2/A	16	1111111110001110
3/1	6	111010
3/2	9	111110111
3/3	12	111111110101
3/4	16	1111111110001111
3/5	16	1111111110010000
3/6	16	1111111110010001
3/7	16	1111111110010010
3/8	16	1111111110010011
3/9	16	1111111110010100
3/A	16	1111111110010101

Example 7.7 (JPEG matrix encoding)

Consider the following quantized luminance DCT matrix. Assuming that the value of the DC coefficient in the preceding block is 40, what is the JPEG symbol sequence

Table 7.4 JPEG coefficient size categories.

Category	Amplitude range
0	0
1	$-1, +1$
2	$-3, -2, +2, +3$
3	$-7 \cdots -4, +4 \cdots +7$
4	$-15 \cdots -8, +8 \cdots +15$
5	$-31 \cdots -16, +16 \cdots +31$
6	$-63 \cdots -32, +32 \cdots +63$
etc.	etc.

for this matrix? Using the default JPEG Huffman tables, compute its Huffman codes and the resulting output bitstream. Assuming the original pixels are 8 bit values, what is the compression ratio for this encoding?

$$
\mathbf{C}_Q =
\begin{bmatrix}
42 & 16 & 0 & 0 & 0 & 0 & 0 & 0 \\
-21 & -15 & 0 & 0 & 0 & 0 & 0 & 0 \\
10 & 3 & -3 & 0 & 0 & 0 & 0 & 0 \\
-2 & 2 & 2 & 0 & 0 & 0 & 0 & 0 \\
0 & -1 & 0 & 0 & 0 & 0 & 0 & 0 \\
0 & 0 & 0 & 0 & 0 & 0 & 0 & 0 \\
0 & 0 & 0 & 0 & 0 & 0 & 0 & 0 \\
0 & 0 & 0 & 0 & 0 & 0 & 0 & 0
\end{bmatrix}
$$

Solution.

1. **DC coefficient**

   The prediction residual for the current DC coefficient is $DC_{i(\text{pred})} = DC_i - DC_{i-1} = 2$. Thus $size = 2$ and $amplitude = 2$.

   The Huffman codeword from Table 7.2 is 011. Hence the bitstream for the DC coefficient in this case is 01110.

2. **AC coefficients**

   The zig-zag scanned vector for the matrix above is:

   $$
   \mathbf{c}_Q = [\; 16 \quad -21 \quad 10 \quad -15 \quad 0 \quad 0 \quad 0 \quad 3 \quad -2 \quad 0 \quad \cdots
   $$
   $$
   \cdots \quad 2 \quad -3 \quad 0 \quad 0 \quad 0 \quad 0 \quad 0 \quad 2 \quad -1 \quad 0 \quad \cdots]
   $$

   For each of these non-zero coefficients, we can now compute the (*run/size*) symbol, its Huffman code (from Table 7.3) and the corresponding ones complement amplitude representation. These are summarized in the table below.

3. **Output bitstream**

   The output bitstream for this matrix is the concatenation of DC and AC codes as follows:

   $$
   (0111011010100001101001010101110101011000011111 \cdots
   $$
   $$
   \cdots 0111110101110111001001111111101111100001010)
   $$

4. **Compression ratio**
   Ignoring any header or side information, the total number of bits for this block is
   $5 + 82 = 87$ bits. In contrast, the original matrix has 64 entries, each of 8 bits.
   Hence it requires 512 bits, giving a compression ratio of 5.9:1 or alternatively
   1.36 bpp.

Coefficients	Run/size	Huffman code	Amplitude	#bits
16	0/5	11010	10000	10
−21	0/5	11010	01010	10
10	0/4	1011	1010	8
−15	0/4	1011	0000	8
3	3/2	111110111	11	11
−2	0/2	01	01	4
2	1/2	11011	10	7
−3	0/2	01	00	4
2	5/2	11111110111	10	13
−1	0/1	00	0	3
EOB	0/0	1010	–	4
Total AC bits				**82**

7.4.3 JPEG performance examples

Some examples that demonstrate the performance of JPEG encoding are shown in
Figures 7.9 and 7.10 for a highly textured image (*maple*) and a more structured image
(*puppet*). Both original images have a wordlength of 8 bits. As can be observed, the
results begin to be psychovisually acceptable beyond 0.25 bpp and become good at
1 bpp. Subjectively the result for *puppet* is more acceptable at 0.22 bpp than that
for *maple*, due to the lower contribution from high frequency textures that are more
heavily quantized during compression.

7.4.4 Lossless JPEG mode

A simple lossless mode is defined in the JPEG standard, which combines predictive
coding with Huffman or arithmetic coding. The prediction framework is shown in
Figure 7.11 and the choices of predictor are shown in Table 7.5. Predictor modes
are selected from the table and signaled as side information in the header—and are
normally kept constant for the whole scan. After prediction, residuals are Huffman
or arithmetically encoded. In practice the reconstructed pixel values, rather than the
actual pixel values, are used in the prediction to eliminate numerical mismatches
between the encoder and decoder.

FIGURE 7.9

JPEG results for various compression ratios for 512×512 *maple* image. Top left: original. Top right: 0.16 bpp. Middle left: 0.22 bpp. Middle right: 0.5 bpp. Bottom left: 1 bpp. Bottom right: 2 bpp.

7.4.5 Context-Adaptive Variable Length Coding (CAVLC) in H.264/AVC

In its baseline and extended profiles, H.264/AVC uses two methods of entropy coding. The first, used for all syntax elements apart from transform coefficients, is based on a single codeword table for all syntax elements (SEs). The mapping of the SE to this

FIGURE 7.10

JPEG results for various compression ratios for 512×512 *puppet* image. Top left: original. Top right: 0.16 bpp. Middle left: 0.22 bpp. Middle right: 0.5 bpp. Bottom left: 1 bpp. Bottom right: 2 bpp.

table is adapted according to the data statistics. An Exp-Golomb code (see Section 7.5) is used which is effective while offering a simple decoding process.

In the case of transform coefficients, Context-Adaptive Variable Length Coding (CAVLC) is used. This provides better performance and supports a number of SE-dependent tables that can be switched between, according to the prevailing data statistics. Although CAVLC is not based on Huffman coding, it is worthy of a brief mention

FIGURE 7.11

Prediction samples for lossless JPEG.

Table 7.5 Prediction modes for lossless JPEG.

Mode	Predictor
0	No prediction
1	$\hat{X} = A$
2	$\hat{X} = B$
3	$\hat{X} = C$
4	$\hat{X} = A + B - C$
5	$\hat{X} = A + (B - C)/2$
6	$\hat{X} = B + (A - C)/2$
7	$\hat{X} = (A + B)/2$

here since it shows the power of context switching. Good examples of CAVLC encoding are given in Ref. [12].

CAVLC exploits the sparsity of the transform matrix as we have seen previously, but also exploits other characteristics of the scanned coefficient vector, most notably: (i) that the scan often ends with a series of ± 1 values and (ii) that the numbers of coefficients in neighboring blocks are correlated. The method uses different VLC tables according to the local context.

CAVLC codes the number of non-zero coefficients and the size and position of these coefficients separately. Based on the characteristics of the data, a number of different coding modes can be invoked. For example, CAVLC can encode:

- **The number of non-zero coefficients and the *Trailing ones* (T1s):** T1s are the number of coefficients with a value of ± 1 at the end of the scan (scans often end with runs of ± 1). Four VLC tables are used for this, based on the number of coefficients in neighboring blocks.
- **The values of non-zero coefficients:** T1s can be represented as signs; other coefficients are coded in reverse scan order because variability is lower at the end of the scan. Six Exp-Golomb code tables are used for adaptation according to the magnitudes of recently coded coefficients.

- **TotalZeroes and RunBefore:** These specify the number of zeros between the start of the scan and its last non-zero coefficient. RunBefore indicates how these zeros are distributed.

A good example of CAVLC coding is provided by Richardson [13].

7.5 Golomb coding

The Golomb family includes Unary, Golomb, Golomb–Rice, and Exponential Golomb codes [14, 15]. These are variable length codes which are generated via a fixed mapping between an index and a codeword. In the same way as with Huffman codes, shorter codewords can be assigned to symbols with higher probabilities and vice versa. Golomb codes have the advantage that they can be simply encoded and decoded algorithmically without the need for look-up tables. They also possess the property that, for any geometric distribution, it is possible to find an encoding that is an optimal prefix code. Adaptive methods have thus been developed and exploited in standards such as the lossless form of JPEG—JPEG-LS.

7.5.1 Unary codes

The simplest type of unary code encodes a non-negative integer, i, by i 0s followed by a single 1 (or the other way around). The first few unary codes are shown in Table 7.6 where it is assumed that the index values correspond to the alphabet symbols ranked in terms of decreasing probability. This encoding is not generally very efficient except when the probabilities are successive powers of 2.

7.5.2 Golomb and Golomb–Rice codes

Golomb codes divide all index values i into equal sized groups of size m. The codeword is then constructed from a unary code that characterizes each group, followed by a

Table 7.6 Golomb codes for index values $i = 0 \cdots 9$.

i	Unary	Golomb–Rice ($m = 4$)	Exp-Golomb
0	1	1|00	1|
1	01	1|01	01|0
2	001	1|10	01|1
3	0001	1|11	001|00
4	00001	01|00	001|01
5	000001	01|01	001|10
6	0000001	01|10	001|11
8	000000001	01|11	0001|000
9	0000000001	001|00	0001|001
⋮	⋮	⋮	⋮

fixed length code v_i that specifies the remainder of the index that has been encoded. The Golomb code residual can be simply generated using equation (7.5), where v_i represents the remainder for index i:

$$v_i = i - \left\lfloor \frac{i}{m} \right\rfloor m \tag{7.5}$$

The Golomb–Rice code is a special case where $m = 2^k$ and is shown in Table 7.6. It can be observerved that Golomb–Rice codes grow more slowly than unary codes and are therefore generally more efficient.

7.5.3 Exponential Golomb codes

These were proposed by Teuhola [16] and, whereas Golomb–Rice codes divide the alphabet into equal sized groups, Exp-Golomb codes divide it into exponentially increasing group sizes. The comparative performance of Golomb–Rice and Exp-Golomb codes will depend on the size of the alphabet and the choice of m, but for larger alphabets, Exp-Golomb coding has significant advantages.

The Exp-Golomb code for a symbol index i comprises a unary code (ζ_i zeros, followed by a single 1), concatenated with a fixed length code, v_i. It can be simply generated as follows:

$$\begin{cases} v_i = 1 + i - 2^{\zeta_i} \\ \zeta_i = \left\lfloor \log_2(i + 1) \right\rfloor \end{cases} \tag{7.6}$$

Exp-Golomb coding has been used in both H.264/AVC and in HEVC.

Example 7.8 (Exp-Golomb encoding and decoding)

(i) What is the Exp-Golomb code for the symbol index 228_{10}?

(ii) Show how the Exp-Golomb codeword 00000100111 would be decoded and compute the value of the corresponding symbol index.

Solution.

(i) The codeword group in this case is given by:

$$\zeta_i = \left\lfloor \log_2(i + 1) \right\rfloor = 7$$

and the residual is given by:

$$v_i = 1 + i - 2^{\zeta_i} = 101_{10} = 1100101_2$$

The complete codeword is thus 000000011100101.

(ii) The algorithm is straightforward in that, for each new codeword, we count the number of leading zeros (ζ_i), ignore the following 1 and then decode the index i from the next ζ_i bits. In this case there are five leading zeros and the encoded residual $v_i = 00111_2 = 7_{10}$. The index can then be calculated from equation (7.6) as $i = v_i - 1 + 2^{\zeta_i} = 134$.

7.6 Arithmetic coding

Although Huffman coding is still widely employed in standards such as JPEG, an alternative, more flexible method has grown in popularity in recent years. Variants of arithmetic coding are employed as the preferred entropy coding mechanism in most video coding standards with *Context-based Adaptive Binary Arithmetic Coding* (CABAC) now forming the basis of H.264/AVC and HEVC [17]. Arithmetic encoding is based on the use of a cumulative probability distribution as a basis for codeword generation for a sequence of symbols. Some of the earliest references to this approach were made by Shannon [18], by Abramson in his book on information theory [2] and by Jelinek [19]. More recent popularization of the technique is due to Rissanen [3] and others. Excellent overviews can be found in Refs. [4,7,8].

7.6.1 The basic arithmetic encoding algorithm

An arithmetic encoder represents a string of input symbols as a binary fraction. In arithmetic coding, a half-open encoding range $[0, 1)$ is subdivided according to the probabilities of the symbols being encoded. The first symbol encoded will fall within a given subdivision; this subdivision forms a new interval which is then subdivided in the same way for encoding the second symbol. This process is repeated iteratively until all symbols in the sequence have been encoded.

Providing some basic additional information is known (i.e. the number of symbols transmitted or the existence of an EOB symbol), the complete sequence of symbols transmitted can thus be uniquely defined by any fractional number (referred to as a *tag*) within the range of the final symbol at the final iteration.

By using binary symbols and approximating the probabilities to powers of 2, an arithmetic encoder can be implemented using only shift and add operations. This achieves a computational complexity which is comparable to or even better than Huffman coding while providing superior compression performance. The basic arithmetic encoding algorithm is illustrated in Algorithm 7.7.

We saw in Chapter 3 that the average codeword length for a source depends highly on how symbols are created and encoded and that, by combining symbols in groups, we could potentially achieve better performance. Put simply, this is what arithmetic coding does, but it does it in a rather elegant and flexible manner.

Let us assume that we have a set of M symbols that comprise our source alphabet, $S = \{s_0, s_1, s_2, \ldots, s_{M-1}\}$, where each symbol maps to an integer x in the range $\{0, 1, 2, 3, \ldots, M - 1\}$ corresponding to its symbol index. The source symbol probabilities are therefore defined as:

$$P(m) = P(x = m); \quad m = 0 \cdots M - 1$$

The cumulative probability distribution can now be defined in terms of these probabilities as follows:

$$P_x(m) = \sum_{k=0}^{m} P(k); \quad m = 0 \cdots M - 1 \tag{7.7}$$

Algorithm 7.7 Arithmetic encoding.

---

1. Subdivide the half open interval $[0, 1)$ according to the symbol cumulative probabilities: equation (7.7);
2. Initialize upper and lower limits: $l(0)=0$; $u(0)=1$;
3. INPUT sequence of symbols $\{x_i\}$;
4. $i=0$;
5. REPEAT
6. $i=i+1$;
7.   Compute lower interval limit: $l(i) = l(i-1) + (u(i-1) - l(i-1))P_x(x_i - 1)$;
8.   Compute upper interval limit: $u(i) = l(i-1) + (u(i-1) - l(i-1))P_x(x_i)$;
9. UNTIL $i=N$;
10. OUTPUT arithmetic codeword, $v \in [l(N), u(N))$;
11. END.

Following our discussion of code optimality in Chapter 3 and in Section 7.3.5, we can see that an optimal coding system will code each symbol in S with an average length of $-\log_2 P_i$ bits.

So now we can proceed to develop the iterative symbol encoding process. Firstly we define a number line as a half open interval in the range $0 \cdots 1$, i.e. $[0, 1)$. We can then define each accumulated probability interval at each iteration by its lower extent and its upper limit (or equivalently, its length). So, if we have a sequence of N input symbols $X = \{x_i\}$; $i = 1 \cdots N$, and we define $l(i)$ to be the lower limit at stage i of the encoding and $u(i)$ to be the corresponding upper limit, then at the start of encoding:

$$l(0) = P(-1) = 0$$
$$u(0) = P_x(M - 1) = 1$$

After the first symbol is encoded, $i = 1$:

$$l(1) = P_x(x_1 - 1)$$
$$u(1) = P_x(x_1)$$

Similarly after the second symbol, $i = 2$, we have:

$$l(2) = l(1) + (u(1) - l(1))P_x(x_2 - 1)$$
$$u(2) = l(1) + (u(1) - l(1))P_x(x_2)$$

and in general after the ith symbol, x_i, we have:

$$l(i) = l(i - 1) + (u(i - 1) - l(i - 1))P_x(x_i - 1)$$
$$u(i) = l(i - 1) + (u(i - 1) - l(i - 1))P_x(x_i)$$

$$(7.8)$$

This is best illustrated by a simple example (Example 7.9).

Example 7.9 (Arithmetic encoding)
Consider a sequence of symbols $\{s_2, s_2, s_3\}$ from an alphabet $\{s_1, s_2, s_3\}$ with probabilities $\{P(1), P(2), P(3)\}$. Show the arithmetic encoding process for the three iterations corresponding to this sequence.

Solution. Using our notation above, the sequence maps on to $X = \{x_0, x_1, x_2\} \Rightarrow$ $\{2, 2, 3\}$. The three iterations of the arithmetic coder for this sequence are illustrated below:

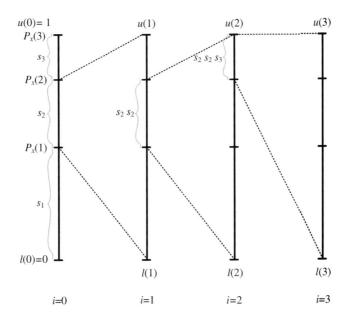

We can see that any value in the range $[l(3), u(3))$ uniquely defines this sequence provided of course that we know when to stop decoding—but more of that later.

Example 7.10 (Arithmetic encoding—numerical example)
Using the same symbols and sequence as in Example 7.9, assume that the symbol probabilities are: $P(1) = 0.5$, $P(2) = 0.25$, $P(3) = 0.25$. Compute the arithmetic code for the sequence $\{s_2, s_2, s_3\}$.

Solution. The three iterations of the arithmetic coder for this sequence are illustrated below, including the tag values for each stage:

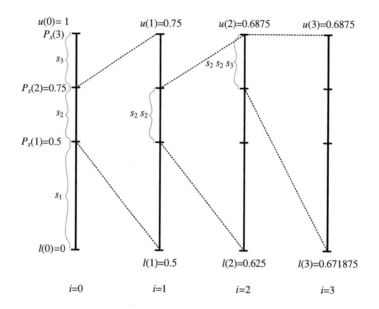

As an example of computing the limits for the interim intervals, consider the case when $i = 2$ ($i = 1$ can be worked out by inspection):

$$l(2) = l(1) + (u(1) - l(1))P_x(1)$$
$$l(2) = 0.5 + (0.75 - 0.5)0.5 = 0.625$$
$$u(2) = l(1) + (u(1) - l(1))P_x(2)$$
$$u(2) = 0.5 + (0.75 - 0.25)0.75 = 0.6875$$

Repeating this for iteration 3 results in an arithmetic code that can assume any value within the interval $v \in [0.671875, 0.6875) = [0.101011_2, 0.1011_2)$. We might take the lower limit for simplicity or take a value midway between the lower and upper limits. In practice it would be preferable to use the shortest binary representation which in this case is also the lower limit, i.e. 0.101011, or, if we drop the leading zero as it is redundant, $\hat{v} = 101011$.

7.6.2 The basic arithmetic decoding algorithm

The encoding algorithm described in Section 7.6.1 creates an arithmetic codeword, $\hat{v} \in [l(N), u(N))$. The purpose of the decoder is to generate the same output sequence, $\hat{x}_i = \{x_1, x_2, \ldots, x_N\}$ as that processed at the encoder, where $\{x_i = k\} \leftarrow \{s_k\}$. We do this by recursively updating the upper and lower limits of the half open interval.

Initializing these to:

$$l(0) = P_x(0) = 0$$
$$u(0) = P_x(M - 1) = 1$$

we then compute the sequence of symbol indices $\hat{x}_i$ using:

$$\hat{x}_1 = \{x : P_x(x_1 - 1) \leq \hat{v} < P_x(x_1)\} \tag{7.9}$$

So in general we need to find the next symbol index such that:

$$\hat{x}_i = \{x : l(i) \leq \hat{v} < u(i)\}; \quad i = 1, 2, \ldots, N \tag{7.10}$$

where, for each value of i we need to update the lower and upper limits:

$$u(i) = l(i - 1) + (u(i - 1) - l(i - 1))P_x(x_i)$$
$$l(i) = l(i - 1) + (u(i - 1) - l(i - 1))P_x(x_i - 1) \tag{7.11}$$

This is formalized in Algorithm 7.8.

Example 7.11 (Arithmetic decoding—numerical example)
Consider decoding the arithmetic code 0.538 which represents a three-symbol sequence from the alphabet $S = \{s_1, s_2, s_3, s_4\}$ where $P(s_1) = 0.6$, $P(s_2) = 0.2$, $P(s_3) = 0.1$, $P(s_4) = 0.1$. What is the symbol sequence represented by this arithmetic code?

Solution. The iterations of the arithmetic decoding process for this codeword are illustrated below:

Algorithm 7.8 Arithmetic decoding.

1. Initialize upper and lower limits: $l(0)=0$; $u(0)=1$;
2. INPUT arithmetic code tag, $\hat{v}$;
3. $i=0$;
4. REPEAT
5. $i=i+1$;
6.   Compute symbol index i: $\hat{x}_i = \{x : l(i) \leq \hat{v} < u(i)\}$ where
        upper interval limit: $u(i) = l(i - 1) + (u(i - 1) - l(i - 1))P_x(x_i)$ and
        lower interval limit: $l(i) = l(i - 1) + (u(i - 1) - l(i - 1))P_x(x_i - 1)$;
7.   OUTPUT symbol s_{x_i};
8. UNTIL $i = N$;
9. END.

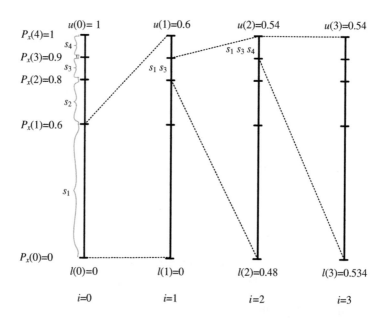

We can see that at stage 1, for decoding symbol 1, the tag lies in the range $[0, 0.6)$ so the symbol output is s_1. Computing the modified cumulative probability values for stage 2 we see that the tag falls in the range $[0.48, 0.54)$ corresponding to the symbol sequence $\{s_1, s_3\}$.

Repeating this for iteration 3 results in the tag falling in the interval $[0.534, 0.54)$ which represents the sequence: $\{s_1, s_3, s_4\}$.

7.6.3 Advantages of arithmetic coding

Arithmetic coding has a number of important properties that make it attractive as a preferred method for entropy encoding. These include:

1. Huffman coding can be efficient with large symbol alphabets, where the highest probability for any individual symbol p_{max} is small and the distribution is relatively uniform. However, for smaller alphabets and where the probability distribution is biased, Huffman coding can be inefficient. In arithmetic coding, codewords represent strings of symbols, hence offering the potential for fractional wordlengths and a corresponding reduction in overall bit rate.

2. Arithmetic coding separates the modeling process from the coding process. Each individual symbol can have a separate probability (usually referred to as conditioning probability) which may be either predetermined or calculated adaptively according to previous data.

3. Symbol statistics can be simply adapted to changing conditions without major recomputation of coding tables.

FIGURE 7.12

A graphical illustration of the arithmetic encoding and decoding processes. Top: probability intervals for three iterations. Bottom: codeword intervals. Shows $P(a) = 0.25$ and $P(b) = 0.75$ with example sequences *abb* and *bbb*. Adapted from Ref. [20].

7.6.4 Binary arithmetic coding

In practical arithmetic coding systems, binary implementations are essential. This poses some constraints on the operation in terms of finite wordlength effects and uniqueness, but also presents opportunities in terms of efficient implementations. To help understanding of this process, consider the illustration in Figure 7.12. This shows three iterations of an arithmetic encoder or decoder for the case of a system with two symbols $\{a, b\}$ where $p(a) = 0.25$ and $p(b) = 0.75$. The top diagram shows probability intervals similar to those we have seen before but visualized slightly differently. The bottom diagram shows the corresponding binary codewords. By scanning vertically from a given sequence, we can easily identify the binary codewords that uniquely define that sequence and select the codeword with the minimum length. For example, consider the sequence $\{a, b, b\}$; scanning down to the bottom diagram, we can see that the shortest codeword that uniquely identifies this sequence is 001. Similarly for $\{b, b, b\}$ it is 11.

Binary encoding can provide advantages in sequential bit-wise processing of tag values, both in encoding and decoding stages. This is demonstrated in the following example.

Example 7.12 (Binary arithmetic encoding)

Consider a source with four symbols with probabilities: $P(s_0) = 0.5$, $P(s_1) = 0.25$, $P(s_2) = 0.125$, $P(s_3) = 0.125$. Perform arithmetic encoding of the sequence: $S = \{s_1, s_0, s_2\}$.

Solution. The encoding proceeds as follows:

Stage 1
The first symbol s_1 defines the interval $I_1 = [0.5, 0.75)_{10} = [0.10, 0.11)_2$. Examining the upper and lower limits of this interval we can see that the first binary digit after the binary point (1) is invariant and hence can be transmitted or appended to the output file.

Stage 2
The second symbol s_0 defines the interval $I_2 = [0.5, 0.625)_{10} = [0.100, 0.101)_2$. Examining the upper and lower limits of this interval we can see that the first two binary digits after the binary point (10) are invariant and hence the 0 can be transmitted or appended to the output file.

Stage 3
The third symbol s_2 defines the interval and arithmetic code: $I_3 = [0.59375, 0.609375)_{10} = [0.10011, 0.100111)_2$. Examining the upper and lower limits of this interval we can see that the first five binary digits after the binary point (10011) are invariant and hence can be transmitted or appended to the output file. We can transmit any value in this range as our arithmetic code, $\hat{v} \in [0.10011, 0.100111)_2$. In this case we can simply send an extra zero to differentiate the lower limit from the upper limit, i.e. $\hat{v} = 0.100110$.

A simplified diagram of this arithmetic encoding process is shown in the figure below.

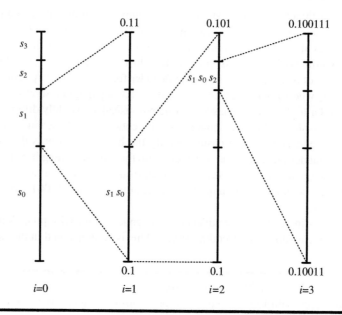

Example 7.13 (Binary arithmetic decoding)
Decode the transmitted sequence from Example 7.12.

Solution. The decoding process is summarized in the following table. As each bit is loaded, it can be processed to further differentiate symbols. When the desired number of symbols or an EOB symbol is received, the decoding terminates. Consider the first row of the table: the first one does not uniquely define a symbol—it could be s_2 or s_3 in this case. However, when the second bit is processed, it uniquely defines s_1. This is repeated until the complete sequence is decoded.

Rx bit	Interval	Symbol
1	$[0.5, 1)_{10} = [0.1, 1.0)_2$	–
0	$[0.5, 0.75)_{10} = [0.10, 0.11)_2$	s_1
0	$[0.5, 0.625)_{10} = [0.100, 0.101)_2$	s_0
1	$[0.5625, 0.625)_{10} = [0.1001, 0.1100)_2$	–
1	$[0.59375, 0.625)_{10} = [0.10011, 0.10100)_2$	–
0	$[0.59375, 0.609375)_{10} = [0.100110, 0.100111)_2$	s_2

7.6.5 Tag generation with scaling

In cases where a transmitted sequence of symbols is large, the arithmetic encoding process will have to iterate a large number of times and at each iteration the size of the intervals reduces (on average by a factor of 2). To cope with this in practical situations, the wordlength must be sufficient to ensure that the intervals and tag can be represented with sufficient numerical accuracy. In practice very long wordlengths are mitigated by a process of scaling the interval each time a bit of the tag is transmitted. Consider stage 1 in Example 7.12; here $l(1) = 0.5$ and $u(1) = 0.75$. Once the first binary 1 is transmitted, this means that the codeword is forever constrained to the upper half of the number line. So, without any loss of generality, we can scale the interval by a factor of 2 as follows:

$$l'(1) = 2(l(1) - 0.5) = 0$$
$$u'(1) = 2(u(1) - 0.5) = 0.5$$

The equations for the next stage are then generated as follows:

$$u(i) = l'(i - 1) + (u'(i - 1) - l'(i - 1))P_x(x_i)$$
$$l(i) = l'(i - 1) + (u'(i - 1) - l'(i - 1))P_x(x_i - 1)$$
$$(7.12)$$

This process of scaling is repeated every time a new symbol constrains the tag to be completely contained in the upper or lower half interval. In cases where the tag interval straddles the center of the number line we can perform a similar process, but this time scaling when the interval falls within [0.25, 0.5). So we have three types of scaling or renormalization that can be used after interval subdivision [4, 7]:

$$E_1 : [0, 0.5) \rightarrow [0, 1); \quad E_1(x) = 2x$$
$$E_2 : [0.5, 1) \rightarrow [0, 1); \quad E_2(x) = 2(x - 0.5) \quad (7.13)$$
$$E_3 : [0.25, 0.75) \rightarrow [0, 1); \quad E_3(x) = 2(x - 0.25)$$

It is not difficult to see how this process can be mimicked at the decoder as the tag bits are processed.

7.6.6 Context-adaptive arithmetic coding

Arithmetic coding has been supported in a number of standards such as JPEG, JPEG2000, H.263, H.264/AVC, and HEVC. However, not until its inclusion in H.264/AVC has it been commonly used in practice. This uptake was due to the fact that Context-Adaptive Binary Arithmetic Coding (CABAC) [17] provided significant performance gains over previous methods. CABAC is now the standard entropy encoding method in HEVC.

It has been shown that, by taking account of prevailing context and adapting conditioning probabilities accordingly, then significant improvements can be obtained with entropy coders. We saw an example of how to do this with Huffman coding earlier in this chapter and also looked at CAVLC. We will consider it briefly here in the context of arithmetic coding. A good review of adaptive coding methods is provided by Tian et al. [12].

Several realizations of adaptive arithmetic encoding engines have been reported in the literature. These include the Q Coder [21], which exploits the idea of renormalization after subdivision (as discussed above) to avoid numerical precision issues (an extension of this, known as the QM Coder, is used in JPEG), and the MQ coder [22]. The latter is used for encoding bit-planes of quantized wavelet coefficient codeblocks in the JPEG2000 EBCOT (Embedded Block Coding with Optimized Truncation) coder.

CABAC has been adopted for the Main and High Profiles of H.264/AVC and HEVC, and employs context-based adaptivity to improve performance. This enables the encode to dynamically adapt to prevailing symbol statistics. In CABAC, conditioning probabilities are based on the statistics of previously encoded Syntax Elements, which are used to switch between a number of probability models. It offers a fast and accurate means of estimating conditioning probabilities over a short interval and excellent computational efficiency.

CABAC comprises three main stages: binarization, context modeling, and binary arithmetic coding. These are illustrated in Figure 7.13. These steps are briefly described below. The reader is referred to Refs. [12, 17] for a more detailed operational explanation.

FIGURE 7.13

Context-Adaptive Binary Arithmetic Coding. Adapted from Ref. [17].

Binarization: Maps non-binary valued SEs to a bin string (a sequence of binary decisions). This reduces the symbol alphabet by producing a set of intermediate codewords (bin string) for each non-binary SE. It also enables context modeling at a subsymbol level, where conditional probabilities can be used for more common bins and simpler schemes for less frequent bins. In CABAC, binarization is implemented using a number of schemes and combinations of them. These include: Unary and Truncated Unary, Fixed Length, kth order Exp-Golomb and combinations of these.

Context modeling: We have seen earlier that one of the primary advantages of arithmetic coding is its ability to separate modeling from coding. CABAC exploits this. The modeler assigns a probability distribution to the symbols being processed and this is used to condition the BAC. In CABAC, the context modeling is restricted to a small number of local symbols, in order to provide a good trade-off between modeling complexity and performance. In cases where SEs exhibit uniform or close to uniform probability distributions, then a bypass mode is invoked.

Binary arithmetic coding: CABAC employs a multiplication-free approach to arithmetic encoding and uses recursive interval subdivision and renormalization based on the approach of Witten [23].

In the H.264/AVC and HEVC standards, CABAC is used to encode coding parameters (e.g. macroblock type, spatial and temporal prediction mode, and slice and macroblock control information) as well as prediction residuals. CABAC has been reported to offer up to 20% improvement over CAVLC, with typical savings between 9% and 14% [17].

7.7 Performance comparisons—Huffman vs arithmetic coding

A comparison of the compression performance achieved for basic arithmetic and Huffman coding is shown in Figure 7.14. These results represent the combination of a DCT codec with both types of entropy coder. It shows that arithmetic coding offers

FIGURE 7.14

Huffman vs arithmetic coding.

advantages (up to 1 dB at lower compression ratios) compared with Huffman coding. As we have seen, further improvements are possible using more sophisticated adaptive and context-based techniques which suit arithmetic coding better than Huffman coding.

7.8 Summary

This chapter has introduced methods for coding images without loss. We have seen the requirement that useful codes need to have prefix properties so as to avoid the need for explicit synchronization. Huffman codes satisfy this property and are able to code individual symbols from a source at a rate close to its entropy. We then described arithmetic coding which offers performance advantages in that it can represent multiple symbols in a group and hence offer fractional symbol wordlengths. It is also more amenable to adaptive encoding—supporting dynamic modification of conditioning probabilities.

References

[1] D.A. Huffman, A method for the construction of minimum-redundancy codes, Proceedings of the IRE 40 (9) (1952) 1098–1101.

[2] N. Abramson, Information Theory and Coding, McGraw-Hill, 1963.

[3] J. Rissanen, Generalized Kraft inequality and arithmetic coding, IBM Journal of Research and Development 20 (1976) 198–203.

[4] A. Said, Introduction to arithmetic coding—theory and practice, in: K. Sayood (Ed.), Lossless Compression Handbook, Academic Press, 2003.

[5] J. Ziv, A. Lempel, A universal algorithm for data compression, IEEE Transactions on Information Theory 23 (3) (1977) 337–343.

[6] T. Welch, A technique for high-performance data compression, IEEE Computer (1984) 8–19.

[7] K. Sayood, Introduction to Data Compression, third ed., Morgan Kaufmann, 2006.

[8] K. Sayood (Ed.), Lossless Compression Handbook, Academic Press, 2003.

[9] R. Gallagher, Variations on a theme by Huffman, IEEE Transactions on Information Theory 24 (6) (1978) 668–674.

[10] J. Vitter, Design and analysis of dynamic Huffman codes, Journal of the ACM 34 (4) (1987) 825–845.

[11] ISO/IEC International Standard 10918-1, Information Technology—Digital and Coding of Continuous-Tone Still Images—Requirements and Guidelines, 1992.

[12] X. Tian, T. Le, Y. Lian, Entropy Coders of the H.264/AVD Standard: Algorithms and VLSI Architectures, Springer, 2011.

[13] I. Richardson, The H.264 Advanced Video Coding Standard, second ed., Wiley, 2010.

[14] S. Golomb, Run-length encodings, IEEE Transactions on Information Theory 12 (3) (1966) 399–401.

[15] D. Taubman, M. Marcelin, JPEG2000 Image Compression Fundamentals, Standards and Practice, Kluwer, 2002.

[16] J. Teuhola, A compression method clustered bit vectors, Information Processing Letters 7 (1978) 308–311.

[17] D. Marpe, H. Schwarz, T. Wiegand, Context-based adaptive binary arithmetic coding in the H.264/AVC video compression standard, IEEE Transactions on Circuits and Systems for Video Technology 13 (7) (2003) 620–636.

[18] C. Shannon, A mathematical theory of communication, Bell System Technical Journal (30) (1948) 623–656.

[19] F. Jelinek, Probabalistic Information Theory, McGraw Hill, 1968.

[20] J.-R. Ohm, Multimedia Communication Technology, Springer, 2003.

[21] J. Pennebaker, G. Mitchell, G. Langdon, R. Arps, An overview of the basic principles of the Q-coder adaptive binary arithmetic coder, IBM Journal of Research and Development 32 (1988) 717–726.

[22] D. Taubman, E. Ordentlich, M. Weinberger, G. Seroussic, Embedded block coding in JPEG 2000, Signal Processing: Image Communication 17 (2002) 49–72.

[23] I. Witten, R. Neal, J. Cleary, Arithmetic coding for data compression, Communications of the ACM 30 (6) (1987) 520–540.

Coding Moving Pictures: Motion Prediction

CHAPTER OUTLINE

8.1 Temporal correlation and exploiting temporal redundancy **256**
   8.1.1 Why motion estimation? 256
   8.1.2 Projected motion and apparent motion 258
   8.1.3 Understanding temporal correlation 259
   8.1.4 How to form the prediction 260
   8.1.5 Approaches to motion estimation 262
8.2 Motion models and motion estimation **263**
   8.2.1 Problem formulation 263
   8.2.2 Affine and high order models 263
   8.2.3 Translation-only models 265
   8.2.4 Pixel-recursive methods 266
   8.2.5 Frequency domain motion estimation using phase
       correlation (PhC) 267
8.3 Block matching motion estimation (BMME) **269**
   8.3.1 Translational block matching 269
   8.3.2 Matching criteria 270
   8.3.3 Full search algorithm 272
   8.3.4 Properties of block motion fields and error surfaces 272
   8.3.5 Motion failure 276
   8.3.6 Restricted and unrestricted vectors 277
8.4 Reduced complexity motion estimation **277**
   8.4.1 Pixel grids and search grids 277
   8.4.2 Complexity of full search 278
   8.4.3 Reducing search complexity 279
   8.4.4 2-D logarithmic search (TDL) 280
   8.4.5 *N*-Step Search (NSS) 281
   8.4.6 Diamond search 282
   8.4.7 Hexagonal search 284
   8.4.8 Initialization and termination criteria 285
   8.4.9 Reduced complexity block matching algorithm
       performance comparisons 286

Communicating Pictures. http://dx.doi.org/10.1016/B978-0-12-405906-1.00008-8
© 2014 Elsevier Ltd. All rights reserved.

8.5 Motion vector coding **287**
   8.5.1 Motion vector prediction 287
   8.5.2 Entropy coding of motion vectors 287
8.6 Summary **288**
References **288**

In previous chapters we have studied various important aspects of image and video coding. We have seen how to create an appropriate input format and color space in Chapter 4, we saw how to decorrelate and quantize images using transforms such as the DCT in Chapter 5 and the DWT in Chapter 6, and in Chapter 7 we described how to represent the quantized output using various forms of entropy coding. This chapter completes "the picture" and addresses the final component in our basic video codec—motion prediction. Motion estimation is an important component in video coding as it enables temporal decorrelation. So rather than code very similar information repeatedly in each frame, we instead code the residual signal after motion prediction, along with the prediction parameters (the motion vectors).

This chapter firstly introduces the idea of temporal correlation and discusses how it can be exploited to provide additional coding gain. It then compares the merits of various motion models. Block matching, based on a translational motion model, is the approach adopted in all standardized video codecs to date and this is studied in detail with its performance characterized in terms of the effects of block size, search range, and motion vector coding. Alternative approaches to motion estimation, such as those based on phase correlation, are also covered. Although such techniques have rarely been adopted for conventional video compression, they are worthy of mention because of their use in video standards conversion.

One of the primary issues associated with motion estimation, even today, is its complexity—if not optimized, motion estimation can account for 60–70% of the processor loading. Methods of reducing motion estimation complexity have thus received significant attention in recent years and selected low complexity methods are compared in Section 8.4 in terms of their accuracy–complexity trade-offs. Finally, Section 8.5 describes how we can efficiently code motion information prior to transmission or storage.

8.1 Temporal correlation and exploiting temporal redundancy

8.1.1 Why motion estimation?

For still natural images, significant spatial redundancies exist and we have seen that these can be exploited by transform, filter bank or other coding methods that provide spatial decorrelation. The simplest approach to encoding a sequence of moving images is thus to apply an *intra-frame* (still image) coding method to each frame (cf. Motion

JPEG or Motion JPEG2000). This, as we will see later, does have some benefits, especially in terms of avoiding error propagation caused by the temporal prediction loop used in most video codecs. It can also be invoked in cases where a motion model fails to capture the prevailing motion in the scene. However, it generally results in limited compression performance.

We have already seen in Chapter 1, for real-time video transmission over low bandwidth channels, that there is often insufficient capacity to code each frame in a video sequence independently (25–30 fps is required to avoid flicker). The solution to this problem lies in taking advantage of the correlation that exists between temporally adjacent frames in a video sequence. This *inter-frame* redundancy can be exploited through motion prediction to further improve coding efficiency. Figure 8.1 demonstrates the effects of this approach, showing the first 53 frames of the *Foreman* CIF sequence, coded with a GOP length of 50 at approximately 500 kb/s. Frames 1 and 51 are coded as intra-frames (I-frames) and the remainder are coded as inter-frames (or predicted frames, also known as P-frames).

A number of issues however arise with this approach and these will be discussed further in the following sections:

- Motion estimation yields a residual signal with reduced correlation. Hence transform coding is less efficient on *displaced frame difference* (DFD) signals than on intra-frames.
- Motion can be a complex combination of translation, rotation, zoom, and shear. Hence model selection is key in balancing motion prediction accuracy with parameter search complexities.
- Lighting levels can change from frame to frame—modifying average luminance levels.
- Occluded objects can become uncovered and visible objects can become occluded, confounding motion models and leading to model failure.

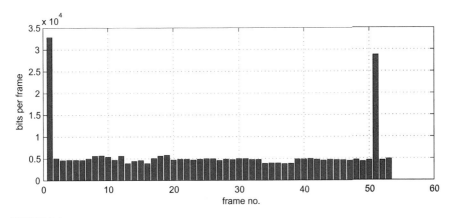

FIGURE 8.1

Variation in the number of bits per frame and frame coding mode.

- Motion estimation and compensation adds complexity to the codec.
- Motion information creates an overhead as it must be transmitted alongside the quantized transform coefficients.

Example 8.1 (Comparing intra- and inter-frame coding)
Consider a sequence with 512×512 spatial samples captured at 30 fps with 8 bits per color channel and a GOP length of 100 frames. Compare uncompressed, intra-compressed and inter-compressed bit rates for this sequence if (for a given quality) the average number of bits per pixel (including motion vector overheads) in intra-frame mode is 0.2 bpp per color channel, and 0.02 bpp in inter-frame mode.

Solution.

1. Uncompressed:

$$\text{Uncompressed bit rate} = 512 \times 512 \times 30 \times 24 = 189 \text{ Mbps}$$

2. Intra-frame compressed:

$$\text{Intra-compressed bit rate} = 512 \times 512 \times 30 \times 0.2 \times 3 = 4.7 \text{ Mbps}$$

3. Inter-frame compressed:

In this case 1 in every 100 frames is an intra-frame @ 0.2 bpp per color channel and 99 out of every 100 frames are inter-frames @ 0.02 bpp per color channel. Thus:

$$\text{Inter-compressed bit rate} = 512 \times 512 \times 30 \times \left(\frac{0.2 \times 3 \times 1 + 0.02 \times 3 \times 99}{100} \right)$$
$$= 514 \text{ kbps}$$

If these sort of figures are achievable without compromising quality, then huge coding gains can be achieved by exploiting temporal redundancy. The goal is thus to exploit both temporal and spatial redundancy when coding the video sequence.

8.1.2 Projected motion and apparent motion

While, in the real world, objects move in three spatial dimensions, we are primarily interested here in the projection of their motion onto a 2-D image plane as captured by a camera. So, when we use the term "motion estimation" in the context of video compression, this normally refers to the *projected motion* of the object. While it could be argued that tracking an object in 3-D might provide a more accurate description of this projected motion, the models and methods associated with doing this are too complex to be of major interest.

In video compression, 2-D motion is conventionally estimated by observing the spatio-temporal variations in signal intensity between frames. This approach measures

FIGURE 8.2

Temporal correlation between two adjacent video frames.

apparent motion rather than true projected motion and can lead to ambiguities. Consider, for example, a disk of uniform luminance rotating about its center. In this case the projected motion is clearly non-zero, whereas the apparent motion is zero. In contrast, an object that is static in the image plane, but experiencing time-varying illumination changes, will have zero projected motion but, because of the intensity changes between frames, will exhibit some apparent motion. The use of apparent motion can be justified since the aim of motion estimation in video compression is not to assess true motion, but instead to minimize the number of bits required to code the prediction residual.

We will see below that there are many possible motion models that could be adopted for motion estimation. However, the most commonly used is the simple 2-D translational model, where the displacement motion vector, $\mathbf{d} = \left[d_x, d_y\right]^{\mathrm{T}}$. This is not normally computed for a whole frame, but instead for a number of smaller blocks—typically of size 16×16 pixels. The set of all vectors collectively represents the *motion field* for the given frame.[1]

8.1.3 Understanding temporal correlation

We showed in Chapter 3 how natural video data is correlated in time, with the temporal autocorrelation function typically remaining high across a wide range of lag values. This property can be exploited to improve coding efficiency by compensating for the movement of objects in a scene between adjacent frames via motion estimation and compensation (MEC). Take, for example, the pair of temporally adjacent frames from a typical 25 fps video sequence (*Table*) shown in Figure 8.2. Although this sequence has relatively high activity—the camera pans from right to left and there are clear movements of the player's hands, bat and ball—the two frames appear visually very similar. We will consider, in the following sections, how best to model this motion in order to improve coding gain.

---

[1] Formally for apparent motion, the displacement field is referred to as the correspondence field.

8.1.4   How to form the prediction

Our aim is to estimate the pixel values in the frame currently being encoded (the *current frame*), based on the values in one or more other frames (*reference frames*) in the sequence. These will often be temporally adjacent to the current frame but need not be.

There are two primary ways to form the prediction:

- **Frame Differencing:** This method does not employ any motion model (although arguably it could be referred to as a zero order prediction). The reference frame is, without any modification, directly used as the prediction for the current frame. The prediction error is thus the difference between the pixel values in the current frame and those co-located in the reference frame and is known as the *frame difference* (FD).
- **Motion-Compensated Prediction:** In this case a motion model is assumed and motion estimation (ME) is used to estimate the motion that occurs between the reference frame and the current frame. Once the motion is estimated, a process known as motion compensation (MC) is invoked to use the motion information from ME to modify the contents of the reference frame, according to the motion model, in order to produce a prediction of the current frame. The prediction is called a *motion-compensated prediction* (MCP) or a *displaced frame* (DF). The prediction error is known as the *displaced frame difference* (DFD) signal.

Figure 8.3 compares the motion residuals from both of these approaches for those frames from *Table* shown in Figure 8.2. It can be seen that the residual energy is significantly reduced for the case of motion estimation. Figure 8.4 reinforces this point, showing how the pdf of pixel values is modified for FD and DFD frames, compared to an original frame from the *Football* sequence.

Three main classes of prediction are used in video compression:

- **Forward Prediction:** The reference frame occurs temporally before the current frame.
- **Backward Prediction:** The reference frame occurs temporally after the current frame.

FIGURE 8.3

Frame differencing (left) vs motion-compensated prediction (right).

FIGURE 8.4

Probability distributions for an original frame and FD and DFD frames from the *Football* sequence.

- **Bidirectional Prediction:** Two (or more) reference frames (forward and backward) are employed and the candidate predictions are combined in some way to form the final prediction.

In *forward prediction*, assuming that the motion model used is perfect and based on translation only, the pixel value at location $\mathbf{p} = [x, y]^T$ in the current frame at time k

is related to a pixel location in the previous frame at time $k - 1$ by:[2]

$$S_k[\mathbf{p}] = S_{k-1}[\mathbf{p} - \mathbf{d}] \qquad (8.1)$$

where $\mathbf{d}$ is the motion vector corresponding to location $\mathbf{p}$ in the current frame. A similar expression can be generated for *backward prediction*. More sophisticated motion models are considered in the next section.

It should be noted that the motion estimation problem is ill-posed and there is not necessarily a unique solution—the number of independent variables exceeds the number of equations. In fact a solution does not necessarily exist at all, for example in the case of uncovered or occluded pixels. The estimate is also sensitive to the presence of noise.

8.1.5   Approaches to motion estimation

The key issues in motion estimation can be summarized as follows:

1. **The motion model:** This can be an accurate, but more complex motion model such as *affine* or *perspective*, or a simple but sufficiently effective model such as *translational*. In practice the translational block matching approach is adopted in most situations. Once the motion model is selected, we must decide how it will be implemented—for example, in the spatial or frequency domain.
2. **The matching criterion:** Once the model is selected, we must measure how good the match is for any candidate parameter set. This can be achieved with techniques such as normalized cross correlation (NCCF), sum squared differences (SSD), or sum of absolute differences (SAD).
3. **The region of support:** This is the group of pixels to which the motion model is applied and can range from a whole frame to a single pixel. The region can be arbitrarily shaped in two or three dimensions so as to best reflect the actual motion characteristics of a scene, but is normally based on rectangular blocks. In simple codecs these are typically 16×16 but variable block sizes have become commonplace.
4. **The search range:** This is the window of pixels in one or more reference frames within which the region of support is compared using the matching criterion, in order to estimate the motion and parameterize the model. The search range will depend on the spatio-temporal resolution of the format used and the activity level of the content coded. Normally a rectangular region is employed with a maximum offset between 7 and 63 pixels.
5. **The search strategy:** Once all of the above choices are made, we need to execute a means of efficiently finding the model parameter that best suit the motion. Because of the complexity associated with performing an exhaustive search, fast methods are commonly employed. In the case of translational block matching these include the *N-Step Search* (NSS), the *Diamond Search* (DS), and the *Hexagon-based Search Method* (HEXBS).

---

[2]Without loss of generality we use the adjacent frame here.

In all situations, there will exist a trade-off between prediction accuracy and coding overhead and appropriate choices have to be made based on processing power available, video format, and content type. The following sections explore these issues in more detail.

8.2 Motion models and motion estimation

8.2.1 Problem formulation

Apparent motion can be attributed to three main causes:

1. Global motion—e.g. camera motion (pan, tilt, zoom, and translation).
2. Local motion—of objects in the scene.
3. Illumination variations—due to changes in lighting, shadows, or noise.

The models and methods described in this chapter address the combined motion due to all three of these reasons. Other methods exist whereby global motion is extracted initially followed by local refinement (e.g. Ref. [1]) but these are not considered further here.

8.2.2 Affine and high order models

The apparent motion in an image or image region can be complex and experience translation, rotation, shear, and dilation as well as suffering from illumination changes between frames and occlusion by other objects (as illustrated in Figure 8.5). In order to represent such motion characteristics, it is necessary to use an appropriate motion model. Examples of such models are the Affine, Bilinear, and Perspective models as described by equations (8.2), (8.3), and (8.4) respectively.

Affine:

$$u = g_x(x, y) = a_1 x + a_2 y + a_3$$
$$v = g_y(x, y) = a_4 x + a_5 y + a_6$$
(8.2)

Bilinear:

$$u = g_x(x, y) = a_1 xy + a_2 x + a_3 y + a_4$$
$$v = g_y(x, y) = a_5 xy + a_6 x + a_7 y + a_8$$
(8.3)

Perspective:

$$u = g_x(x, y) = \frac{a_1 x + a_2 y + a_3}{a_7 x + a_8 y + 1}$$
$$v = g_y(x, y) = \frac{a_4 x + a_5 y + a_6}{a_7 x + a_8 y + 1}$$
(8.4)

where location (x, y) is mapped to (u, v) using appropriate warping parameters $a_1 \cdots a_8$. An excellent review of image warping transforms is provided by Glasbey and Mardia in Ref. [2].

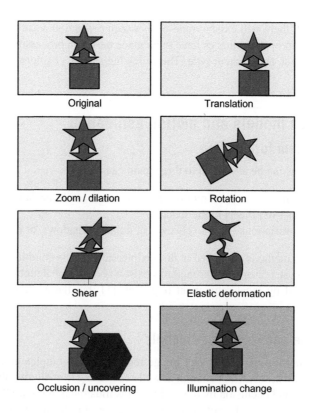

FIGURE 8.5

Motion-based image deformations.

Node-based warping

The use of higher order models has been widely reported in the research literature, but none have been adopted for standardization. There has however been recent renewed interest through the work of Zhang and Bull [3] and Vigars et al. [4] and several others. Furthermore, some authors propose the practice of using affine or eight-parameter models to model camera motion, while using a translational model for local motion based on small blocks.

In most cases, the warping parameters are obtained through the use of a 2-D deformable mesh formed from an interconnected set of node points, as shown in Figure 8.6. A method such as *block matching motion estimation* (BMME) is used to estimate the motion of each node through comparisons between a current frame and one or more reference frames, and this in turn is used to find the parameters $\{a_i\}$ of the motion model. The model is then used to transform each pixel in the current block region. Meshes can be continuous or discrete, fixed or adaptive and can adopt a range of approaches to node tracking. A good overview and comparison of these variations is provided in Ref. [5].

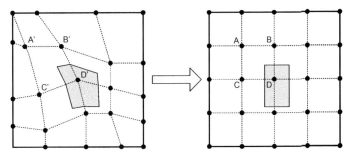

FIGURE 8.6

Mesh-based warping.

Once the nodes of the mesh have been tracked, the motion parameters can be obtained through solving a set of simultaneous equations, such as those shown in equation (8.5) for the bilinear model. Subsequently the transformation can be applied to each pixel in the patch using the appropriate model equation, in this case equation (8.3).

$$
\begin{bmatrix} u_A & u_B & u_C & u_D \\ v_A & v_B & v_C & v_D \end{bmatrix} = \begin{bmatrix} a_1 & a_2 & a_3 & a_4 \\ a_5 & a_6 & a_7 & a_8 \end{bmatrix} \begin{bmatrix} x_A y_A & x_B y_B & x_C y_C & x_D y_D \\ x_A & x_B & x_C & x_D \\ y_A & y_B & y_C & y_D \\ 1 & 1 & 1 & 1 \end{bmatrix} \quad (8.5)
$$

8.2.3 Translation-only models

The translational model is widely used in practice because it is simple to parameterize, it fits well into a regular block-based coding architecture and, importantly, it provides a good fit to the apparent motion in most scenarios. The latter point is worth explaining since it is obvious that true motion, projected motion, and apparent motion are not usually purely translational.

Recall that our primary purpose is to optimize rate–distortion performance, and this can be achieved with a simple motion model with low coding overhead (i.e. few parameters) and which delivers a low energy prediction residual. The translational model succeeds in this respect since, at typical video frame rates (25–50 fps) and spatial resolutions, most motion is approximately translational. This can be observed in Figures 8.2 and 8.3, which show two temporally adjacent frames together with their residual image after block-based translational motion estimation. While obvious small areas exist where the model fails, in most of the frame it works exceptionally well, providing a very low residual signal. Some further examples, where this assumption breaks down, are shown later.

The translational model is a subset of the perspective model, and is given by equations (8.6):

$$
\begin{aligned}
u &= g_x(x, y) = x + a_1 = x + d_x \\
v &= g_y(x, y) = y + a_2 = y + d_y
\end{aligned} \quad (8.6)
$$

Table 8.1 Comparison of warping and translational models for three test sequences [5].

PSNR (dB)	Akiyo	Foreman	Table
BMA	39.88	27.81	29.06
WBA	41.45	29.09	29.22
BMA-HO	41.77	29.51	29.87

It should also be noted that models embracing translation and rotation or translation and dilation are also possible. For example, the latter is given by:

$$u = g_x(x, y) = cx + a_1$$
$$v = g_y(x, y) = cy + a_2$$
(8.7)

Al-Mualla et al. [5] compared warping methods with translational block matching, both based on 16×16 blocks, for low bit rate coding applications. The results are summarized in Table 8.1. Three algorithms were compared: BMA—a basic integer-pixel block matching algorithm, WBA—a warping-based algorithm based on 16×16 blocks with node warping using an eight-parameter model, and BMA-HO—BMA enhanced with half-pixel accuracy and overlapped block motion estimation. It can be seen that the warping approach does offer gains in prediction quality over the basic BMME algorithm but that this can be offset through enhancements to BMME such as sub-pixel estimation and the use of overlapped blocks. In terms of complexity (processor execution time), the BMA-HO was found to be 50% more complex than BMA, while WBA was found to be 50 times slower than BMA.

As mentioned above, the translational motion model is very widely used in practical video compression systems. We will return to it and describe it fully in Section 8.3.

8.2.4 Pixel-recursive methods

Pixel-recursive motion estimation methods were first introduced by Netravali and Robbins [6] and are based on the idea of iteratively minimizing the prediction error using a gradient descent approach. Given a displaced frame difference signal for a given frame (or block):

$$\text{DFD}[\mathbf{p}, \mathbf{d}] = S_k[\mathbf{p}] - S_{k-1}[\mathbf{p} - \mathbf{d}]$$
(8.8)

the estimate for the motion vector at each location is then formed iteratively using a steepest descent approach, hence:

$$\hat{\mathbf{d}}^{i+1} = \hat{\mathbf{d}}^i - \varepsilon \text{DFD}[\mathbf{p}, \hat{\mathbf{d}}^i] \nabla S_{k-1}[\mathbf{p} - \hat{\mathbf{d}}^i]$$
(8.9)

It has been observed that the convergence characteristics of pixel-recursive methods can be highly dependent both on the content (i.e. DFD characteristics) and on the step-size parameter ε. They struggle with smooth intensity regions and large displacements; they also inherently produce dense motion fields and this can lead to a high motion overhead unless parameter prediction methods are used. For these reasons, pixel-recursive methods have not been widely adopted in practice.

8.2.5 Frequency domain motion estimation using phase correlation (PhC)

Principles

The basic idea of phase correlation (PhC) was introduced by Kuglin and Hines [7] and refined by Thomas for broadcast applications [8]; it is illustrated in Figure 8.7. This method employs the 2-D Fourier transform to estimate the correlation between blocks in adjacent frames. Peaks in the correlation surface correspond to the cross correlation lags which exhibit the highest correlation between the current and reference blocks. One advantage of this approach is that the surface may possess multiple peaks, corresponding to multiple candidate motions. Some form of peak detection would then be required to establish the dominant motion or motion vector. Advantages of the PhC approach include:

- It can detect multiple motions in each block, facilitating object-based scene decomposition.
- Block size reduction and/or optimum vector selection for each pixel is possible.
- It can easily cope with large displacements.
- It offers robustness to noise and illumination variations as the Fourier phase is insensitive to mean shifts.
- It facilitates sub-pixel resolution through increasing the length of the DFT.

Assuming perfect translational block motion, for a block or picture, **S**, each location is related as follows:

$$S_k[x, y] = S_{k-1}[x + d_x, y + d_y] \tag{8.10}$$

where x and y represent the center pixel locations in block **S** of frame k, and d_x and d_y are the offset parameters which globally translate the block in frame k to align it with that in frame $k - 1$. From the 2-D DFT:

$$\mathcal{S}_k(f_x, f_y) = \mathcal{S}_{k-1}\left(f_x, f_y\right) e^{j2\pi(d_x f_x + d_y f_y)} \tag{8.11}$$

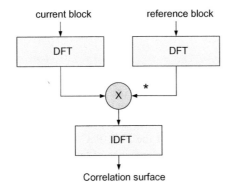

FIGURE 8.7

Phase correlation-based motion estimation.

In the case of translational motion, the difference of the 2-D Fourier phases in the respective blocks 'defines a plane in the spatial frequency variables f_x, f_y. The motion vector can be estimated from the orientation of this plane, thus:

$$\arg(\mathcal{S}_k(f_x, f_y)) - \arg(\mathcal{S}_{k-1}(f_x, f_y)) = 2\pi(d_x f_x + d_y f_y) \quad (8.12)$$

However, this requires phase unwrapping and it is not easy to identify multiple motions. Instead, the phase correlation method is used to estimate the relative shift between two image blocks by means of a normalized cross correlation function in the 2-D Fourier domain. In this case the correlation between frames k and $k - 1$ is defined by:

$$C_{k,k-1}[x, y] = S_k[x, y] \star S_{k-1}[-x, -y] \quad (8.13)$$

Taking the Fourier transform of both sides, we obtain a complex cross-power spectrum:

$$\mathcal{C}_{k,k-1}[f_x, f_y] = \mathcal{S}_k(f_x, f_y)\mathcal{S}_{k-1}^*(f_x, f_y) \quad (8.14)$$

and normalizing this by its magnitude gives the phase of the cross-power spectrum:

$$\angle(\mathcal{C}_{k,k-1}[f_x, f_y]) = \frac{\mathcal{S}_k(f_x, f_y)\mathcal{S}_{k-1}^*(f_x, f_y)}{|\mathcal{S}_k(f_x, f_y)\mathcal{S}_{k-1}^*(f_x, f_y)|} \quad (8.15)$$

Assuming ideal translational motion by an amount $[d_x, d_y]$ we can obtain from equation (8.15):

$$\angle(\mathcal{C}_{k,k-1}[f_x, f_y]) = e^{2\pi(d_x f_x + d_y f_y)} \quad (8.16)$$

and taking the inverse Fourier transform of equation (8.16) yields the phase correlation function:

$$C_{k,k-1}[x, y] = \delta[x - d_x, y - d_y] \quad (8.17)$$

i.e. the phase correlation function yields an impulse whose location represents the displacement vector.

Applications and performance

Phase correlation is the preferred method for TV standards conversion (e.g. frame rate conversion between US and EU frame rates), primarily due the fact that it produces a much smoother motion field with lower entropy than BMME, significantly reducing motion artifacts. Due to the use of the FFT, PhC can be computationally efficient. Phase correlation works well in the presence of large scale translational motion and it can handle multiple motions. However, the use of variable block size BM is now considered a more flexible approach and is favored for conventional motion estimation in modern video coding standards. An example of PhC is given in Figure 8.8.

Δy=13 pixels

Δ x=65 pixels

FIGURE 8.8

Example of phase correlation-based motion estimation. Top: input frames of a tarmac road surface taken with an approximately translating camera. Bottom: phase correlation surface.

8.3 Block matching motion estimation (BMME)

8.3.1 Translational block matching

Motion vector orientation

Figure 8.9 illustrates the convention we will adopt for motion vector orientation. The motion vector represents the displacement of the closest matching block in the reference frame with respect to the current block location (reversing the polarity of true motion). We will assume the [0,0] vector corresponds to the center of the search grid and offsets in x and y are positive right and down respectively.

Region of support—size of the search window

Intuitively, we would wish to search the whole of the reference picture for the best match to the current block. However, this is both impractical and unnecessary in most cases. Consider a 1080p25 sequence with an object traversing the screen in 1 s; with 1920 horizontal pixels, this means that the object moves horizontally by $1920/25 = 77$ pixels per frame. One could thus argue that such motion would dictate a search window of ±77 pixels in order to predict it. In practice, a number of other factors come into play. Firstly, an object traversing a screen in 1 s is very fast and would rarely occur (not in directed content at least); secondly, if it did happen, there would be significant motion blur so the object could never be faithfully depicted anyway; and

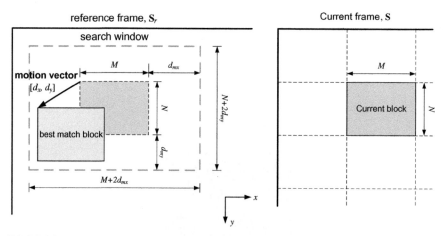

FIGURE 8.9

Translational block matching motion estimation (BMME).

thirdly, motion masking effects will exist which further diminish the need for faithful reproduction. As such, a compromise window is normally employed where the region of support caters for most motion scenarios while limiting search complexity.

The search grid is normally restricted, as shown in Figure 8.9, to a $[\pm d_{mx}, \pm d_{my}]$ window and, in most cases, $d_{mx} = d_{my} = d_m$. The value of d_m depends on image resolution and activity; for example, in broadcast applications, d_m typically ranges from 15 for low motion to 63 for high (sports) motion.

8.3.2 Matching criteria

The block distortion measure (BDM)

For block matching motion estimation, uniform motion and a translational motion model are normally assumed. We need to select the best motion vector for the current block and this will correspond to the lowest residual distortion. This is assessed with some form of block distortion measure (BDM), formulated as follows:

$$\text{BDM}(i, j) = \sum_{x=0}^{N-1} \sum_{y=0}^{M-1} g(s(x, y) - s_r(x + i, y + j)) \qquad (8.18)$$

where $g(\cdot)$ is the function used to compute the BDM, i and j are the offsets for the current motion vector candidate, and x, y are the local coordinates within the current block.

We choose the motion vector that minimizes the BDM for the current block:

$$\mathbf{d} = [d_x, d_y]^{\text{T}} = \arg\left(\min_{\forall(i,j)}(\text{BDM}(i, j))\right) \qquad (8.19)$$

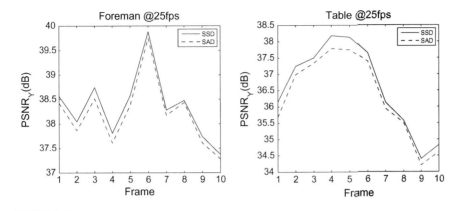

FIGURE 8.10

Effect of matching function: SAD vs SSD.

Absolute difference vs squared difference measures

There are many possible block distortion functions that could be used to assess the error residual after motion estimation. These include normalized cross correlation (NCCF), mean squared error (or sum of squared differences (SSD)), and sum of absolute differences (SAD). The choice of BDM function is key as it influences both the search complexity and the prediction accuracy.

The two most common functions are SAD (equation (8.20)) and SSD (equation (8.21)).[3]

$$\text{SAD}(i, j) = \sum_{x=0}^{N-1} \sum_{y=0}^{M-1} |(s(x, y) - s_r(x + i, y + j))| \qquad (8.20)$$

$$\text{SSD}(i, j) = \sum_{x=0}^{N-1} \sum_{y=0}^{M-1} (s(x, y) - s_r(x + i, y + j))^2 \qquad (8.21)$$

Figure 8.10 illustrates the difference in prediction accuracy between these two functions for two different sequences. It can be observed that, although the SSD metric consistently outperforms SAD in terms of quality, there is actually very little difference between the two methods. Furthermore, the SAD metric requires only $3NM$ basic operations (Add, Sub, Abs) per block to compute whereas the SSD requires NM multiplications plus $NM - 1$ additions. For this reason, the SAD function is frequently chosen over SSD.

---

[3]*Note:* SAD is normally used in preference to mean absolute difference (MAD).

8.3.3 **Full search algorithm**

The simplest BMME algorithm is based on *full search* or *exhaustive search*. In this case the global minimum of the error surface is found by exhaustively checking all possible $(2d_m + 1)^2$ motion vector candidates within the search grid.

8.3.4 **Properties of block motion fields and error surfaces**

The block motion field

For natural scenes, the block motion field is generally smooth and slowly varying. This is because there is normally high correlation between the motion vectors of neighboring blocks. Figure 8.11 illustrates this for the example of the *Foreman* sequence captured at 25 fps.

The effect of block size

Figure 8.12 shows the effect of different block sizes (8×8 vs 16×16) in terms of prediction quality and motion overhead. It can be seen that smaller block sizes normally result in better prediction accuracy, but only at the expense of significantly increased bit rates. The improved prediction accuracy is due to the fact that smaller blocks can deal better with multiple motions, complex motions (i.e. provide a better fit to the model), and the different properties of textured and non-textured regions. Because of the conflict between rate and distortion, many modern compression standards employ variable block sizes for motion estimation alongside rate–distortion optimization (RDO) techniques that select the block size that optimizes a rate–distortion trade-off. This is discussed further in Chapters 9 and 10.

$\rho_X = 0.56$ $\rho_Y = 0.33$	$\rho_X = 0.69$ $\rho_Y = 0.49$	$\rho_X = 0.64$ $\rho_Y = 0.46$
$\rho_X = 0.66$ $\rho_Y = 0.48$	**Current Block**	$\rho_X = 0.72$ $\rho_Y = 0.63$
$\rho_X = 0.64$ $\rho_Y = 0.43$	$\rho_X = 0.76$ $\rho_Y = 0.61$	$\rho_X = 0.68$ $\rho_Y = 0.56$

FIGURE 8.11

Autocorrelation values for typical block motion fields (*Foreman* at 25 fps). (Adapted from Ref. [5].)

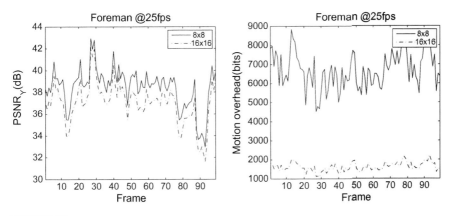

FIGURE 8.12

The effect of block size on motion estimation performance.

Example 8.2 (Full search motion estimation)

Given the following 2×2 block in the current frame **S**, and the corresponding search window **W**, determine the motion vector, **d**, for this block. Use a full search strategy with an SAD error metric.

$$
\mathbf{W} = \begin{bmatrix} 1 & 5 & 4 & 9 \\ 6 & 1 & 3 & 8 \\ 5 & 7 & 1 & 3 \\ 2 & 4 & 1 & 7 \end{bmatrix}; \quad \mathbf{S} = \begin{bmatrix} 3 & 9 \\ 1 & 4 \end{bmatrix}
$$

Solution. Consider, for example, the candidate motion vector $\mathbf{d} = [0, 0]$. In this case the displaced frame difference signal is given by:

$$
\mathbf{E} = \begin{bmatrix} 3 & 9 \\ 1 & 4 \end{bmatrix} - \begin{bmatrix} 1 & 3 \\ 7 & 1 \end{bmatrix} = \begin{bmatrix} 2 & 6 \\ -6 & 3 \end{bmatrix}
$$

and the SAD = 17. Similarly for all other $[d_x, d_y]$ offsets:

i	j	SAD(i,j)	i	j	SAD(i,j)
-1	-1	14	0	1	18
-1	0	18	1	-1	7
-1	1	5	1	0	2
0	-1	8	1	1	11
0	0	17			

Hence the motion vector for this block is $\mathbf{d} = [1, 0]$.

The effect of search range

The *search grid* size is dictated by the maximum motion vector offset from the current block location, denoted as d_m. The value of d_m also has an influence on complexity and accuracy, since full search complexity will grow with its square (there are $(2d_m + 1)^2$ candidates to be evaluated) and most fast search methods grow with it linearly. A small value of d_m will mean that fast moving objects will not be captured in the search widow, resulting in poor prediction quality and motion failure. In contrast, a large value of d_m may be inefficient and lead to excessive computational complexity.

The choice of search window size is dictated by content type and video format. For example, sports content will normally require a larger window than talk shows, and HDTV will require a larger window than SDTV for the same frame rate. Figure 8.13 shows the effect of search window size for the CIF Foreman sequence at 25 fps. It can be seen that a small ±5 window performs satisfactorily most of the time in this case, except for a few temporal regions (around frames 12, 50, 80, 90) where higher motion causes a significant drop in quality. The larger search window will cope with these fast motions, but does require a small increase in bit rate in order to deal with the increased average motion vector size.

The motion residual error surface

The error surface for each block (i.e. the BDM figures for each candidate vector in the search grid) will normally be multimodal, containing one or more local minima. This can be caused by periodicities in local textures, the presence of multiple motions within the block, or just a poor fit between the motion model and the apparent motion. In addition, the global minimum and maximum SAD values can vary enormously, even across a single frame in a sequence. An example of an error surface is shown in Figure 8.14 together with its corresponding current block (top right) and search

FIGURE 8.13

The influence of search range on prediction accuracy and coding bit rate.

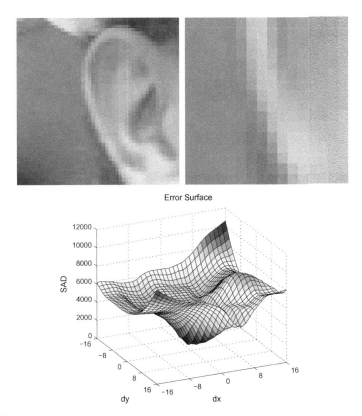

FIGURE 8.14

Typical motion residual error surface for a 16×16 block from the *Foreman* sequence (CIF at 25 fps, $\pm 16 \times \pm 16$ search window). Top left: search window. Top right: current block.

window (top left). We can observe that in the example shown, the SAD surface is indeed multimodal and extends in value from below 1000 to over 11,000.

Motion vector probabilities

The distribution of the block motion field is center-biased, implying that smaller displacements are more probable than larger ones and that the [0,0] vector is (in the absence of global motion) the most likely. This property is illustrated in Figure 8.15, for a range of sequences coded at 25 fps. The peak in the pdf at [0,0] is clear for the *Akiyo* and *Football* sequences, whereas there is an obvious offset due to the camera pan in *Coastguard*. The low activity sequence, *Akiyo*, shows a large peak in the pdf at [0,0] with very small values elsewhere. The *Football* sequence on the other hand shows a much lower peak together with a clear element due to camera panning and a significant pdf energy away from [0,0] due to the complex nature of the scene. The *Coastguard* sequence shows a wider main peak, biased due to the camera pan,

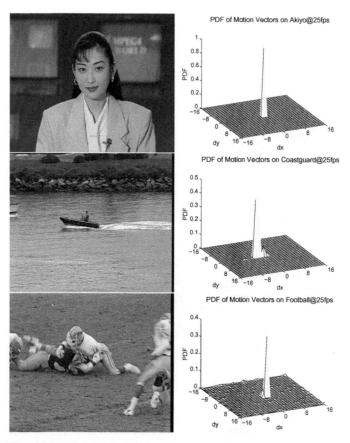

FIGURE 8.15

Motion vector pdfs for *Akiyo*, *Coastguard* and *Football* sequences.

but also energy away from [0,0], largely due to the difficulty of estimating the random movements of the dynamic water texture.

8.3.5 Motion failure

Although the block-based translational motion model works well in most cases and can provide excellent rate–distortion performance with low complexity, there are situations where the motion characteristics are such that it fails badly. Such a situation is shown in Figure 8.16. These two frames, from *Football*, capture the high activity typical in such content and the DFD, even after sub-pixel motion estimation, shows significant artifacts. These artifacts will cause the DFD to have a very low autocorrelation coefficient and hence the decorrelating transform (DCT or otherwise) will perform poorly.

So how do we deal with this situation? Basically two approaches can be used, independently or in combination. The first compares the rate–distortion performance

FIGURE 8.16

Example of motion model failure for *Football* sequence. Top left: current frame, right: reference frame. Bottom left: full-pixel estimation, right: half-pixel estimation.

of intra- and inter-frame coding and chooses the optimum coding mode. The second uses an in-loop deblocking filter to remove introduced edge artifacts in the reference frame. These methods are discussed further in Chapters 9 and 10.

8.3.6 Restricted and unrestricted vectors

In the case of blocks situated at or near the edge of a frame, a portion of the search area will fall beyond the frame border. In standard motion estimation, vectors are restricted to fall within the frame. However, benefits can be obtained, especially in the case of camera motion, by artificially extending the search window through border pixel extrapolation, thus supporting the use of *unrestricted motion vectors*. External pixels are normally based on a zero order or symmetric extrapolation of the border values.

8.4 Reduced complexity motion estimation

8.4.1 Pixel grids and search grids

Before we look at advanced search techniques, it is worth spending a little time differentiating between *pixel grids* and *search grids*. Both represent the region of support for motion search but we will use the latter as the basis for describing the operation of search methods from now on. The pixel grid is simply the image matrix where each element represents the pixel value within the search window. In contrast, the search grid represents the search window in terms of its offsets. Without loss of generality,

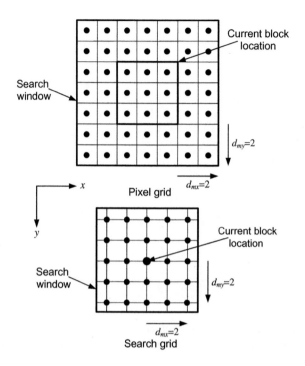

FIGURE 8.17

The pixel grid and the search grid.

assuming a square search window, the dimensions of the search grid are $2d_m + 1$ by $2d_m + 1$. This corresponds, for an N by N block size, to a search window with dimensions $2d_m + N$ by $2d_m + N$. These two representations are illustrated in Figure 8.17.

8.4.2 Complexity of full search

For each motion vector, there are $(2d_m + 1)^2$ motion vector candidates within the search grid. At each search location MN pixels are compared, where each comparison requires three operations (Sub, Abs, Add). The total complexity (in terms of basic arithmetic operations) per macroblock is thus $3MN(2d_m + 1)^2$.

Thus, for a picture resolution of $I \times J$ and a frame rate of F fps, the overall complexity is (in operations per second) given by equation (8.22):

$$C_{FS} = 3IJF(2d_m + 1)^2 \qquad (8.22)$$

Example 8.3 (Full search complexity)
Consider a video sequence with the following specifications: $M = N = 16, I = 720, J = 480, F = 30, d_m = 15$. Compute the complexity of a full search motion estimation algorithm for this sequence.

Solution. As indicated above the full search complexity is given by $3IJF(2d_m+1)^2$ operations per second. Substituting the values for the given sequence gives a processor loading of 29.89×10^9 operations per second!

8.4.3 Reducing search complexity

As we will see shortly, search complexity can be reduced considerably but usually at the expense of sub-optimum performance. Many approaches exist and these can be based on:

1. **A reduced complexity BDM:** In practice it is difficult to improve upon the use of SAD which already offers a very good trade-off between complexity and performance. One interesting approach in this category is that of Chan and Siu [9] where the number of pixels used in the BDM evaluation is varied according to the block activity.

2. **A sub-sampled block motion field:** This approach exploits the smoothness properties of the block motion field. Only a fraction of the block candidates are evaluated and the remainder are interpolated before selecting the minimum BDM. For example, Liu and Zaccarin [10] use a checkerboard pattern to evaluate the motion vectors for half of the blocks. This reduces search complexity by a factor of 2 with little decrease in prediction quality.

3. **Hierarchical search techniques:** This class of method uses a multiresolution approach to progressively refine motion vector estimates across finer scales. An example of such methods, using a mean pyramid, is presented by Nam et al. [11]. Such approaches are also reported to provide more homogeneous block motion fields with a better representation of the true motion.

4. **A reduced set of motion vector candidates:** This method restricts the number of blocks evaluated using some form of directed search pattern. Most fast ME algorithms in this category are based on the assumption of a smooth and unimodal error surface (i.e. the error surface increases monotonically with distance from the minimum BDM point). We have already seen that this assumption is not always valid and algorithms can get trapped in local minima (see Figure 8.14) so the search pattern used must take this into account. The earliest approach of this type was the Two-Dimensional Logarithmic (TDL) search proposed by Jain and Jain [12], and many variants have followed. This class of method is widely considered to offer an excellent compromise between complexity and prediction quality and is commonly used in practice. Selected approaches from this category are described in the following subsections.

5. **Intelligent initialization and termination:** Again, these exploit the properties of the block motion field. Intelligent initialization criteria are used to start the search at the location of a predicted motion vector rather than at the center of the search grid. Such methods can be used with (4) to help avoid local minima.

The reader is directed to Ref. [5] for a detailed comparison of these methods. In the following sections we examine some of the more popular techniques in more detail.

8.4.4 2-D logarithmic search (TDL)

The TDL search was introduced by Jain and Jain in 1981 [12] as a method of reducing the number of motion vector candidates evaluated. The search patterns used in this method are shown in Figure 8.18 and are in the shape of a + with five points, except for the final stage where a nine-point square pattern is checked. The largest pattern is initially located at the center of the search grid, all five points are evaluated and the one with the lowest BDM is selected as the center of the next search. The process is repeated until the minimum point remains at the center, in which case the search pattern radius is halved. This process continues until the radius is equal to 1, when the nine-point square pattern is invoked. This process is described in Algorithm 8.1.

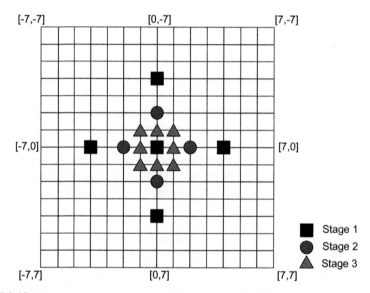

FIGURE 8.18

2-D logarithmic search patterns.

Algorithm 8.1 TDL search.

---

1. Set initial search radius $r = 2^{k-1}$ where $k = \lceil \log_2(d_m) \rceil$;
2. Set checking points to: $\mathbf{\Gamma} = \{[0, 0], [0, \pm r][\pm r, 0]\}$;
3. Evaluate the BDM at the five candidate locations on the cross and select the grid point with the lowest BDM: $\mathbf{d}' = \arg\left(\min_{(i,j)\in\Gamma} (\text{BDM}(i, j)) \right)$;
4. IF the minimum remains at the center of the pattern THEN $r = \frac{r}{2}$;
5. IF $r > 1$ THEN $\mathbf{\Gamma} = \mathbf{d}' + \{[0, 0], [0, \pm r][\pm r, 0]\}$, GO TO (3);
6. IF $r = 1$ THEN invoke nine-point square pattern with $\mathbf{\Gamma} = \mathbf{d}' + \{[0, 0], [\pm r, \pm r], [0, \pm r][\pm r, 0]\}$, $\mathbf{d} = \arg\left(\min_{(i,j)\in\Gamma} (\text{BDM}(i, j)) \right)$, STOP.

8.4.5 *N*-Step Search (NSS)

The *N*-Step Search (NSS), introduced by Koga et al. [13], is similar to the TDL search. It uses a uniform pattern comprising nine evaluation points as shown in Figure 8.19, where the search radius decreases by a factor of 2 at each iteration. As for the TDL search, the process terminates when the search radius is equal to 1. The advantage of this approach is that it requires a fixed number of iterations, so offers regularity in terms of hardware or software implementation. The method is described in Algorithm 8.2. A worked example of a Three-Step Search (TSS) is given in Example 8.4.

The complexity of the NSS algorithm is easy to compute and is:

$$C_{NSS} = 3IJF(8k + 1) \tag{8.23}$$

where *k* is the number of steps as indicated in Algorithm 8.2.

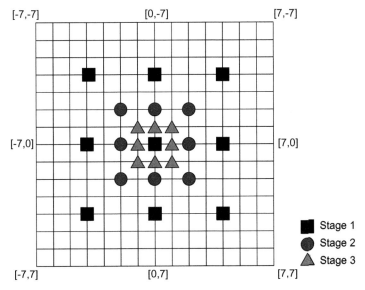

FIGURE 8.19

N-Step Search patterns.

Algorithm 8.2 NSS search.

1. Set initial search radius $r = 2^{k-1}$ where $k = \lceil \log_2(d_m) \rceil$;
2. Set checking points to: $\mathbf{\Gamma} = \{[0, 0], [\pm r, \pm r], [0, \pm r], [\pm r, 0]\}$;
3. Evaluate the BDM at the nine candidate locations and select the grid point with the lowest BDM: $\mathbf{d'} = \arg\left(\min_{(i,j) \in \Gamma} (BDM(i, j)) \right)$;
4. Set $r = \frac{r}{2}$;
5. IF $r < 1$ THEN $\mathbf{d} = \mathbf{d'}$, STOP; ELSE set new search locations: $\mathbf{\Gamma} = \mathbf{d'} + \{[0, 0], [\pm r, \pm r], [0, \pm r], [\pm r, 0]\}$; GO TO (3).

Example 8.4 (The three-step search)
Assuming a 15×15 search grid and a monotonically decreasing error surface with a single minimum, demonstrate how the TSS search algorithm would converge to a motion vector of $\mathbf{d} = [-3, 2]$.

Solution. The NSS algorithm will always converge in N steps, but the convergence profile will depend on the actual shape of the error surface. One possible solution is shown in the figure below.

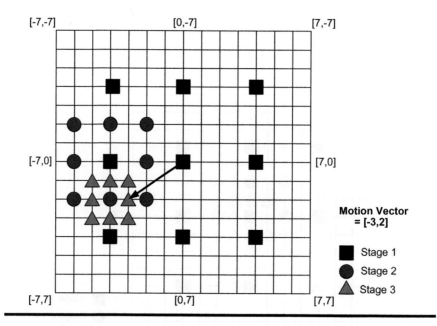

8.4.6 Diamond search

The Diamond Search (DS) was first introduced by Zhu and Ma in 1997 [14] and has demonstrated faster processing, with similar distortion, in comparison with TSS and NTSS approaches. DS adopts two search patterns—one large diamond for general gradient search and a second smaller diamond for final stage refinement. The approach is again similar to the TDL search in that the large diamond pattern at stage k is centered on the point with minimum BDM from stage $k - 1$. When the search minimum remains at the center of the pattern, the small diamond pattern is invoked to refine the motion vector before termination. The two diamond patterns are shown in Figure 8.20 and the process is described in Algorithm 8.3. An illustration of the diamond search process is given in Example 8.5. It should be noted that the complexity of DS is dependent on the orientation of the optimum motion vector and so is content dependent.

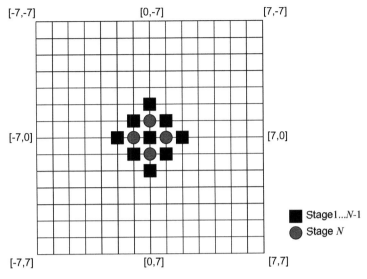

FIGURE 8.20

Diamond search patterns.

Algorithm 8.3 Diamond search.

1. Initialize motion vector estimate: $\mathbf{d'} = [0, 0]$;
2. Set checking points to: $\mathbf{\Gamma} = \mathbf{d'} + \{[0, 0], [0, \pm2][\pm2, 0][\pm1, \pm1]\}$;
3. Evaluate the BDM at the nine candidate locations on the diamond and select the grid point with the lowest BDM: $\mathbf{d'} = \arg\left(\min_{(i,j)\in\Gamma}(\text{BDM}(i, j))\right)$;
4. IF the minimum remains at the center of the pattern THEN $\mathbf{\Gamma} = \mathbf{d'} + \{[0, 0], [0, \pm1], [\pm1, 0]\}$, $\mathbf{d} = \arg\left(\min_{(i,j)\in\Gamma}(\text{BDM}(i, j))\right)$, STOP;
5. GO TO (2).

Example 8.5 (The diamond search)

Assuming a 15×15 search grid and a monotonically decreasing error surface with a single minimum, demonstrate how the DS search algorithm would converge to a motion vector of $\mathbf{d} = [-1, -4]$.

Solution. The DS algorithm will not converge in a specific number of steps but will track the gradient of the error surface using the large diamond shape until the minimum stays at the center of the diamond (stages 1–3 in this example). The small diamond shape is then introduced at stage 4 to identify the final motion vector. The convergence profile will of course depend on the actual shape of the error surface. One possible solution is shown in the following figures:

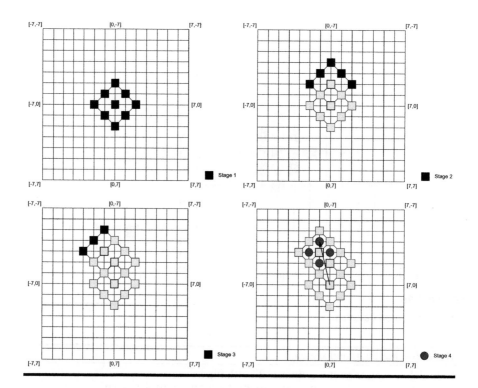

8.4.7 Hexagonal search

There are many variants on the methods presented above. One worthy of note is the hexagonal search (HEXBS), which again uses two pattern types, but this time in the shape of a hexagon (Figure 8.21). HEXBS was introduced by Zhu et al. [15] and, like DS, the smaller pattern is used only once for refinement at the final iteration. HEXBS has the advantage over diamond search that it offers greater orientation flexibility with the number of new checking points introduced at each iteration of the gradient descent stage being constant (three new points).

The advancing speed for DS is 2 pixels per step horizontally and vertically, and $\sqrt{2}$ pixels per step diagonally, whereas the HEXBS has horizontal and vertical advancements of 2 pixels per step horizontally and vertically and $\sqrt{5}$ pixels per step in the direction of the off-axis search points. This means that HEXBS is more consistent in its search performance and can outperform DS by up to 80% for some motion vector locations.

The complexity of the HEXBS algorithm can easily be seen to be:

$$C_{\text{HEXBS}} = 7 + 3k + 4 \tag{8.24}$$

where k is the number of iterations required to reach the minimum of the error surface.

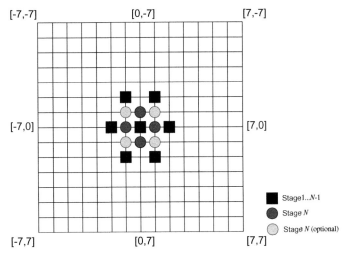

FIGURE 8.21

Hexagonal search patterns.

8.4.8 Initialization and termination criteria

Initialization

The use of intelligent initialization and termination criteria can, in many cases, improve search performance and help to avoid the problems of the search becoming trapped in local minima. Referring to Figure 8.22, which shows a current block P and its near causal neighbors $A-D$, a range of candidate initialization vectors can be formulated, based on the local smoothness property of the block motion field. These include:

$$\hat{\mathbf{d}}_P = \{[0, 0], \mathbf{d}_A, \mathbf{d}_B, \mathbf{d}_C, \mathbf{d}_D, \text{med}(\mathbf{d}_A, \mathbf{d}_C, \mathbf{d}_D), \text{med}(\mathbf{d}_A, \mathbf{d}_B, \mathbf{d}_C)\} \quad (8.25)$$

An initialization algorithm would typically evaluate some or all of the candidate vectors from equation (8.25), choose that with the lowest BDM, and start the search at $\hat{\mathbf{d}}_P$ rather than always at [0,0].

B	C	D	
A	P		

FIGURE 8.22

Predicted motion vectors for initialization.

FIGURE 8.23

Fast motion estimation performance comparison for the CIF Bream sequence at 25 fps: 4PMV+0 vs ADZS-ER vs FS.

Termination

The use of termination criteria is largely dictated by the implementation method used. For example, in the case of a software implementation, a search could be terminated by setting an acceptable block distortion threshold BDM_{Th} and exiting the search when $BDM \leq BDM_{Th}$. Furthermore, inside each SAD evaluation loop, execution could be terminated when the distortion measure for the current offset $[i, j]$ exceeds the current minimum. i.e. when $BDM_{i.j} \geq BDM_{opt}$, or when the rate–distortion cost is sufficiently small.

Several methods have been reported that provide significant speed-up using intelligent initialization and termination. These include the work of Tourapis et al. [16,17] on PVMFast and EPZS and the work in the author's lab by Sohm [18]. For example, Figure 8.23 shows complexity results including the 4PMV+0 algorithm, Diamond Search and Full Search.

8.4.9 Reduced complexity block matching algorithm performance comparisons

Several authors have published tables comparing motion estimation methods in terms of complexity and prediction accuracy (see, for example, Ref. [16]). Table 8.2 compares some of the algorithms presented in this chapter. It can be seen that methods such as the Simplex Search [19] provide the most accurate results (closest to FS) with a competitive complexity. Techniques adopting intelligent initialization and termination criteria such as 4PMV+0 on the other hand require extremely low numbers of checking points with only a modest reduction in prediction accuracy.

Table 8.2 Reduced complexity BMME comparisons for the Bream CIF sequence at 25 fps. Top: checking points per macroblock. Bottom: SAD error relative to full search.

Algorithm	4PMV + 0	ADZS	DS	Simplex	NSS	FS
Minimum	1	1	13	9	32.9	1002.4
Average	4.1	5.9	16.1	14.2	33.0	1018.3
Maximum	9.6	15.3	22.2	21.8	33.1	1024.0
Average	4.248	4.492	4.311	4.078	4.494	3.975
% vs FS	106.9	113.0	108.5	102.6	113.1	100.0

8.5 Motion vector coding

8.5.1 Motion vector prediction

Standards such as H.264/AVC and HEVC have made substantial progress in improving rate–distortion performance. However, as a consequence, the motion coding overhead has proportionally increased, with over 50% of the bits being dedicated to motion for some sequences with high QP values.

Coding of motion vectors is normally performed predictively to exploit the smoothness property of the block motion field. Consider coding the current block P in Figure 8.22. A predicted motion vector can be generated using equation (8.26). If block D is not available then block B can be used instead.

$$\hat{\mathbf{d}}_P = \{\mathrm{med}(\mathbf{d}_A, \mathbf{d}_C, \mathbf{d}_D)\} \qquad (8.26)$$

The motion vector difference or residual is then computed and entropy coded for transmission:

$$\mathbf{e}_P = \mathbf{d} - \hat{\mathbf{d}}_P \qquad (8.27)$$

8.5.2 Entropy coding of motion vectors

The motion vector prediction process will result in a decorrelated residual signal that will mainly comprise small values. In the same way as we used entropy coding based on Huffman or arithmetic methods in Chapter 7 to code quantized transform coefficients, so too can we apply this to motion vector residuals. Clearly, different VLC tables will be required for Huffman coding and different probability distributions for arithmetic coding.

Example 8.6 (Motion vector coding)
Motion vectors for four adjacent blocks in two partial rows of a frame are shown below, outlined in bold. Using the prediction scheme: $\hat{\mathbf{d}}_P = \mathrm{med}(\mathbf{d}_A, \mathbf{d}_C, \mathbf{d}_D)$, compute the predicted motion vectors for each of these blocks together with their residuals.

[1,3]	[1,3]	[2,4]	[2,3]
[1,3]	[1,3]	[2,4]	[2,3]
[1,5]	[1,5]	[2,4]	[2,4]

B	C	D
A	P	

Solution. The predicted motion vectors and residuals for each of the four blocks outlined (labeled top left to bottom right) are given in the table below:

Block, P	1	2	3	4
$\hat{\mathbf{d}}_P$	[1,3]	[2,3]	[1,4]	[2,4]
$\mathbf{e}_P$	[0,0]	[0,1]	[0,1]	[0,0]

Note that the prediction process described here must be modified for boundary blocks. This is described further in Chapter 12.

8.6 Summary

This chapter has addressed the important topic of motion estimation, the primary technique for reducing temporal correlation in an image sequence. We saw how temporal correlations can be exploited to provide additional coding gain. Most of the chapter focused on block matching and block-based motion estimation using a translational motion model. This approach is in widespread use and has been adopted in all standardized video codecs to date. The chapter also described a number of methods for reducing the complexity of motion estimation, comparing a range of fast search methods alongside descriptions of intelligent initialization and termination techniques.

In the next chapter we will show how to integrate the motion estimation algorithm into a DPCM prediction loop at the encoder. We also examine a range of methods for improving still further the coding gain of this block-based hybrid codec.

References

[1] T. Ebrahimi, M. Kunt, Visual data compression for multimedia applications, Proceedings of the IEEE 86 (6) (1998) 1109–1125.

[2] C. Glasbey, K. Mardia, A review of image warping methods, Journal of Applied Statistics 25 (1998) 155–171.

[3] F. Zhang, D. Bull, A parametric framework for video compression using region-based texture models, IEEE Journal of Selected Topics in Signal Processing 6 (7) (2011) 1378–1392.

[4] R. Vigars, A. Calway, D. Bull, Context-based video coding, in: Proceedings of the IEEE International Conference on Image Processing, 2013, pp. 1953–1957.

[5] M. Al-Mualla, C. Canagarajah, D. Bull, Video Coding for Mobile Communications, Academic Press, 2002.

[6] A. Netravali, J. Robins, Motion compensated television coding: part 1, Bell Systems Technical Journal 58 (1979) 631–670.

[7] D. Kuglin, D. Hines, The phase correlation image alignment method, in: Proceedings of the IEEE International Conference on Cybernetics and Society, 1975, pp. 163–165.

[8] G. Thomas, Television Motion Measurement for DATV and Other Applications, BBC Research Technical Report 1987/11, 1987.

[9] Y. Chan, W. Siu, New adaptive pixel decimation for block vector motion estimation, IEEE Transactions on Circuits and Systems for Video Technology 6 (1) (1996) 113–118.

[10] B. Liu, A. Zaccarin, New fast algorithms for the estimation of motion vectors, IEEE Transactions on Circuits and Systems for Video Technology 3 (2) (1993) 148–157.

[11] K. Nam, J. Kim, R. Park, Y. Shim, A fast hierarchical motion vector estimation algorithm using mean pyramid, IEEE Transactions on Circuits and Systems for Video Technology 5 (4) (1995) 344–351.

[12] J. Jain, A. Jain, Displacement measurement and its application in interframe image coding, IEEE Transactions on Communications 29 (12) (1981) 1799–1808.

[13] T. Koga, K. Linuma, A. Hirano, T. Ishiguro, Motion compensated interframe coding for video conferencing, in: Proceedings of the National Telecommunications Conference, 1981, pp. G5.3.1–G5.3.5.

[14] S. Zhu, K. Ma, A new diamond search algorithm for fast block matching motion estimation, IEEE Transactions on Image Processing 9 (2) (2000) 287–290.

[15] C. Zhu, X. Lin, L. Chau, Hexagon-based search pattern for fast block motion estimation, IEEE Transactions on Circuits and Systems for Video Technology 12 (5) (2002) 255–349.

[16] A. Tourapis, O. Au, M. Liou, Predictive motion vector field adaptive search technique (PMVFAST): enhancing block-based motion estimation, in: Proceedings of the SPIE Visual Communications and Image Processing, vol. 4310, 2001, pp. 883–893.

[17] A. Tourapis, Enhanced predictive zonal search for single and multiple frame motion estimation, in: Proceedings of the SPIE Visual Communications and Image Processing, vol. 4671, 2002, pp. 1069–1078.

[18] O. Sohm, Method for Motion Vector Estimation, US Patent 7,260,148, 2007.

[19] M. Al-Mualla, C. Canagarajah, D. Bull, Simplex minimization for single- and multiple-reference motion estimation, IEEE Transactions on Circuits and Systems for Video Technology 11 (12) (2001) 1209–1220.

The Block-Based Hybrid Video Codec

CHAPTER OUTLINE

9.1 **The block-based hybrid model for video compression** **292**
    9.1.1 Picture types and prediction modes 292
    9.1.2 Properties of the DFD signal 293
    9.1.3 Operation of the video encoding loop 294
    9.1.4 Operation of the video decoder 295
9.2 **Intra-frame prediction** **298**
    9.2.1 Intra-prediction for small luminance blocks 298
    9.2.2 Intra-prediction for larger blocks 300
9.3 **Sub-pixel motion estimation** **301**
    9.3.1 Sub-pixel matching 301
    9.3.2 Interpolation methods 303
    9.3.3 Performance 305
    9.3.4 Interpolation-free methods 306
9.4 **Multiple reference frame motion estimation** **307**
    9.4.1 Justification 307
    9.4.2 Properties, complexity, and performance of MRF-ME 307
    9.4.3 Reduced complexity MRF-ME 308
9.5 **Variable block sizes for motion estimation** **309**
    9.5.1 Influence of block size 309
    9.5.2 Variable block sizes in practice 310
9.6 **Variable sized transforms** **310**
    9.6.1 Integer transforms 310
    9.6.2 DC coefficient transforms 313
9.7 **In-loop deblocking operations** **313**
9.8 **Summary** **315**
References **315**

In Chapter 8 we saw how motion estimation formed the basis of an efficient video coding system. This chapter describes how it can be integrated into a practical video compression framework based on a hybrid structure, combining block-based

For colour and a higher quality version of Figure 9.17 please refer to the electronic version or the website.

Communicating Pictures. http://dx.doi.org/10.1016/B978-0-12-405906-1.00009-X

motion prediction with block-based transform coding. The architecture of a generic video encoder is presented and the operation of the encoding and decoding stages is described in detail.

The limitations of the basic hybrid architecture are discussed and a collection of practical codec enhancements that have been introduced into recent standards are described. These techniques serve to significantly enhance the performance of the basic video coding structure and have enabled the basic hybrid architecture to provide a doubling of coding gain every 8–10 years over the past 40 years or so. The techniques described are not an exhaustive list of all the features found in modern codecs, but are some of the most significant ones.

We first of all focus on intra-prediction—a means of enhancing performance through spatial prediction of blocks in each frame. We then extend the motion prediction methods introduced in Chapter 8 through sub-pixel estimation and the use of multiple reference frames. We also examine the influence of variable block sizes and their exploitation in the coding process, both in terms of transforms and motion estimation. Finally we study the important area of in-loop deblocking as an approach to reducing edge artifacts.

9.1 The block-based hybrid model for video compression

9.1.1 Picture types and prediction modes

Prediction modes

As we saw in Chapter 8, three main classes of prediction are used in video compression:

- **Forward prediction:** The reference picture occurs temporally before the current picture.
- **Backward prediction:** The reference picture occurs temporally after the current picture.
- **Bidirectional prediction:** Two (or more) reference pictures (forward and backward) are employed and the candidate predictions are combined in some way to form the final prediction.

Picture types

Three major types of picture (or frame) are employed in most video codecs:

- **I-pictures:** These are intra-coded (coded without reference to any other pictures).
- **P-pictures:** These are inter-coded with forward (or backward) prediction from another I- or P-picture.
- **B-pictures:** These are inter-coded with bidirectional prediction from more than one I- and/or P-picture.

Coded pictures are arranged in a sequence known as a *Group of Pictures* (GOP). A typical GOP structure comprising 12 frames is shown in Figure 9.1. A GOP will

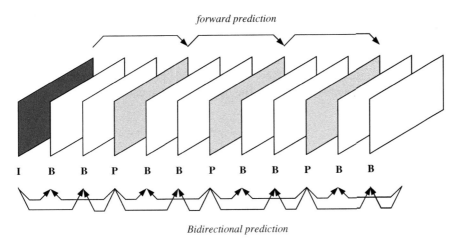

FIGURE 9.1

Typical group of pictures structure.

contain one I-picture and zero or more P- and B-pictures. The 12-frame GOP in Figure 9.1 is sometimes referred to as an IBBPBBPBBPBB structure and it is clear, for reasons of causality, that the encoding order is different to that shown in the figure since the P-pictures must be encoded prior to the preceding B-pictures.

9.1.2 Properties of the DFD signal

The DCT is a special case of the Karhunen–Loeve expansion for stationary and first order Markov processes whose autocorrelation coefficient approaches unity. Most still images possess good correlation properties, typically with $\rho \geq 0.9$; this justifies the use of the DCT for intra-frame coding as it will provide close to optimum decorrelation. However, in the case of inter-frame coding, where motion is predicted through motion estimation in a DPCM loop, the autocorrelation coefficient for the DFD signal typically falls to 0.3–0.5. Strobach [1] showed that, in many practical cases, the DCT coding gain for a DFD signal is small compared to the intra-frame case and that most of the gain is associated with temporal decorrelation. The edges introduced in the DFD signal due to motion failure can also cause ringing artifacts after quantization.

This discussion implies that the hybrid codec, which uses a decorrelating transform such as the DCT to code the DFD residual, is not always optimum. In practice, however, the DFD can be made smoother (i.e. exhibit higher autocorrelation values) if the motion model is more accurate and if any artificial edges are filtered out. These characteristics are promoted by codec enhancements such as sub-pixel motion estimation, variable block sizes, multiple reference frames, and the use of a deblocking filter in the motion prediction loop. Such enhancements, alongside efficient rate–distortion optimization methods, are largely responsible for the performance, and universal acceptance, of the block-based hybrid codec. They are all discussed in more detail later in this chapter.

9.1.3 Operation of the video encoding loop

Following the brief overview of the video encoding architecture given in Chapter 1, we describe here the operation of the encoding loop more formally and in more detail. A generic structure is given in Figure 9.2 for the case of intra-frame coding, and in Figure 9.3 for the inter-frame mode. The operation of the encoder in intra-mode is

FIGURE 9.2

Video encoder structure (intra-frame mode). Dotted lines depict control signals relating to quantizer step size, etc.

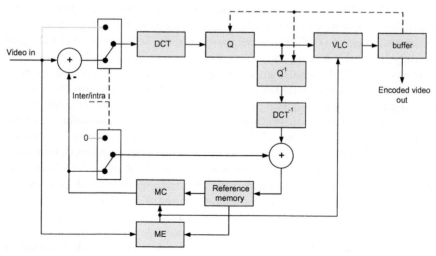

FIGURE 9.3

Video encoder structure (inter-frame mode).

Algorithm 9.1 Intra-mode encoder operation.

1. Inter/intra switch is in the intra position;
2. Compute displaced frame difference (DFD) signal (equal to the video input frame in this case): $\mathbf{E}_k = \mathbf{S}_k$;
3. Perform a forward decorrelating transform (typically a DCT or variant) on the input frame and quantize according to the prevailing rate–distortion criteria: $\mathbf{C}_k = Q\left(\mathrm{DCT}\left(\mathbf{E}_k\right)\right)$;
4. Entropy code the transformed frame and transmit to the channel;
5. Inverse quantize $\mathbf{C}_k$ and perform an inverse DCT to produce the same decoded frame pixel values as at the decoder: $\mathbf{E}'_k = \mathrm{DCT}^{-1}\left(Q^{-1}\left(\mathbf{C}_k\right)\right)$;
6. Update the reference memory with the reconstructed frame: $\mathbf{S}'_k = \mathbf{E}'_k + \mathbf{0}$.

Algorithm 9.2 Inter-mode encoder operation.

1. Inter/intra switch is in the inter position;
2. Estimate motion vector for current frame: $\mathbf{d} = \mathrm{ME}\left(\mathbf{S}_k, \mathbf{S}_{k-1}\right)$;
3. Form the motion compensated prediction frame, $\mathbf{P}_k$: $\mathbf{P}_k = \mathbf{S}'_{k-1}\left[\mathbf{p} + \mathbf{d}\right]$;
4. Compute the DFD signal: $\mathbf{E}_k = \mathbf{S}_k - \mathbf{P}_k$;
5. Perform a forward decorrelating transform on the DFD and quantize according to the prevailing rate–distortion criteria: $\mathbf{C}_k = Q\left(\mathrm{DCT}\left(\mathbf{E}_k\right)\right)$;
6. Entropy code the transformed DFD, motion vectors and control parameters, and transmit to the channel;
7. Inverse quantize $\mathbf{C}_k$ and perform an inverse DCT to produce the same decoded frame pixel values as at the decoder: $\mathbf{E}'_k = \mathrm{DCT}^{-1}\left(Q^{-1}\left(\mathbf{C}_k\right)\right)$;
8. Update the reference memory with the reconstructed frame: $\mathbf{S}'_k = \mathbf{E}'_k + \mathbf{P}_k$.

---

described in Algorithm 9.1. Similarly the operation in inter-mode is described in Algorithm 9.2.

9.1.4 Operation of the video decoder

The operation of the video decoder is illustrated in Figure 9.4 and described formally in Algorithms 9.3 and 9.4. We can see, by comparing the encoder and decoder architectures, that the encoder contains a complete replica of the decoder in its prediction feedback loop. This ensures that (in the absence of channel errors) there is no drift between the encoder and decoder operations.

Example 9.1 (Operation of a basic video encoder)
Based on the encoder diagrams in Figures 9.1 and 9.3 together with the following data at time k, compute:

1. The current integer-pixel motion vector, $\mathbf{d}$, at time k. Assume the search window is the whole reference frame and that vectors are restricted to point within the frame.

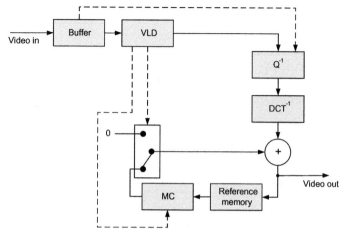

FIGURE 9.4

Video decoder structure.

Algorithm 9.3 Intra-mode decoder operation.

---

1. Inter/intra switch is in the intra position;
2. Perform entropy decoding of control parameters and quantized DFD coefficients;
3. Inverse quantize $\mathbf{C}_k$ and perform an inverse DCT to produce the decoded frame pixel values: $\mathbf{E}_k' = \mathrm{DCT}^{-1}\left(\mathrm{Q}^{-1}\left(\mathbf{C}_k\right)\right)$;
4. Update the reference memory with the reconstructed frame and output to file or display: $\mathbf{S}_k' = \mathbf{E}_k' + \mathbf{0}$.

Algorithm 9.4 Inter-mode decoder operation.

---

1. Inter/intra switch is in the inter position;
2. Perform entropy decoding of control parameters, quantized DFD coefficients and motion vector;
3. Inverse quantize $\mathbf{C}_k$ and perform an inverse DCT to produce the decoded DFD pixel values: $\mathbf{E}_k' = \mathrm{DCT}^{-1}\left(\mathrm{Q}^{-1}\left(\mathbf{C}_k\right)\right)$;
4. Form the motion compensated prediction frame, $\mathbf{P}_k$: $\mathbf{P}_k = \mathbf{S}_{k-1}'\left[\mathbf{p}+\mathbf{d}\right]$;
5. Update the reference memory with the reconstructed frame and output to file or display: $\mathbf{S}_k' = \mathbf{E}_k' + \mathbf{P}_k$.

---

2. The motion compensated output, $\mathbf{P}_k$, at time k.

3. The DFD for the current input frame, $\mathbf{E}_k$.

4. The transformed and quantized DFD output, $\mathbf{C}_k$.

Assume an image frame size of 12×12 pixels and a macroblock size of 4×4 pixels. The codec is operating in its inter-frame mode, using block-based translational motion estimation with a block size of 4×4 pixels:

Transform matrix: $\mathbf{A} = \frac{1}{2} \begin{bmatrix} 1 & 1 & 1 & 1 \\ 1 & 1 & -1 & -1 \\ 1 & -1 & -1 & 1 \\ 1 & -1 & 1 & -1 \end{bmatrix}$

Quantization matrix: $\mathbf{Q} = \begin{bmatrix} 2 & 2 & 2 & 2 \\ 2 & 2 & 2 & 2 \\ 2 & 2 & 4 & 4 \\ 2 & 2 & 2 & 4 \end{bmatrix}$

Current input block: $\mathbf{S}_k = \begin{bmatrix} 6 & 6 & 6 & 6 \\ 8 & 8 & 8 & 8 \\ 6 & 6 & 6 & 6 \\ 8 & 8 & 8 & 8 \end{bmatrix}$; assume top left of frame

Reference memory: $\mathbf{S}'_{k-1} = \begin{bmatrix} 0 & 0 & 5 & 6 & 2 & 6 & 0 & 0 & 0 & 0 & 0 & 0 \\ 0 & 0 & 8 & 8 & 8 & 8 & 0 & 0 & 0 & 0 & 0 & 0 \\ 0 & 0 & 6 & 6 & 6 & 6 & 0 & 0 & 0 & 0 & 0 & 0 \\ 0 & 0 & 8 & 8 & 12 & 8 & 0 & 0 & 0 & 0 & 0 & 0 \\ 0 & 0 & 0 & 0 & 0 & 0 & 0 & 0 & 0 & 0 & 0 & 0 \\ 0 & 0 & 0 & 0 & 0 & 0 & 0 & 0 & 0 & 0 & 0 & 0 \\ 0 & 0 & 0 & 0 & 0 & 0 & 0 & 0 & 7 & 0 & 0 \\ 0 & 0 & 0 & 0 & 0 & 0 & 0 & 3 & 4 & 2 & 3 & 4 \\ 0 & 0 & 0 & 0 & 0 & 0 & 0 & 4 & 4 & 5 & 8 & 0 \\ 7 & 7 & 8 & 9 & 0 & 0 & 0 & 0 & 6 & 7 & 0 & 0 \\ 0 & 0 & 0 & 10 & 3 & 0 & 0 & 0 & 0 & 7 & 0 & 0 \\ 0 & 0 & 2 & 2 & 2 & 0 & 0 & 0 & 0 & 0 & 0 & 0 \end{bmatrix}$

Solution.

1. Current motion vector:
By inspection:

$$\mathbf{d} = \begin{bmatrix} 2, 0 \end{bmatrix}$$

2. Motion compensated output:

$$\mathbf{P}_k = \begin{bmatrix} 5 & 6 & 2 & 6 \\ 8 & 8 & 8 & 8 \\ 6 & 6 & 6 & 6 \\ 8 & 8 & 12 & 8 \end{bmatrix}$$

3. DFD for current input frame:

$$\mathbf{E}_k = \mathbf{S}_k - \mathbf{P}_k = \begin{bmatrix} 1 & 0 & 4 & 0 \\ 0 & 0 & 0 & 0 \\ 0 & 0 & 0 & 0 \\ 0 & 0 & -4 & 0 \end{bmatrix}$$

4. Transformed and quantized output:
The transformed output is:

$$\mathbf{C}_k = \mathbf{A}\mathbf{E}_k\mathbf{A}^{\mathrm{T}} = \frac{1}{4} \begin{bmatrix} 1 & 1 & 1 & 1 \\ 9 & -7 & -7 & 9 \\ 1 & 1 & 1 & 1 \\ 9 & -7 & -7 & 9 \end{bmatrix}$$

and the quantized output is:

$$\mathbf{C}_{k(\mathrm{Q})} = \frac{1}{4} \begin{bmatrix} 1 & 1 & 1 & 1 \\ 9 & -7 & -7 & 9 \\ 1 & 1 & 1 & 1 \\ 9 & -7 & -7 & 9 \end{bmatrix} / \mathbf{Q} = \begin{bmatrix} 0 & 0 & 0 & 0 \\ 1 & -1 & -1 & 1 \\ 0 & 0 & 0 & 0 \\ 1 & -1 & -1 & 1 \end{bmatrix}$$

9.2 Intra-frame prediction

In Chapter 8 we saw the benefits of temporal prediction as a means of decorrelating between frames prior to transformation. H.264/AVC and HEVC also support intra-prediction in the spatial domain as well as an I-PCM mode that enables direct encoding of pixel values. Intra-coding has been found to be especially beneficial in regions which have certain oriented textures or directional edges. There are two main prediction modes supported in H.264/AVC—intra_4×4 and intra_16×16. These are described below.

9.2.1 Intra-prediction for small luminance blocks

This mode performs well when there is a high level of local detail, particularly with oriented features such as edges. Coding of a given block (4×4 in H.264/AVC) is performed with reference to a subset of samples from previously coded blocks that lie to the left and above the current block. As we will see in Chapter 11, intra-coding can lead to the propagation of errors between frames when it is based on data predicted from previously inter-coded macroblocks. To avoid this a *constrained intra-coding mode* can be used that restricts intra-prediction only to be based on intra-coded neighbors.

Figure 9.5 shows, for example, the intra-coding relationships and orientations used in H.264/AVC. The prediction for each 4×4 block (pixels *a–p* here) is based on its neighboring pixels (*A–M*). In H.264/AVC, the predictor can select from nine

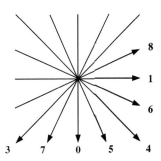

FIGURE 9.5

Intra-prediction modes in H.264/AVC.

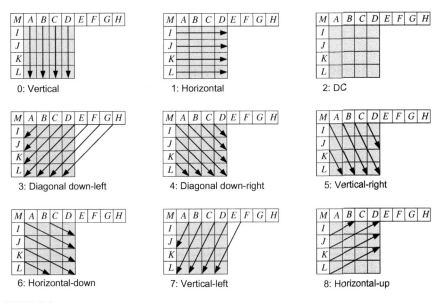

FIGURE 9.6

4×4 intra-prediction modes.

prediction orientations plus a DC mode where the average of the neighboring pixels is used as the prediction. These nine modes are shown in Figure 9.6. Some examples of how the predictions are computed are given below:

Mode 0 (Vertical)

$$\{a, e, i, m\} = A \tag{9.1}$$

Mode 1 (Horizontal)

$$\{a, b, c, d\} = I \tag{9.2}$$

Mode 2 (DC)

$$\{a, b, \cdots, p\} = \left\lfloor \left(\frac{A + B + C + D + I + J + K + L}{8} \right) + 0.5 \right\rfloor \qquad (9.3)$$

Mode 4 (Down-right)

$$\{a, f, k, p\} = \left\lfloor \left(\frac{A + 2M + I}{4} \right) + 0.5 \right\rfloor \qquad (9.4)$$

Other modes are formed by a similar extrapolation approach and a full description can be found in Ref. [2].

9.2.2 Intra-prediction for larger blocks

For larger smoother image regions it can be beneficial to perform intra-prediction on larger blocks. In the H.264 intra_16 × 16 mode, the entire 16 × 16 luma block is predicted simultaneously. The mode supports only four prediction types: mode 0 (vertical), mode 1 (horizontal), mode 2 (DC), and mode 3 (plane). The first three modes are similar to the 4 × 4 modes described above except they reference 16 pixels above and to the left of the current block, rather than just four. Details of the plane mode can be found in Ref. [2].

Chroma samples can be predicted in a similar manner to that used for luma blocks. Since chroma signals normally exhibit smoother characteristics than luma signals, only larger block sizes tend to be used.

Example 9.2 (Intra-prediction)
Compute the SAD values using the intra_4 × 4 vertical, horizontal, and DC modes, for the 4 × 4 block of pixels shown below:

1	2	3	4	5	6	7	8	9
1	2	3	4	5				
1	2	2	4	6				
2	2	3	4	3				
2	2	3	4	5				

Solution. The prediction blocks based on vertical, horizontal, and DC modes are given in the figure below:

2	3	4	5
2	3	4	5
2	3	4	5
2	3	4	5

vertical

1	1	1	1
1	1	1	1
2	2	2	2
2	2	2	2

horizontal

3	3	3	3
3	3	3	3
3	3	3	3
3	3	3	3

DC

and the corresponding SAD values are: SAD(vertical) = 4; SAD(horizontal) = 30; SAD(DC) = 16. Hence, based on this limited subset, the best match in terms of prediction residual energy is mode 0 (vertical). In this case the residual prior to coding would be as shown below:

0	0	0	0
0	−1	0	1
0	0	0	−2
0	0	0	0

9.3 Sub-pixel motion estimation

9.3.1 Sub-pixel matching

In Chapter 8 we introduced the concepts and realization of motion estimation as the basis for efficient video encoding. We limited the resolution of our search and matching to that of the sampling grid used to acquire the images. In reality, the motion in a scene is not related to the sampling grid and the most accurate match might occur with a sub-pixel displacement. Ericsson [3] introduced the concept of sub-pixel estimation in 1985 demonstrating, for the sequences used, that up to 2 dB coding gain could be achieved when changing from full-pixel to 1/8 pixel accuracy. This was formalized by Girod [4] who showed that limited gains were achieved by decreasing the estimation accuracy beyond 1/8 pixel resolution.

It is clear that the benefits of sub-pixel estimation will depend on a number of factors, including video format (spatial and temporal resolution) and type of content. However, it is acknowledged that sub-pixel estimation offers significant performance gains and modern coding standards all include the option to predict motion with fractional-pixel accuracy. MPEG-1 was the first international standard to use half-pixel estimation, closely followed by MPEG-2 in 1994. H.263 introduced more flexible block sizes and quarter-pixel estimation in 1996 and its improvement in performance over H.261 has been largely attributed to this innovation [5].

It is worth spending a few moments discussing the interpolation needed to support sub-pixel motion estimation. We could adopt a number of approaches:

1. Interpolate all of the current frame block and all of the search window and perform the entire motion estimation process at sub-pixel resolution.
2. Interpolate just the search window and proceed as in (1).
3. Perform integer-pixel search initially, then refine this locally using an interpolated search window and an interpolated current block.
4. Perform integer search initially, then refine this locally just interpolating the search window.

In practice method (4) is adopted, as it offers the lowest complexity with only a minor fall-off in performance (see Figure 9.7). This approach is illustrated in Figure 9.8, which shows the black square as the best full-pixel solution. The grid is then interpolated locally at half-pixel resolution and a refinement search is performed to obtain the

FIGURE 9.7

Sub-pixel search with local refinement.

FIGURE 9.8

Motion estimation search grid with half-pixel refinement.

best half-pixel solution (black circle). The refinement search is normally performed exhaustively and the same matching criterion used as for the integer-pixel case.

9.3.2 Interpolation methods

Figure 9.9 illustrates the interpolation method used for half-pixel and quarter-pixel interpolation in H.264/AVC. In the case of half-pixel estimation, the search window is interpolated locally around the optimum integer-pixel result. If we assume that the pixels in the frame's search window are locally upsampled by a factor of 2 (see Chapter 6 for a more detailed discussion of upsampling), then the upsampled result in the x direction is given by:

$$s_u[x] = \begin{cases} s[x/2]; & x = 2m \\ 0; & \text{otherwise} \end{cases} \tag{9.5}$$

where m is an integer. We can then filter the result to interpolate the half-pixel locations. Normally the filter used is symmetrical and separable to reduce complexity. In the case of H.264/AVC [6], the filter used to create an interpolated half-pixel value, $g[x]$, is as follows:

$$g[x] = \sum_{i=-5}^{+5} h[i] s_u[x-i] \tag{9.6}$$

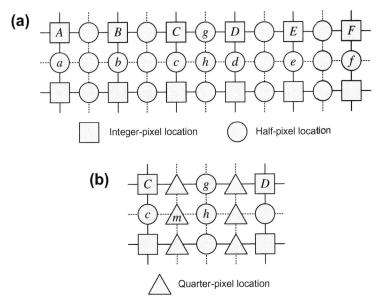

FIGURE 9.9

H.264 sub-pixel interpolation. (a) Half-pixel interpolation. (b) Quarter-pixel interpolation.

where the filter coefficients are:

$$h[i] = \{1, 0, -5, 0, 20, 0, 20, 0, -5, 0, 1\}/32 \tag{9.7}$$

In simple terms this means, with reference to the labeling in Figure 9.9, that:

$$g = \frac{(A - 5B + 20C + 20D - 5E + F)}{32} \tag{9.8}$$

and a similar approach is used vertically and on intermediate pixels such as h, based on pixels $\{a, b, c, d, e, f\}$.

In the case of quarter-pixel refinement, a simpler interpolation is employed; for example, to obtain the interpolated quarter-pixel value for location m in Figure 9.9, we use the following equation:

$$m = \frac{(c + h)}{2} \tag{9.9}$$

All interpolated results are rounded into the range $0 \cdots 255$, prior to the execution of the motion search phase.

It should be noted that the choice of interpolation filter is important as it can introduce bias (see for example the work of Bellers and de Haan [7]).

Example 9.3 (Sub-pixel motion estimation)
Implement the full search block matching motion estimation algorithm on the 6×6 search window $\mathbf{S}_{k-1}$, using the current frame template $\mathbf{S}_k$, given below. Use a simple two-tap interpolation filter to refine the integer-pixel result to half-pixel accuracy:

$$\mathbf{S} = \begin{bmatrix} 1 & 5 & 4 & 9 & 6 & 1 \\ 6 & 1 & 3 & 8 & 5 & 1 \\ 5 & 7 & 1 & 3 & 4 & 1 \\ 2 & 4 & 1 & 7 & 6 & 1 \\ 2 & 4 & 1 & 7 & 8 & 1 \\ 1 & 1 & 1 & 1 & 1 & 1 \end{bmatrix} ; \mathbf{M} = \begin{bmatrix} 3 & 9 \\ 1 & 4 \end{bmatrix}$$

Solution.   SAD values for selected integer-pixel $[d_x, d_y]$ candidates are:

i	j	SAD(i,j)	i	j	SAD(i,j)	i	j	SAD(i,j)
−1	−1	17	0	1	7	−1	−2	8
−1	0	18	1	−1	11	−2	−2	14
−1	1	15	1	0	13	−2	−1	18
0	−1	2	1	1	17	−2	0	5
0	0	11	0	−2	7	2	−2	18

Clearly by inspection, all other candidate vectors give SAD values greater than 2. Hence the integer-pixel motion vector for this block is $\mathbf{d} = [0, -1]$.

Next compute the half-pixel accurate motion vector. Use a simple two-tap filter to interpolate the values of the reference frame only. For example, in the horizontal direction:

$$s[x + 0.5, y] = \frac{s[x, y] + s[x + 1, y]}{2}$$

Once all values have been interpolated it is conventional to round the result prior to computing the sub-pixel SAD. We can use: $\lceil s[x + 0.5] - 0.5 \rceil$ to round down values ending in 0.5. Using this approach, the interpolated half-pixel values in the vicinity of the integer-pixel result are given below:

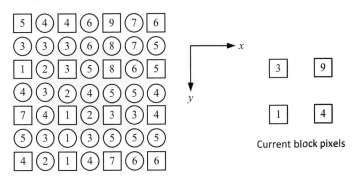

Half-pixel interpolation

Current block pixels

and the sub-pixel SAD values are given in the following table:

i	j	SAD(i,j)	i	j	SAD(i,j)
0	−1	2	0.5	−1.5	>2
−0.5	−1	10	0	−0.5	>2
0.5	−1	7	−0.5	−0.5	>2
0	−1.5	>2	0.5	−0.5	>2
−0.5	−1.5	>2	–	–	–

Hence the half-pixel motion vector for this block is the same as before, i.e. $\mathbf{d} = [0, -1]$.

9.3.3 Performance

The benefits of fractional-pixel search are shown in Figures 9.10 and 9.11. Figure 9.10 shows the PSNR improvements obtained from using half- and quarter-pixel estimation for ten frames of the *Foreman* sequence (CIF at 25 fps). This is based on using the original frame as the reference in each case. It can be seen that up to 1.5 dB gain is achieved in this case. It can also be observed that, while quarter-pixel search does provide additional benefits, the gain is reduced. In most cases search at one-eighth pixel resolution will provide some small additional gain, but going beyond this provides diminishing returns.

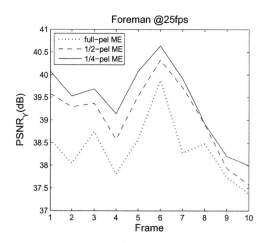

FIGURE 9.10

Comparison of full, half-, and quarter-pixel motion estimation.

FIGURE 9.11

Influence of sub-pixel motion estimation on residual error. Left: integer-pixel ME. Right: half-pixel ME.

Figure 9.11 shows an example DFD signal for *Coastguard*, comparing integer- and half-pixel search methods. The reduction in residual energy is clear, especially in the region of the boat and the riverbank in the background. However, while there is some improvement in the region of the river, the dynamic textures of the water cause severe problems for the translational motion model. Techniques which model such textures using dynamic models will be discussed further in Chapter 13. It has been found that the use of quarter-pixel accurate motion prediction in H.264/AVC contributes between 10% and 20% saving in bit rate, compared with the integer-pixel case.

9.3.4 Interpolation-free methods

We have seen that quarter-pixel motion estimation can provide a significant contribution to the rate–distortion performance of a video codec. By using sub-pixel estimation, based on interpolated search values, the residual energy for each predicted macroblock can be significantly reduced, but only at the cost of increased computational complexity. This latter point is particularly relevant for codecs such as H.264/AVC and HEVC, where the cost of an exhaustive set of macroblock segmentations needs to be estimated for optimal mode selection.

To mitigate the cost of interpolation, novel schemes for sub-pixel motion estimation based solely on the whole-pixel SAD distribution have been proposed. The approach of Hill et al. [8] enables both half-pixel and quarter-pixel searches to be guided by an intelligent model-free estimation of the SAD surface. They use a method based on a 2-D parameterized parabolic interpolation between the estimated ambiguity samples, approximating the actual behavior in the presence of pure translations. They also include a fallback strategy that improves the rate–distortion performance to a level close to full interpolation. This approach is faster (in terms of frames per second) than a fully interpolated system by a factor of between 12% and 20%.

Hill and Bull [9] extended the method of Ref. [8] using a two-dimensional kernel approach. This reduces the number of quarter-pixel search positions by a factor of 30% giving an overall speed-up of approximately 10% compared to the EPZS quarter-pixel method.

9.4 Multiple reference frame motion estimation

9.4.1 Justification

In conventional motion estimation, only one reference frame is employed. However, it was observed by Wiegand et al. [10] that such approaches do not exploit long term statistical dependencies in video sequences and that *multiple reference frame motion estimation* (MRF-ME) can provide many benefits. Multiple reference frames were first standardized in H.263 and have been a feature in most codecs since that time [6]. Normally the frames employed as references are simply past decoded frames; however, they may also include warped versions to account for various camera motions.

Reference frames are assembled in a reference picture buffer at both the encoder and decoder, and synchrony of contents must be maintained at both ends. This is illustrated in Figure 9.12. This is simple when a sliding window update is employed but requires signaling when other, more intelligent updates are used.

9.4.2 Properties, complexity, and performance of MRF-ME

Properties

A good description of the properties of long term memory is provided by Al-Mualla et al. [12] and the reader is referred there for more details. To summarize (assuming a sliding window update method):

1. The distribution of long term memory spatial displacements is center-biased. This means that $\mathbf{d} = [0, 0, d_z]$ has the highest relative frequency of occurrence.
2. The distribution of long term memory temporal displacements is zero-biased. This means that the most recent reference frame is the most commonly adopted candidate. i.e. $\mathbf{d} = [0, 0, 0]$ is the most common vector.

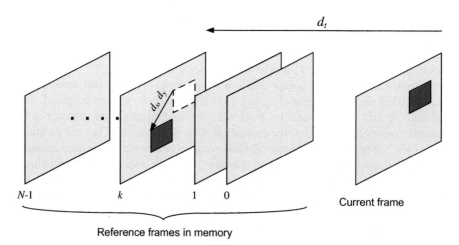

FIGURE 9.12

Multiple reference frame motion estimation.

3. The long term memory motion field is smooth and slowly varying. Hence search methods that work well for single-frame search can also offer solutions for the multiple-frame case.

Performance and complexity

MRF-ME exploits the long term statistical dependencies in an image sequence to provide coding gain. However, in order to achieve this, the motion estimation algorithm must not only search across spatial locations in the search grid, but also temporally across multiple candidate frames. With increases in processor power and the increasing integration and lowering cost of memory, this approach is now feasible. Nonetheless it can still present a major computational overhead and reduced complexity approaches are desirable. Figure 9.13 (left) illustrates the performance gains possible for the case of *Foreman* (QCIF @25 fps). The figure shows the benefits of long term memory up to 50 reference frames, indicating an increase in PSNR from 33 dB for one frame, up to 35.5 dB for 50 frames.

This improvement in coding performance is, however, offset by the additional complexity required to search for the optimum motion vector. This is shown in Figure 9.13 (right) which demonstrates, for the same sequence, that the number of locations searched per frame (for full search) increases from 80,000 for the case of one reference frame to over 4,000,000 for 50 reference frames. Such an increase in complexity, even with today's technology, could make MRF-ME intractable.

9.4.3 Reduced complexity MRF-ME

Several methods for improving the efficiency of MRF-ME without compromising its associated coding gain have been proposed. Su and Sun [11] exploit the smoothness

FIGURE 9.13

MRF-ME performance. Left: PSNR. Right: complexity.

Table 9.1 Performance of simplex-based (SMS) multiple reference frame motion estimation. Results from Ref. [13].

Algorithm:	FS	MR-FS	MR-SMS
PSNR (dB):	32.2	33.97	33.87
Search locations/frame	77,439	3,554,700	106,830

and correlations within the block motion field to provide similar coding performance to the H.264/AVC JM but at around 20% of the complexity. Al-Mualla et al. [13] showed that their simplex method can provide benefits for long term memory motion estimation. The simplex-based approach has a number of useful properties that make it attractive in such situations. For example, it is not based inherently on a unimodal error surface and it offers diversity in its candidate search locations. Table 9.1 presents coding performance and complexity results for this algorithm; this is based on 300 frames of *Foreman* (QCIF @25 fps), using an SAD distortion measure with a block size of 16 × 16 pixels and a maximum motion displacement of 15 pixels. The search was performed to full-pixel accuracy over 50 reference frames.

It can be seen that significant gains in performance are possible using this approach; a coding gain decrease of only 0.1 dB is accompanied by a speed-up of some 33 times.

9.5 Variable block sizes for motion estimation

9.5.1 Influence of block size

One of the main reasons for the performance gains of recent standards has been the introduction of variable block sizes, both for transforms and for motion estimation.

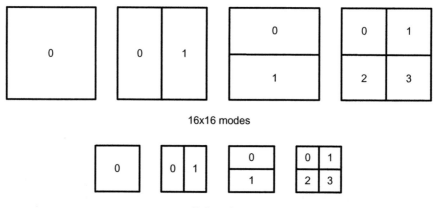

16x16 modes

8x8 modes

FIGURE 9.14

Variable block sizes supported by H.264/AVC. Top: 16 × 16 modes 1–4. Bottom: 8 × 8 modes 1–4.

This enables the shape and size of the block to be optimized in a rate–distortion sense, taking account of content type, spatial and temporal characteristics, and bit rate constraints. We saw in Chapter 8 that block size can influence both DFD accuracy and motion bit rate overhead. It thus seems perfectly sensible that block sizes should not be fixed, but instead should be allowed to adapt to content type within a rate–distortion framework.

9.5.2 Variable block sizes in practice

The block sizes supported by H.264/AVC are:

- **16 × 16 modes:** $16 \times 16, 16 \times 8, 8 \times 16, 8 \times 8$.
- **8 × 8 modes:** $8 \times 8, 8 \times 4, 4 \times 8, 4 \times 4$.

These are shown in Figure 9.14 and have been found to offer approximately 15% saving over the use of fixed block sizes. The distribution of block sizes in an actual coded frame is shown in Figure 9.15 for the *Foreman* sequence, produced using the Elecard Streameye analyzer [14].

9.6 Variable sized transforms

9.6.1 Integer transforms

Recent standards such as H.264/AVC [6] and HEVC [15] use block transforms to code the intra-frames or prediction residuals, but, rather than employ a single transform type, variable sizes are employed. For example, in H.264/AVC, the smallest block

FIGURE 9.15

Block size distribution (Foreman sequence coded using H.264/AVC). Produced using Ref. [14].

size is 4×4 and a separable integer transform which is closely related to the 4×4 DCT is used. This integer transform was introduced by Malvar et al. [16] and offers benefits:

- The transform block size is matched to the basic motion estimation block, so it performs better in the presence of complex local motions and regions with random textures.
- Because of the improvements resulting from variable block sizes in HEVC and H.264/AVC, the residual signal after motion compensation has reduced spatial correlation. As discussed previously, this implies that the transform is less efficient and hence a 4×4 integer transform can often be as good as a larger transform in terms of decorrelation.
- It offers the benefits of reducing noise around edges.
- The inverse transform is matched to the forward transform with exact integer operations, hence encoder–decoder mismatches are eliminated.
- The integer transform is computationally efficient requiring a smaller processing wordlength, and can be implemented with just addition and shift operations.

In Chapter 5 we saw that the 1-D 4×4 DCT is given by:

$$
\mathbf{A} = \begin{bmatrix}
\frac{1}{2} & \frac{1}{2} & \frac{1}{2} & \frac{1}{2} \\
\sqrt{\frac{1}{2}}\cos\left(\frac{\pi}{8}\right) & \sqrt{\frac{1}{2}}\cos\left(\frac{3\pi}{8}\right) & -\sqrt{\frac{1}{2}}\cos\left(\frac{3\pi}{8}\right) & -\sqrt{\frac{1}{2}}\cos\left(\frac{\pi}{8}\right) \\
\frac{1}{2} & -\frac{1}{2} & -\frac{1}{2} & \frac{1}{2} \\
\sqrt{\frac{1}{2}}\cos\left(\frac{3\pi}{8}\right) & -\sqrt{\frac{1}{2}}\cos\left(\frac{\pi}{8}\right) & \sqrt{\frac{1}{2}}\cos\left(\frac{\pi}{8}\right) & -\sqrt{\frac{1}{2}}\cos\left(\frac{3\pi}{8}\right)
\end{bmatrix}
$$

$$= \begin{bmatrix} a & a & a & a \\ b & c & -c & -b \\ a & -a & -a & a \\ c & -b & b & -c \end{bmatrix} \quad \text{where:} \quad \begin{aligned} a &= \frac{1}{2} \\ b &= \sqrt{\frac{1}{2}} \cos\left(\frac{\pi}{8}\right) \\ c &= \sqrt{\frac{1}{2}} \cos\left(\frac{3\pi}{8}\right) \end{aligned} \tag{9.10}$$

Letting $d = \frac{c}{b}$, the integer version of the transform can be written as follows:

$$\mathbf{A} = \begin{bmatrix} 1 & 1 & 1 & 1 \\ 1 & d & -d & -1 \\ 1 & -1 & -1 & 1 \\ d & -1 & 1 & -d \end{bmatrix} \otimes \begin{bmatrix} a & a & a & a \\ b & b & b & b \\ a & a & a & a \\ b & b & b & b \end{bmatrix}$$

where $\otimes$ represents a scalar multiplication. Since $d \approx 0.414$, the next step is to approximate d by a value of 0.5, to give:

$$\mathbf{A} = \begin{bmatrix} 1 & 1 & 1 & 1 \\ 2 & 1 & -1 & -2 \\ 1 & -1 & -1 & 1 \\ 1 & -2 & 2 & -1 \end{bmatrix} \otimes \begin{bmatrix} a & a & a & a \\ \frac{b}{2} & \frac{b}{2} & \frac{b}{2} & \frac{b}{2} \\ a & a & a & a \\ \frac{b}{2} & \frac{b}{2} & \frac{b}{2} & \frac{b}{2} \end{bmatrix} \tag{9.11}$$

Since for all row vectors, $\mathbf{a}_i \mathbf{a}_j^T = 0$, the basis vectors remain orthogonal. To make the matrix orthonormal, we must ensure that $\|\mathbf{a}_i\| = 1$. Hence:

$$a = \frac{1}{2}; \quad b = \sqrt{\frac{2}{5}}; \quad d = \frac{1}{2}$$

Extending this separable transform to process 2-D signals gives:

$$\begin{aligned} \mathbf{C} &= \mathbf{ASA}^T \\ &= \begin{bmatrix} 1 & 1 & 1 & 1 \\ 2 & 1 & -1 & -2 \\ 1 & -1 & -1 & 1 \\ 1 & -2 & 2 & -1 \end{bmatrix} [\mathbf{S}] \begin{bmatrix} 1 & 2 & 1 & 1 \\ 1 & 1 & -1 & -2 \\ 1 & -1 & -1 & 2 \\ 1 & -2 & 1 & -1 \end{bmatrix} \otimes \mathbf{E} \end{aligned} \tag{9.12}$$

where:

$$\mathbf{E} = \begin{bmatrix} a^2 & \frac{ab}{2} & a^2 & \frac{ab}{2} \\ \frac{ab}{2} & \frac{b^2}{4} & \frac{ab}{2} & \frac{b^2}{4} \\ a^2 & \frac{ab}{2} & a^2 & \frac{ab}{2} \\ \frac{ab}{2} & \frac{b^2}{4} & \frac{ab}{2} & \frac{b^2}{4} \end{bmatrix}$$

Thus we can perform the forward and inverse transform operations as: $\mathbf{S} = \mathbf{ACA}^T$ with the scalar multiplication by $\mathbf{E}$ absorbed into the quantization process.

9.6.2 DC coefficient transforms

We saw earlier that intra-prediction for luma and chroma signals can provide significant benefits for coding non-textured regions. If a smoothly varying region extends across the whole coding block, then additional benefits can be obtained by performing a further transform on the DC coefficients from the 4×4 transforms. In H.264, a 2×2 transform is also applied to the DC coefficients of the four 4×4 blocks of each chroma component. This provides benefits because, for smoothly varying regions, the autocorrelation coefficient will approach unity and the reconstruction accuracy is proportional to the inverse of the size of the transform. For non-textured regions the reconstruction error for an 8×8 transform is therefore half of that for a 4×4 transform.

Example 9.4 (Reduced complexity integer transform)
Consider the forward integer 4×4 transform given in Eq. (9.11). Show that this can be implemented using only shift and add operations and compute its complexity.

Solution. The architecture for the fast integer transform is given below:

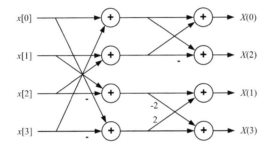

It can be observed that the transform requires only eight additions/subtractions and two 1-bit shifts.

9.7 In-loop deblocking operations

Blocking artifacts can arise in the hybrid video codec because of motion estimation and transform coding of the DFD signal. Because of this, deblocking filters have been used in various forms in standardized video codecs since H.261. They have been proven to improve both objective and subjective performance and have become progressively more sophisticated over time. Wiegand et al. [6] and List et al. [17] provide excellent overviews of the deblocking filter used in H.264/AVC; a summary is given below.

Deblocking filters perform best when they are integrated into the motion prediction loop, as this eliminates drift and enables benefits for all frame types. The idea of the deblocking operation is illustrated in Figure 9.16. The edge samples are filtered according to a set of criteria that relate to the prevailing quantization conditions coupled with estimates of whether the edge is real or induced by coding. The absolute differences of pixels near a block edge are first computed and then referenced against

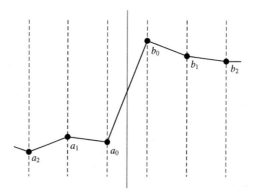

FIGURE 9.16

One-dimensional edge example.

Algorithm 5 Deblocking operation.

1. Set thresholds $\alpha\left(QP\right)$ and $\beta\left(QP\right)$;
2. Filter a_0 and b_0 iff: $|a_0 - b_0| < \alpha\left(QP\right)$ AND $|a_1 - a_0| < \beta\left(QP\right)$ AND $|b_1 - b_0| < \beta\left(QP\right)$;
3. Filter a_1 and b_1 iff: $|a_2 - a_0| < \beta\left(QP\right)$ OR $|b_2 b_0| < \beta\left(QP\right)$.

---

the likely size of any quantization-induced artifact. If the edge is larger than any that might have been induced by coding, then it is most likely to be an actual edge and should not be filtered. It has been reported [6] that the incorporation of the deblocking filter saves as much as 10% in bit rate for equivalent quality.

The conditions for filtering to be invoked depend on two thresholds $\alpha\left(QP\right)$ and $\beta\left(QP\right)$ where the second threshold is much smaller than the first. The deblocking process is described in Algorithm 5.

Example frames from *Foreman* with and without a deblocking filter applied are shown in Figure 9.17. The subjective benefits of the filtering operation can be clearly

FIGURE 9.17

Illustration of the effect of a deblocking filter. Left: without deblocking. Right: with deblocking filter applied.

seen. What cannot be seen directly are the benefits of the smoothed and better correlated DFD signal which will enable more efficient transform coding and hence reduced bit rate.

9.8 Summary

This chapter has integrated the concepts from Chapters 4, 5, 7, and 8 to form what is referred to as a block-based hybrid coder. This architecture has been universally adopted through standardization processes and is the workhorse of all of our TV, storage and streaming applications. This basic architecture has benefited from several enhancements by way of advancing international standards. Some of the key enhancements have been described in this chapter and it is these that are responsible for providing the excellent performance offered by recent standards such as H.264/AVC and HEVC.

For interactive demonstrations of video codec operation, the reader is referred to the Matlab code associated with this text, which can be downloaded from www.bristol. ac.uk/vi-lab/demos.

References

[1] P. Strobach, Tree-structured scene adaptive coder, IEEE Transactions on Communications 38 (4) (1990) 477–486.

[2] Recommendation ITU-T H.264 I International Standard ISO/IEC 14496-10, ITU-T, 2013.

[3] S. Ericsson, Fixed and adaptive predictors for hybrid predictive/transform coding, IEEE Transactions on Communications 33 (12) (1985) 1291–1302.

[4] B. Girod, Motion-compensated prediction with fractional-pel accuracy, IEEE Transactions on Communications 41 (4) (1993) 604–612.

[5] B. Girod, E. Steinbach, N. Farber, Performance of the H.263 compression standard, Journal of VLSI Signal Processing 17 (1997) 101–111.

[6] T. Wiegand, G. Sullivan, G. Bjøntegaard, A. Luthra, Overview of the H.264/AVC video coding standard, IEEE Transactions on Circuits and Systems for Video Technology 13 (7) (2003) 560–576.

[7] E. Bellers, G. de Haan, Analysis of sub-pixel motion estimation, in: Proceedings of SPIE Visual Communications and Image Processing, 1999, pp. 1452–1463.

[8] P. Hill, T. Chiew, D. Bull, C. Canagarajah, Interpolation-free subpixel accuracy motion estimation, IEEE Transactions on Circuits and Systems for Video Technology 16 (12) (2006) 1519–1526.

[9] P. Hill, D. Bull, Sub-pixel motion estimation using kernel methods, Signal Processing: Image Communication 25 (4) (2010) 268–275.

[10] T. Wiegand, X. Zhang, B. Girod, Long term memory motion-compensated prediction, IEEE Transactions on Circuits and Systems for Video Technology 9 (1) (1999) 70–84.

[11] Y. Su, M.-T. Sun, Fast multiple reference frame motion estimation for H.264/AVC, IEEE Transactions on Circuits and Systems for Video Technology 16 (3) (2006) 447–452.

[12] M. Al-Mualla, C. Canagarajah, D. Bull, Video Coding for Mobile Communications, Academic Press, 2002.

[13] M. Al-Mualla, C. Canagarajah, D. Bull, Simplex minimization for single- and multiple-reference motion estimation, IEEE Transactions on Circuits and Systems for Video Technology 11 (12) (2001) 1209–1220.

[14] <http://www.elecard.com/en/products/professional/analysis/streameye.html>.

[15] G. Sullivan, J.-R. Ohm, W.J. Han, T. Wiegand, Overview of the high efficiency video coding (HEVC) standard, IEEE Transactions on Circuits and Systems for Video Technology 22 (12) (2012) 1649–1668.

[16] H. Malvar, A. Hallapuro, M. Karczewicz, L. Kerofsky, Low-complexity transform and quantization in H.264/AVC, IEEE Transactions on Circuits and Systems for Video Technology 13 (2003) 598–603.

[17] P.A. List, J. Lainema, G. Bjøntegaard, M. Karczewicz, Adaptive deblocking filter, IEEE Transactions on Circuits and Systems for Video Technology 13 (2003) 614–619.

Measuring and Managing Picture Quality

10

CHAPTER OUTLINE

10.1 **General considerations and influences** **318**
    10.1.1 What do we want to assess? 319
    10.1.2 Influences on perceived quality 319
10.2 **Subjective testing** **321**
    10.2.1 Justification 321
    10.2.2 Test sequences and conditions 321
    10.2.3 Choosing subjects 323
    10.2.4 Testing environment 324
    10.2.5 Testing methodology and recording of results 324
    10.2.6 Statistical analysis and significance testing 328
    10.2.7 The complete evaluation process 329
10.3 **Test data sets and how to use them** **330**
    10.3.1 Databases 330
    10.3.2 The relationship between Mean Opinion Score
         and an objective metric 331
    10.3.3 Evaluating metrics using public (or private) databases 332
10.4 **Objective quality metrics** **333**
    10.4.1 Why do we need quality metrics? 333
    10.4.2 A characterization of PSNR 335
    10.4.3 A perceptual basis for metric development 337
    10.4.4 Perception-based image and video quality metrics 338
    10.4.5 The future of metrics 341
10.5 **Rate–distortion optimization** **343**
    10.5.1 Classical rate–distortion theory 343
    10.5.2 Practical rate–distortion optimization 345
    10.5.3 The influence of additional coding modes and parameters . . . 348
    10.5.4 From Rate–Distortion to Rate–Quality Optimization 350
10.6 **Rate control** **350**
    10.6.1 Buffering and HRD 352
    10.6.2 Rate control in practice 354
    10.6.3 Regions of interest and rate control 357

Communicating Pictures. http://dx.doi.org/10.1016/B978-0-12-405906-1.00010-6
© 2014 Elsevier Ltd. All rights reserved.

10.7   Summary **357**

References **358**

Assessing perceptual quality is one of the most critical yet challenging tasks in image and video processing. Visual perception is highly complex, influenced by many confounding factors, not fully understood and difficult to model. For these reasons, the characterization of compression algorithm performance has invariably been based on subjective assessments where a group of viewers are asked their opinions on quality under a range of test conditions. Methods for conducting such trials and for analyzing the results from them are described in Section 10.2. A discussion of the properties of some publicly available subjective test databases is given in Section 10.3.

Objective measures of video quality have conventionally been computed using the absolute or squared difference between the coded version of a picture and its original version. It is however well known that the perceptual distortion experienced by the human viewer cannot be fully characterized using such simple mathematical differences. Because of the limitations of distortion-based measures, perception-based metrics have begun to replace them. These offer the potential for enhanced correlation with subjective opinions, thus enabling more accurate estimates of visual quality. The most promising of these are reviewed in Section 10.4, including Structural Similarity (SSIM), the Video Quality Metric (VQM), Visual Signal-to-Noise Ratio (VSNR), Spatial and Temporal Most Apparent Distortion (MAD), and Motion Tuned Spatio-Temporal Quality Assessment (aka MOVIE). Although these approaches have demonstrated various degrees of improved performance compared to conventional metrics such as Peak Signal-to-Noise Ratio (PSNR), a number of problems including high computational complexity, latency and flexibility of integration into the coding process still need to be resolved. In this context we introduce an alternative, the Perception-based Video Metric (PVM), that successfully addresses some of these issues.

Video metrics are not just used for comparing different coding algorithms. They also play an important role in the picture compression and delivery processes, for example enabling in-loop Rate–Quality Optimization (RQO). Section 10.5 addresses this issue, describing some of the most common techniques that enable us to select the optimum coding parameters for each spatio-temporal region of a video. Finally, methods for controlling the bit rate generated by a codec in the context of time-varying channel conditions (rate control) are described in Section 10.6.

10.1   General considerations and influences

Firstly it is worth stating that assessing the quality of impaired image or video content, whether due to transmission losses or compression artifacts, is not straightforward. It can be achieved in one of two ways:

For colour higher quality versions of Figures 10.14 and 10.15 please refer to the electronic version or the website.

1. **Subjectively:** Requiring many observers and many presentations of a representative range of impairment conditions and content types. Subjective testing conditions must be closely controlled, with appropriate screening of observers and post-processing of the results to ensure consistency and statistical significance. They are costly and time consuming, but generally effective.

2. **Objectively:** Using metrics that attempt to capture the perceptual mechanisms of the human visual system. The main issue here is that simple metrics bear little relevance to the HVS and generally do not correlate well with subjective results, especially at lower bit rates when distortions are higher. More complex, perceptually inspired metrics, although improving significantly in recent years, can still be inconsistent under certain test conditions. The outcome of this is that mean squared error (MSE) based metrics are still the most commonly used assessment methods, both for in-loop optimization and for external performance comparisons.

10.1.1 What do we want to assess?

Good overviews of the reasons and processes for evaluating visual quality are provided in Refs. [1,2]. The primary motivations are summarized below:

1. To compare the performance of different video codecs across a range of bit rates and content types.
2. To compare the performance of different video codecs across a range of channel impairments.
3. To compare the influence of various parameters and coding options for a given type of codec.

The latter is of particular relevance to in-loop Rate–Distortion (or Quality) Optimization RDO (RQO) as we will see later.

10.1.2 Influences on perceived quality

Human visual perception

More than a century ago, vision scientists began to pay attention to our perceptual sensitivity to image and video distortions. We saw in Chapter 2 that this sensitivity varies with screen brightness, local spatial and temporal frequency characteristics, types of motion, eye movements, various types of artifact, and of course the viewing environment. In order to ensure validity of subjective tests and consistency in the performance of objective metrics, the influence of these sensitivities must be, as far as possible, represented in the content used for evaluations and captured in the structure of any metric employed.

It should also be noted that the performance of the HVS varies significantly across subjects, depending on age, illness, fatigue, or visual system defects. It is also possible that biases can influence the opinions of viewers through personal preferences or even boredom.

Viewing environment

In order to provide consistent subjective assessments of picture quality it is essential that the details and parameters of the viewing conditions are recorded and kept as consistent as possible between tests. This is particularly important when comparing results across different laboratories. Several factors about the viewing environment will influence the perception of picture quality. These include:

- **Display size:** The more the display fills peripheral vision, the more engaging it will appear.
- **Display brightness and dynamic range:** Flicker and temporal CSF will depend on display brightness and higher dynamic range displays impart a greater sense of depth.
- **Display resolution:** The spatial and temporal update rates will influence the types of artifacts perceived.
- **Ambient lighting:** High ambient lighting levels and reflections from the display will reduce perceived contrast levels and introduce additional artifacts.
- **Viewing distance:** This parameter interacts with spatio-temporal resolution and will have a significant impact on perceived quality.
- **Audio quality:** Finally it has been clearly demonstrated that variations in audio quality will influence the perception of video quality.

Content type

We have seen in Chapter 2 how content type can influence perceived quality. For example, a plain scene with little motion will be much easier to code than a spatio-temporally busy sequence with complex textures and motions. If codec A is used to code easy content and a similar codec, B, is used to code more difficult content, then it would appear that codec A is superior in terms of rate–distortion performance to codec B, when in reality this is not the case.

Tests should therefore be based on representative and reasonably challenging content offering a range of spatial and temporal activity levels (see Section 10.2.2). To ensure that the focus of the assessor remains on picture quality, rather than on the narrative, short sequences (typically 10 s) are normally selected and these are not usually accompanied by an audio track. All codecs under test should be evaluated on all sequences and at all impairment levels.

Artifact types

We will not list all of the artifact types produced by all types of codec here, but hopefully the reader is by now aware of these from previous chapters. In general they can be classified as follows:

- **Edges:** Due to blocking from transforms or motion estimation, or to contouring effects associated with quantization.
- **Blurring:** Due to the loss of high frequency detail during quantization of block transform or wavelet coefficients.

- **Ringing:** Quantization artifacts in filter bank synthesis stages can cause ringing effects.
- **Dissimilarity:** Due for example to inconsistencies in geometric transformations of planar regions in certain codecs, errors in warping rigid textures or synthesizing dynamic textures (see Chapter 13 for further details).

10.2 Subjective testing

10.2.1 Justification

Despite recent advances in the performance of objective metrics and the large number of them available, none are yet universally accepted as a definitive measure of quality.[1] As a consequence it is necessary to use controlled subjective testing. Subjective assessment methods are employed widely to characterize, compare, or validate the performance of video compression algorithms. Based on a representative set of test content and impairment conditions, they are intended to provide a robust indication of the reactions of those who might view the systems tested.

Most subjective testing experiments conducted are based on recommendations from the ITU and other organizations that have been developed over many years, based on the collective experiences of many organizations. The primary reference documents are ITU-R rec. BT.500 [4] and ITU-T rec. P.910 [5]. The discussion that follows is primarily based on these recommendations.

10.2.2 Test sequences and conditions

Test material

In general, test material should be selected that is appropriate to the problem or application area being addressed. It will normally be defined in terms of its fixed parameters, which include: the number of sequences used, sequence duration, sequence content, spatial resolution, temporal resolution, and the bit depth.

For most general assessments, the test material will be "critical, but not unduly so" [4]. This means that it should contain content that is difficult to code but should remain representative of typical viewing. The sequences selected should always include critical material since results for non-critical material cannot usually be extrapolated. However, if the tests are intended to characterize performance for specific difficult cases, then it would be expected that the material used would be selected to reflect these cases.

In general, it is usual to select at least four types of sequence for testing. This number will provide a minimum coverage of activity levels while not boring the assessors with an over-long test session.

---

[1]It could be argued that VQM [34], as standardized by ANSI, is more widely accepted than others although it is not the best performing.

Activity or information levels

The amount of spatial and temporal activity present in a clip has a high impact on the degree of compression possible and the quality of the resulting reconstruction. In general, test sequences should be selected that are consistent with the range of channel conditions prevailing in the application areas of interest. It is useful, prior to final selection of test material, to formally assess the spatio-temporal activity levels of the candidate clips to ensure that these cover the appropriate spatio-temporal information space [3,5].

Following the approach recommended in recommendation ITU-T rec. BT.910 [5], the spatial information measure is based on the standard deviation of each frame S_z in the sequence after Sobel filtering. The maximum value of the standard deviation, over all frames, is selected as the SI metric:

$$\text{SI} = \max_{\forall z} \left\{ \sigma_{\forall(x,y)} \left(\text{Sobel} \left(\mathbf{S}_z(x, y) \right) \right) \right\} \tag{10.1}$$

The temporal information measure (TI) used in BT.910 is based on the standard deviation of the difference, over consecutive frames, between co-located luma pixel values. The TI measure is computed, as above, as the maximum value of the standard deviation over all frames in the sequence. Hence:

$$\text{TI} = \max_{\forall z} \left\{ \sigma_{\forall(x,y)} \left(\mathbf{S}_z(x, y) - \mathbf{S}_{z-1}(x, y) \right) \right\} \tag{10.2}$$

It is usual to produce an SI vs TI plot to ensure that the sequences selected for testing provide adequate coverage of the SI–TI space. An indicative plot, with a good coverage, is shown in Figure 10.1. It should be noted that a slightly different approach to that described above is used by Winkler [3], based on motion vectors. He also includes an information measure based on color.

Test conditions

The *test conditions* refer to those parameters that will change during the test session. These might include codec types, codec parameters, and bit rates, and a typical test would compare two or more codecs or codec configurations for a range of sequences and coding rates.

It is normal to evaluate codec performance over a representative, well-distributed set of test conditions. Most assessment methods are sensitive to the range and distribution of conditions. The sensitivity can be reduced by restricting the range of conditions but also by including some explicit (direct anchoring) extreme cases or by distributing these throughout the test without explicit identification (indirect anchoring). Furthermore, the larger the number of test conditions, the longer the test session becomes. Tests must avoid viewer fatigue and this imposes constraints on the number of sequences employed and the number of test conditions evaluated.

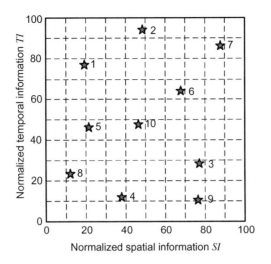

FIGURE 10.1

Spatial and temporal information coverage for a range of test sequences.

The number of presentations is equal to $N = N_s \times N_c \times N_r$ where N_s is the number of source sequences, N_c is the number of test conditions ($N_c = N_{c_1} N_{c_2}$ where N_{c_1} is the number of codecs under test and N_{c_2} is the number of bit rates tested (usually including uncompressed)), and N_r is the redundancy factor (i.e. the number of times each condition is repeated). If each presentation takes T seconds then the time for each test is $N \times T$ seconds and if there are K observers, then the total time for the complete trial is $N \times T \times K$ seconds (not including dead time for change-overs and breaks).

10.2.3 Choosing subjects

Depending on the nature of the test, observers may be expert or non-expert. Studies have found that systematic differences can occur between different laboratories conducting similar tests [4]. The reasons for this are not fully understood, but it is clear that expert observers will view material differently to non-experts. Other explanations that have been suggested include gender, age, and occupation [4]. In most cases, for consumer applications, it is expected that the majority of observers should be non-expert and that none should have been directly involved in the development of the system under test.

Before final selection of the assessors, all candidates should be screened to ensure that they possess normal visual acuity (with or without corrective lenses). This can be done using a Snellen Chart for visual acuity and an Ishihara Chart to check for color vision. The number of assessors used depends on the scale of the test and the sensitivity and reliability of the methodology adopted. Normally it is recommended that at least 15 subjects are employed. This author recommends that the number is slightly higher to allow for outlier removal during results processing.

Table 10.1 Selected subjective testing viewing environment parameters. From Refs. [4, 5].

Viewing condition	Home	Laboratory
Ratio of inactive to peak screen luminance	≤ 0.02	≤ 0.02
Display peak luminance	200 cd/m^2	See recommendations BT.814 and BT.815
Background chromaticity	–	D65
Maximum observer angle	30°	30°
Viewing distance	3–6H	3–6H
Room illumination	200 lx	Low ($\leq$ 20 lx)

10.2.4 Testing environment

Testing environments are normally specified according to the requirements of the test—usually meaning either a realistic consumer environment or laboratory conditions:

- **Laboratory conditions:** These are intended to provide test conditions that enable maximum detectability of impairments.
- **Home conditions:** These are intended to enable assessment in viewing conditions closer to typical consumer-end environments.

Selected parameters associated with these two environments are listed in Table 10.1.

10.2.5 Testing methodology and recording of results

A generic diagram of a subjective testing methodology is shown in Figure 10.2. The source delivers the presentation clips, either directly to the subject or via a system under test that introduces impairments dependent on the test conditions set at the time. The subject or subjects will view the content on an assessment display which shows the impaired video clip and in some cases also the unimpaired clip, either simultaneously or sequentially. The controller will control the timing and switching between different test conditions. In some cases the user may be able to influence the order, repetition, and timing of clips.

Testing methodologies broadly fall into two categories—based on their use of single or multiple stimuli. Before we consider these, there are a few general points that are common to all methods.

General principles of subjective testing

- **Duration of test:** To avoid fatigue, BT.500 recommends that the duration of the test session should be limited to 30 min. In practice, it has been the author's experience that tests often do consume more time and 40–45 min is not uncommon, especially in cases where the observer is permitted to control the display and revisit sequences

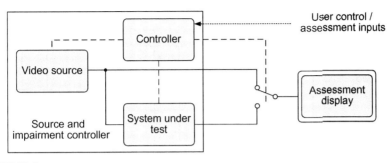

FIGURE 10.2

Generic arrangement for subjective video evaluation.

multiple times (e.g. the Subjective Assessment Methodology for Video Quality (SAMVIQ) [2]).

- **Preparing assessors:** Prior to the test, assessors should receive full instructions on the reason for the test, its content, the types of impairment that will occur and the method of recording results. The first part of any test should include dummy runs which familiarize the assessor(s) with the methodology. These also help to stabilize the observer's opinions.
- **Presentation order:** It is normal during the test session to randomize the presentations—within and across tests. This ensures that any influences of fatigue or adaptation are balanced out. It is also normal to include anchors (extreme cases). In some cases the assessor will know which presentations contain anchors, but this is usually not the case.
- **Recording opinions:** The opinions of each assessor must be recorded for post-test analysis. This is often done using a line bisection process based on gradings such as those shown for the Double Stimulus Continuous Quality Scale test (DSCQS) in Figure 10.3. The observer places a mark on the line for each presentation assessed and this typically falls into one of five quality scale bins as shown in the figure. The final grade is normally scaled to the range 0–100. In other cases, an automated user interface will be provided, such as that shown in Figure 10.4 for the SAMVIQ methodology [2].
- **Recording test conditions:** Because of the inconsistencies that can exist between subjective trials, it is vital to record all salient information about the test and the parameters and equipment used. Good examples are provided by Ebrahimi's group in the context of HEVC—for example, see Refs. [6,7].

Double stimulus methods

Double stimulus evaluations remain the most popular means of evaluating compressed video quality. The most commonly used procedure is ITU-R rec. BT.500 which, although originally intended for television applications, is now used more widely. ITU-T rec. P.910 was targeted specifically at multimedia applications, but shares

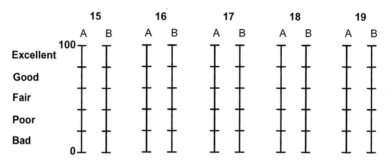

FIGURE 10.3

A section of the DSCQS assessment form.

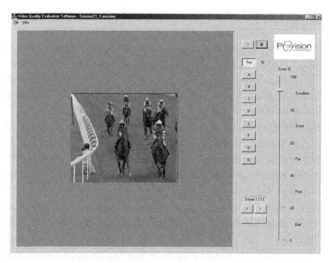

FIGURE 10.4

Interface for SAMVIQ testing. Courtesy: ProVision Communications.

many similarities with BT.500. ITU-R rec. BT.500 describes two main double stimulus methods:

- **Double Stimulus Continuous Quality Scale (DSCQS):** The DSCQS methodology is best suited for cases where the qualities of the test material and the original are similar and the aim is to assess how well the system under test performs relative to the original. The test is arranged as a sequence of paired clips (A and B), one of which is the original (anchor) and the other is the original impaired by the system under test. The assessor does not know which clip is the original as the order is randomized. This is shown in Figure 10.5. The pair is shown twice, the first time for familiarization and the second for voting.

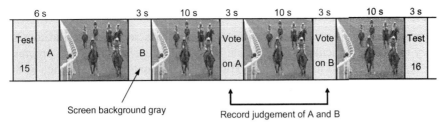

FIGURE 10.5

Typical timing of a DSCQS subjective test.

- **Double Stimulus Impairment Scale (DSIS):** DSIS is similar to DSCQS except that the pair is presented only once and the assessor knows which clip is the anchor as it is always shown before the impaired version. DSIS is generally better suited to assessing the robustness of a system or the effects of more noticeable distortions such as those imparted by transmission errors. This method is very similar to the Degradation Category Rating method (DCR) in ITU-T rec. P.910. The DSIS grading scale again comprises five grades but this time they are labeled:

  - Imperceptible; Perceptible but not annoying; Slightly annoying; Annoying; and Very annoying.

Single stimulus methods

In single stimulus methods, there is no explicit anchor and the assessor is presented with a randomized sequence of test conditions, normally including the original. The Absolute Category Rating (ACR) in ITU-T rec.P.910 does this using 10 s clips across a range of test conditions, with voting on the same scale as in Figure 10.3. A Single Stimulus Continuous Quality Evaluation method is proposed as part of ITU-R rec. BT.500. This is intended to better capture the time and content variations that happen in practical content delivery. Here it is recommended that the test session is organized with longer programme segments (e.g. sport) of around 5 min duration. These are concatenated into a test session of duration 30–60 min comprising a range of program segments with various quality parameters. One of the main advantages of single stimulus methods is reduced testing time.

SAMVIQ [2] is a single stimulus method where the assessor has some control over viewing order and repetitions. The test is organized as a series of test sequences which are assessed in a given order and where the assessor cannot proceed to the next sequence until the previous one has been completely assessed. Within each sequence a number of algorithms and test conditions can be presented in any order as selected by the assessor (however, the selection buttons are randomized for each new sequence evaluated). An example interface with a slider for score entering is shown in Figure 10.4.

Triple stimulus methods

Some authors have proposed that triple stimulus methods provide increased consistency in results, particularly for example when comparing interlaced and progressive scanning methods for HDTV. The Triple Stimulus Continuous Evaluation Scale methodology (TSCES) was proposed by Hoffmann et al. [8], and simultaneously displays the video processed by the system under test alongside (usually they are stacked vertically) two extreme anchors—the original and a low quality coded version. The conventional five-grade scale is also used and the assessor is asked to score the system under test in the context of both anchors.

Pair comparison methods

Pair comparison methods are also double stimulus, but this time are based on comparisons between two systems under test for the same test conditions.

10.2.6 Statistical analysis and significance testing

It is essential that the data collected is analyzed and presented in a robust and consistent fashion. Most test results are scaled to the range 0–100 and that is assumed here. Generally, following the process described in BT.500 [4], let us assume that there are N presentations, where each consists of one of N_c test conditions.

Calculation of mean scores

The mean score across all observers for each presentation is given by:

$$\bar{u}_{csr} = \frac{1}{K} \sum_{k=1}^{K} u_{kcsr} \tag{10.3}$$

where u_{kcsr} is the score for observer k in response to sequence s under test condition c for repetition r, and K is the total number of observers. This is then similarly processed after screening of observers to produce $\bar{u}_{sc}$, the Mean Opinion Score (MOS) for each sequence under a given test condition.

It is common in many tests to compensate scores relative to the reference content. This is particularly important for single stimulus methods where a process of hidden reference removal is applied to produce Difference Mean Opinion Scores (DMOS).

Confidence interval

When the results of a study are presented, it is good practice to also include the confidence interval [9]. The confidence interval for a mean is the interval that will contain the population mean a specified proportion of the time, typically 95%. Confidence intervals are based on the size of each sample and its standard deviation and provide more information than point estimates. The 95% confidence interval is given by:

$$\left[\bar{u}_{csr} - \delta_{csr}, \bar{u}_{csr} + \delta_{csr} \right] \tag{10.4}$$

Algorithm 10.1 Screening of observers in subjective trials using the β_2 test [4].
1. REPEAT for each observer, k:
2. $k = k + 1$;
3. FOR $c,s,r = 1,1,1$ to C,S,R DO:
4.   IF $2 \leq \beta_{2csr} \leq 4$ THEN: IF $u_{kcsr} \geq \bar{u}_{csr} + 2\sigma_{csr}$ THEN $P_k = P_k + 1$;   IF $u_{kcsr} \leq \bar{u}_{csr} - 2\sigma_{csr}$ THEN $Q_k = Q_k + 1$;
5.   ELSE: IF $u_{kcsr} \geq \bar{u}_{csr} + \sqrt{20}\sigma_{csr}$ THEN $P_k = P_k + 1$; IF $u_{kcsr} \leq \bar{u}_{csr} -$ $\sqrt{20}\sigma_{csr}$ THEN $Q_k = Q_k + 1$;
6. END FOR;
7. IF $\frac{P_k + Q_k}{N_c N_s N_r} > 0.5$ and $\left\lvert \frac{P_k - Q_k}{P_k + Q_k} \right\rvert < 0.3$ THEN reject observer k;
8. UNTIL $k = K$.

where

$$\delta_{csr} = 1.96 \frac{\sigma_{csr}}{\sqrt{K}}$$

The standard deviation for each presentation is given by:

$$\sigma_{csr} = \sqrt{\sum_{k=1}^{K} \frac{(\bar{u}_{csr} - u_{kcsr})^2}{(K-1)}} \tag{10.5}$$

The absolute difference between the experimental mean and the true mean (based on an infinite number of observers) is smaller than the 95% confidence interval of equation (10.4) with a probability of 95% (on the condition that the scores are normally distributed).

Screening of observers

The screening (used in DSIS and DSCQS) for a normally distributed set of scores conventionally uses the β_2 test, based on the kurtosis coefficient. When β_2 is between 2 and 4, the distribution can be assumed to be normal. The kurtosis coefficient is defined as the ratio of the fourth order moment to the square of the second order moment. Thus:

$$\beta_{2csr} = \frac{m_4}{(m_2)^2} \quad \text{with} \quad m_x = \frac{\sum_{k=1}^{K} (u_{kcsr} - \bar{u}_{kcsr})^x}{K} \tag{10.6}$$

We now compute P_k and Q_k, as given in Algorithm 10.1, in order to determine whether any observers should be rejected in forming the distribution.

10.2.7 The complete evaluation process

The preceding sections have provided a general overview of subjective testing methods. Both double and single stimulus methods have been formalized through recommendations such as ITU-R rec. BT.500-13, ITU-T rec. P.910 and SAMVIQ [2,4,5].

Algorithm 10.2 Subjective testing methodology.

---

1. Select the codecs under test and the parameter settings to be used (GOP structures, filtering, RDO modes, ME modes, etc.);
2. Define the coding conditions to be evaluated (spatial and temporal resolutions, bit rates, etc.);
3. Identify or acquire the test sequences that will be used, ensuring that they are sufficiently varied and challenging, representative of the applications of interest and of appropriate duration;
4. Select an appropriate evaluation methodology (dependent on conditions being tested, impairment levels, etc.);
5. Design the test session and environment (complying with recommendations of whichever methodology is being adopted);
6. Dry run the session to remove any bugs or inconsistencies;
7. Organize the full test, inviting sufficient subjects (assessors), with pre-screening to eliminate those with incompatible attributes;
8. Run the tests (including assessor briefing and dummy runs) and collect the results;
9. Perform post-processing to remove outliers and assess significance;
10. Produce test report fully documenting laboratory conditions, test conditions, and analysis/screening methods employed.

The reader is advised to refer to the relevant standard or procedure documents before implementing any actual subjective testing experiments. As a guide, however, a general methodology for subjective testing is described in Algorithm 10.2.

10.3 Test data sets and how to use them

A reliable subjective database has three primary uses. Firstly it will produce a robust comparison of the selected compression methods in the context of the test conditions employed. Secondly it provides a very useful basis for validating objective metrics, based on well-established statistical methods. Finally it can be utilized to characterize HVS properties in the context of compression and hence provides a valuable tool for refining objective quality metrics.

10.3.1 Databases

A brief overview of the primary publicly available databases is presented below. The reader is referred to the appropriate reports associated with each trial for further details. An excellent critique and comparison of currently available databases is provided by Winkler [3].

VQEG FRTV

The earliest major subjective database for objective video quality assessment was generated via the Video Quality Experts Group (VQEG) FRTV Phase 1 program [10] (followed by Phase 2 in 2003 [11]). The Phase I database was constructed in 2000 to address quality issues associated with the introduction of digital TV standards worldwide. The database used 287 assessors (in four groups) and 20 video sequences with 16 distortion types to generate over 26,000 subjective opinion scores. FRTV-I employed two resolutions, 720×486 and 720×576, all interlaced. The testing methodology used was the Double Stimulus Continuous Quality Scale (DSCQS) method as described earlier.

Although the FRTV-I database exhibits good coverage and uniformity, it does have limitations in the context of contemporary coding requirements:

1. The artifacts presented do not fully reflect recent advances in video compression as coding is primarily based on MPEG-2.
2. Its assessments are based on standard definition formats. It does not include high definition material.
3. Its content is interlaced, whereas the current trend is toward progressive formats, especially for streaming applications.

However, despite these limitations, because of the large number of sequences, the large number of subjects, its good coverage and its good uniformity [3], VQEG FRTV Phase I is still one of the more commonly used databases for objective quality metric validation.

LIVE

A more recent and equally important subjective test database is that developed in the Laboratory for Image and Video Engineering (LIVE) [12–14]. The LIVE database presents impairments based on both MPEG-2 and H.264/AVC compression algorithms and includes results for simulated wired and wireless transmission errors. Ten reference sequences are used with a resolution of 768×432, where each reference has 15 distorted versions. A single stimulus continuous test procedure was adopted for this data set and final scores were again scaled between 0 and 100. Compared to the VQEG FRTV-I database, only 38 assessors were employed in the LIVE viewing trials. According to the analysis of Winkler [3], LIVE also provides a narrower range of source content (in terms of SI and TI) and distortions (in terms of PSNR and MOS values).

Others

Other databases that can be used for assessing metrics include the IRCCyN/IVC [15], the VQEG HDTV database [16], and the VQEG multimedia Phase I database [17]. These are often used for specific research purposes and are in general less suitable for comprehensively validating image and video quality metrics. A further comparison of these can be found in Ref. [3].

10.3.2 The relationship between Mean Opinion Score and an objective metric

Subjective results are frequently used as a benchmark for establishing a relationship between DMOS (or MOS) scores and a specific objective picture quality metric.

The scores produced by the objective video quality metric must be correlated with the viewer scores in a predictable and repeatable fashion. The relationship between predicted and DMOS need not be linear as subjective testing can have non-linear quality rating compression at the extremes of the test range. The linearity of the relationship is thus not so critical, but rather the stability of the relationship and a data sets error-variance that determine predictive usefulness.

BT.500 [4] describes a method of finding a simple continuous relationship between $\bar{u}$ (the mean score) and the metric based on a logistic function. Firstly the range of mean score values is normalized as follows (after screening):

$$p = \frac{(\bar{u} - u_{min})}{(u_{max} - u_{min})} \tag{10.7}$$

Typical relationships between p and a given distortion measure D generally exhibit a skew-symmetric sigmoid form. Hence the function $p = f(D)$ can be approximated by a logistic function of the form:

$$\tilde{p} = \frac{1}{1 + e^{(D - D_M)G}} \tag{10.8}$$

where D_M and G are constants that can be quite simply derived from the experimental data [4].

10.3.3 Evaluating metrics using public (or private) databases

Judgements of the performance of a particular objective metric relative to a body of MOS values associated with a specific database are normally based on certain statistical attributes. These are conventionally related to measures of prediction accuracy, monotonicity, and consistency of its fit. The following measures are commonly used:

Linear correlation

The Pearson Linear Correlation Coefficient (LCC) is used as a measure of the accuracy of fit of the metric to the subjective scores. It characterizes how well the metric under test can predict the subjective quality ratings. The general form is defined for a set of N measurement–prediction pairs (x_i, y_i) as:

$$r_p = \frac{\sum_{i=0}^{N-1}(x_i - \bar{x})(y_i - \bar{y})}{\sqrt{\sum_{i=0}^{N-1}(x_i - \bar{x})^2}\sqrt{\sum_{i=0}^{N-1}(y_i - \bar{y})^2}} \tag{10.9}$$

where x_i would represent the actual MOS (or DMOS) score and y_i would represent the predicted MOS (or DMOS) score.

Rank-order correlation

The degree to which the model's predictions correlate with the relative magnitudes of subjective quality ratings is assessed using a rank-order metric. This characterizes the prediction monotonicity, i.e. to what degree the sign of differences across tests correlate between the subjective scores and the metric's prediction of them. Conventionally the Spearman Rank-Order Correlation Coefficient is used for this purpose:

$$r_s = \frac{\sum_{i=0}^{N-1} (\mathcal{X}_i - \bar{\mathcal{X}})(\mathcal{Y}_i - \bar{\mathcal{Y}})}{\sqrt{\sum_{i=0}^{N-1}(\mathcal{X}_i - \bar{\mathcal{X}})^2}\sqrt{\sum_{i=0}^{N-1}(\mathcal{Y}_i - \bar{\mathcal{Y}})^2}} \qquad (10.10)$$

where $\mathcal{X}_i$ is the rank order of x_i, $\mathcal{Y}_i$ is the rank order of x_i, and $\bar{\mathcal{X}}$ and $\bar{\mathcal{Y}}$ are their median values.

Outlier ratio

The outlier ratio, r_o, effectively measures prediction consistency—how well the metric predicts the subjective scores over the range of content and impairments. An outlier is normally classed as a predicted data point that is greater than a threshold distance from the corresponding MOS point. Conventionally a threshold of twice the standard deviation error of the MOS values is used. So if the number of data points that satisfy equation (10.11) is N_o then the outlier ratio is simply given in equation (10.12).

$$|x_i - y_i| > 2\sigma_{y_i} \qquad (10.11)$$

$$r_o = \frac{N_o}{N} \qquad (10.12)$$

10.4 Objective quality metrics

10.4.1 Why do we need quality metrics?

Image and video quality assessment plays a crucial role in many aspects of image and video processing, in particular related to video coding and communications. In the context of communicating pictures, they have three primary uses:

1. **Algorithm development and benchmarking:** We have seen previously that subjective evaluations, while very effective in characterizing the strengths and weaknesses of competing coding algorithms, are expensive and time consuming. The existence of reliable and consistent objective metrics provides a much simpler means of comparison. Furthermore they can provide guidance regarding the benefits or otherwise of various algorithmic modifications to a given codec (for example, the benefits of adding a loop filter or using multiple reference frames).
2. **Rate–Quality Optimization:** Quality assessments are increasingly needed in the encoding loop to make instantaneous RQO decisions about which coding modes and parameter settings to use for optimum performance given certain content and rate constraints.

FIGURE 10.6

Reduced reference video quality assessment.

3. **Streaming control:** In the case of network delivery of video content, it is beneficial for the encoder and transmitter to be aware of the quality of the signal at the receiver after decoding. This enables the encoder to be informed of the prevailing channel conditions and hence to make appropriate decisions in terms of rate- and error-control.

Objective quality assessment methods are generally classified as either full-reference (FR), reduced-reference (RR), or no-reference (NR).

- **FR methods** are widely used in applications where the original material is available, such as when assessing image and video coding algorithm performance, or when making Rate–Quality Optimization decisions.
- **NR methods** are normally only employed where reference content is not available [18]. Scenarios could include when evaluating the influence of a lossy communication system at the receiver. It is extremely difficult to produce "blind" metrics and hence the use of NR methods is generally restricted to specific operating scenarios and distortion types. They do not generalize well and reduced reference metrics are preferable if possible.
- **RR methods** [19] use only partial information about the source for predicting quality. They find application in lossy communication networks where quality predictions can be informed by partial knowledge of the original content and possibly also the channel state. Figure 10.6 depicts a possible scenario for an RR metric, where sparse features are extracted at the encoder and transmitted through a protected (low data rate) side channel to the decoder. At the decoder, similar features are extracted from the reconstructed signal and compared in the RR metric. An indication of the reconstruction quality at the decoder can then be fed back to the encoder so that it can make informed coding decisions based on the prevailing channel state. Clearly any additional side information places an overhead on the bit rate of the coded information and this must be assessed in the context of the quality gains achieved.

10.4.2 **A characterization of PSNR**

For certain types of visual signal under certain test conditions, PSNR can provide a simple and efficient approach to distortion estimation. For example, Huynh-Thu and Ghanbari [20] showed that PSNR can offer consistent results when used to compare between similar codecs or codec enhancements based on the same test data. However, it is equally clear that MSE measures will fail badly for certain impairments that in reality have little perceptual impact. For example, a small spatial or temporal shift, an illumination change or a small variation in a contextual texture [21] will all cause a large increase in MSE. Girod [22] provides a comprehensive overview of the limitations of MSE-based measures and an excellent analysis is also presented by Wang and Bovik [21]. They sum up the situation nicely, listing the assumptions that underpin the use of MSE:

1. Signal quality is independent of temporal or spatial relationships between samples.
2. Signal quality is independent of any relationship between the original signal and the error signal.
3. Signal quality is independent of the signs of the error signal.
4. All samples contribute equally to signal quality.

Based on results from the VQEG FRTV Phase I database, Zhang and Bull [42] analyzed the correlation between PSNR quality predictions (after non-linear regression) and subjective Differential Mean Opinion Scores (DMOS) for a range of different conditions. The primary results are illustrated in Figure 10.7 where it can be observed that in general the points fit only loosely and that the fit worsens at lower quality levels (higher DMOS).

Zhang and Bull investigated how content influences PSNR-based predictions by sub-dividing the FRTV-I dataset into five groups based on the dominant content type (spatial texture, structural elements, plain luminance, dynamic motion or mixed). This grouping was determined subjectively because conventional spatial information (SI) and temporal information (TI) measures do not provide sufficient granularity. The correlations for these content classes are presented in Figure 10.7. It can be observed that, for videos with significant texture content, most subjective DMOS scatter points fall below the PSNR predictions. This illustrates the existence of HVS masking effects for static and dynamic textured content. In contrast, for sequences with significant structural content, the PSNR-predicted DMOS values tend to fall below the subjective scores, indicating that (despite edge masking effects) the HVS is very sensitive to errors around structural elements. The scatter plot for plainer content with little high frequency energy similarly indicates little masking protection.

Zhang and Bull [42] also presented PSNR-predicted DMOS vs DMOS scatter plots based on different types of coding distortion: compressed content, compressed with transmission errors, and compressed with interpolation. These results are shown in the final three sub-plots in Figure 10.7. PSNR can be seen to perform reasonably well for pure coding distortion, where errors are widely distributed, but provides much poorer predictions for the cases with interpolation errors and transmission failures.

FIGURE 10.7

Scatter plots of DMOS vs PSNR predicted DMOS for various classes of sequence from the VQEG database [42].

It is clear that in the case of highly distorted content, PSNR is not effective and alternative methods are needed.

This analysis confirms that:

1. Visual masking exists and is more evident for content with spatial and temporal textures than for plain luminance areas.
2. Highly visible artifacts such as those in plain areas or due to transmission loss tend to cause the HVS to overestimate distortion.
3. Two distinct perceptual strategies are utilized by the HVS—*near-threshold* and *supra-threshold*. It is clear that artifact detection is more important for cases of lower perceptual quality.

Example 10.1 (Texture masking and PSNR)
Consider the following original 256 × 256 image (left) and its two distorted versions. In the middle image, the foreground is corrupted by additive Gaussian white noise (AGWN) of variance 0.003. The right-hand image has the background corrupted, again with AGWN of variance 0.003. The PSNR values for the two corrupted images are identical and the foreground and background regions are of equal area.

Compare these images visually and comment on the visibility of the noise (Readers are referred to the electronic version of this figure on the website).

Solution. The PSNR value for the distorted images is 28 dB, so both have identical distortions in the mean squared error sense. Most observers will find the distortions to the background more noticeable than those to the foreground and this is particularly the case where textured content is contextual in the scene and where the content is viewed in the absence of a reference image. This shows that distortions with different perceptual impacts are not always differentiated by PSNR measures.

10.4.3 A perceptual basis for metric development

The principal aim of any video quality metric is to correlate well with visual perception under a wide range of conditions and impairments. Secondary requirements might also include metric complexity and metric latency, particularly relevant for on-line decision making within a codec or transmission system.

HVS characteristics such as contrast sensitivity and visual masking have been exploited, both in coding and in quality assessment. For example, CSF weighting has been employed to reflect the HVS sensitivity across the range of spatial and tempo-ral frequencies, and visual masking—the reduction of perceived distortions at certain luminances and in spatio-temporal textures—has also been exploited. Masking effects are more evident in spatially textured regions, where the HVS can tolerate greater dis-tortions than in smoother areas. A similar phenomenon exists with dynamic textures.

When HVS properties are exploited, it is generally found that the metric provides enhanced correlation with subjective judgements, compared to conventional distortion measures such as MSE. It has been noted that this improvement is more significant when the distorted content is similar to the original [24].

When distortion becomes significant, visual attention is more likely to be attracted by visible artifacts. Under these circumstances, it is often more efficient to characterize the artifacts rather than to purely compute distortions. Based on this concept, quality metrics have been developed which emulate the perception process using a two-step strategy: *near-threshold* and *supra-threshold*. Examples of this class of metric include those reported by Kayargadde et al. [25], Karunasekera et al. [26], Carnec et al. [27], Chandler and Hemami (VSNR) [28], and Larson and Chandler (MAD) [29].

10.4.4 Perception-based image and video quality metrics

Over the past decade or so, several perceptually inspired objective quality assessment metrics have been proposed. These often exploit visual statistics and features in both frequency and spatial domains. Some of them are described below. For further details the reader is referred to Refs. [1, 30–33].

VQM

Pinson and Wolf's VQM [34] is an objective method for video quality assessment that closely predicts subjective quality ratings. It is based on impairment filters that combine measures of blurring, jerkiness, global noise, block distortion and color distortion. VQM computes seven parameters based on different quality features. These are obtained by filtering the impaired and reference videos to extract the property of interest. Spatio-temporal features are then extracted and a quality parameter is obtained for the feature by comparing the statistics of the filtered original and impaired video regions.

The parameters used in VQM are *si_loss* (blurring), *hv_loss* (edge shifts), *hv_gain* (blocking), *si_gain* (edge sharpening), *chroma_spread* (color distribution), *chroma_extreme* (localized color impairments), and *ct_ati_gain* (spatio-temporal impairments). VQM forms a linear combination of these to yield the final quality assessment. In VQEG tests in 2004, VQM demonstrated superior performance in predicting MOS scores compared to all other algorithms evaluated. It has subsequently been adopted as an ANSI and ITU standard.

SSIM

The integrity of structural information in an image or video is an important cue for visual perception. Wang et al. [35] developed an image quality assessment approach,

SSIM (Structural Similarity Image Metric), which estimates the degradation of structural similarity based on the statistical properties of local information between a reference and a distorted image. This is an improved version of the previous Universal Image Quality Index (UIQI) and combines three local similarity measures based on luminance, contrast, and structure. After some simplification, these three terms can be rearranged as in equation (10.13) to give the structural similarity between an impaired image **s** and its reference $\mathbf{s}_R$:

$$\text{SSIM} = \frac{(2\mu_s\mu_{s_R} + C_1)\,(2\sigma_{ss_R} + C_2)}{(\mu_s^2 + \mu_{s_R}^2 + C_1)\,(\sigma_s^2 + \sigma_{s_R}^2 + C_2)} \tag{10.13}$$

where μ and σ are the local mean and standard deviation of the two images, C_1 and C_2 are constants used to stabilize the equation in the presence of weak denominators, and σ_{ss_R} is the sample cross correlation (zero mean). Conventionally $C_1 = (k_1 L)^2$ and $C_2 = (k_2 L)^2$ where $k_1 = 0.01$, $k_2 = 0.03$, and L is the dynamic range of the pixel values.

The range of SSIM values extends between -1 and $+1$ and only equals 1 if the two images are identical. It is conventional for SSIM to be calculated using a sliding window, typically of size 11×11, using a circularly symmetric Gaussian weighting function. The SSIM value for the whole image is then computed as the average across all individual windowed results.

The advantage of SSIM is that it offers superior performance to PSNR in many cases and that it is relatively simple to implement. As such it is probably, at the time of writing, the most commonly used non-MSE metric. It has however been recognized that SSIM suffers from a number of problems, particularly that it is sensitive to relative scalings, translations, and rotations. A complex wavelet-based approach, CW-SSIM, has been developed to address these issues [36] as well as an enhanced multiscale version (MS-SSIM) [37]. A further extension to SSIM called V-SSIM which also takes account of temporal information [38] weights the SSIM indices of all frames. This metric has demonstrated improved performance compared to PSNR on the VQEG FRTV Phase I database.

It is interesting to note in the context of Example 10.1 that, although the PSNR values are identical for both distorted images, the SSIM scores are 0.81 for the case of the right-hand image and 0.95 for the middle image. This indicates that SSIM better reflects subjective opinions in the context of texture masking.

MOVIE

Scene motion is a key component in influencing visual perception and it can provide a considerable degree of artifact masking. However, perceived quality is also dependent on viewing behavior; for example, whether or not a moving object is being tracked by the observer. Seshadrinathan et al. [13] introduced a motion tuned spatio-temporal quality assessment method (MOVIE). MOVIE analyzes both distorted and reference content using a spatio-temporal Gabor filter family, and the quality index consists of a spatial quality component (inspired by SSIM) and a temporal quality component based on motion information.

The accuracy of MOVIE-based predictions has been evaluated using the VQEG FRTV Phase I database, where it was demonstrated to offer significant correlation improvements compared to both PSNR and SSIM. It also demonstrates excellent performance on the LIVE video database. One issue with MOVIE however is its high computational complexity, due to the large number of Gabor filters used and the temporal extent required for its calculation.

VSNR

The near-threshold and supra-threshold properties of the HVS were exploited by Chandler and Hemami in their Visual Signal-to-Noise Ratio (VSNR) still image metric [28]. This emulates the cortical decomposition of the HVS using a wavelet transform. A two-stage approach is then applied to assess the detectability of distortions and determine a final measure of visual SNR. VSNR has been evaluated using the LIVE image database with very good results.

MAD and STMAD

Building on the approach used in VSNR, Larson and Chandler [29] developed the Most Apparent Distortion model (MAD). MAD models both near-threshold distortions and appearance-based distortions. It employs different approaches for high quality images (near threshold distortions) and low quality images (supra-threshold distortion). These are combined using a non-linear model to obtain a final quality index.

MAD was extended to cope with temporal components in Ref. [39] where spatio-temporal slices are processed based on the spatial MAD and the results are then weighted using motion information. The temporal MAD index is combined with spatial MAD (computed from individual frames) to obtain spatio-temporal MAD (ST-MAD) for video. Excellent correlation performance with subjective results is reported based on the LIVE video database.

VDP and VDP-2

In VDP-2 Mantiuk et al. [40] built on the previous VDP metric to create a metric that correlates well with subjective opinions for content with extended or diminished intensity ranges. The VDP-2 metric predicts both error visibility and quality (MOS) and was based on new contrast sensitivity measurements. The model was validated using LIVE (image) and TID2008 image databases and demonstrated improved performance compared to its predecessor HDR-VDP and VDP metrics, especially for low luminance conditions. It also compared favorably with the MS-SSIM metric.

Reduced complexity metrics and in-loop assessment

Current image and video quality assessment methods have been developed based on various HVS characteristics in order to achieve quality prediction close to subjective opinions. These approaches provide enhanced correlation with subjective scores but are often disadvantaged by high complexity or high latency, and hence are not appropriate for real-time operation.

Recently, Soundararajan and Bovik presented a reduced reference video quality metric [19] based on spatio-temporal entropic differencing. This performs well and supports low complexity operation through the use of reduced spatio-temporal content.

In the context of perceptual video coding, Zhang and Bull proposed an Artifact-based Video Metric (AVM) [23] using the DT-CWT as the basis for assessment of both conventionally compressed and synthesized content. AVM correlates well with subjective VQEG scores and has the advantage that it can be easily integrated into a synthesis-based framework because of its high flexibility and low complexity due to extensive parameter reuse.

PVM

A Perception-based Video quality Metric (PVM) was recently proposed by Zhang and Bull [41,42], building on their AVM model. PVM simulates the HVS perception processes by adaptively combining noticeable distortion and blurring artifacts (both exploiting the shift-invariance and orientation selectivity properties of the Dual-Tree Complex Wavelet Transform) using an enhanced non-linear model. Noticeable distortion is defined by thresholding absolute differences using spatial and temporal masks which characterize texture masking effects, and this makes a significant contribution to quality assessment when the distorted video is similar to the original. Blurring artifacts, estimated by computing high frequency energy variations and weighted with motion speed, are found to improve the overall metric performance in low quality cases when it is combined with noticeable distortion.

Importantly PVM, as with its predecessor, AVM, is intended to be used with synthesized as well as conventionally coded material. Early results indicate that these are the only metrics capable of robust performance in this respect.

Comparing results

The performances of PVM, MOVIE, and ST-MAD are compared for both LIVE video and VQEG FRTV Phase I databases as scatter plots in Figure 10.8. Values for correlation coefficients, Outlier ratio and RMSE are given on each graph. It can be seen that PVM is more consistent across the two databases and offers superior performance in the context of the VQEG database.

10.4.5 The future of metrics

While it is clear that perceptual metrics have advanced significantly in the past decade or so, they are still not widely accepted for general picture quality assessment. There are many complex reasons for this. For example, they mostly measure video fidelity (the closeness of the impaired version to the original), they generally measure degradation (i.e. the impaired version is assumed to be worse than the original), and they

FIGURE 10.8

Comparison of MOVIE, ST-MAD, and PVM on LIVE and VQEG FRTV database [42].

do not generally take account of viewing patterns (viewers tend to focus on regions of interest).

One of the most challenging problems for the future is to create a reliable, yet low complexity, in-loop quality assessment measure with which to precisely estimate subjective quality and detect any noticeable coding artifacts. MSE methods are used extensively for this at present, but will be inappropriate as synthesis-based coding becomes more commonplace.

For real-time or in-loop processing, quality metrics should be able to:

1. Perform assessment at different spatial levels (such as GOP, picture, region, or coding unit).
2. Offer manageable computational complexity.

3. Differentiate the effects of an extended video parameter space, including higher dynamic range, frame rate and resolution, and take account of different environmental conditions.
4. Ideally provide compatibility with emerging perceptual coding and analysis–synthesis compression methods (see Chapter 13).

Few if any existing metrics meet these criteria.

10.5 Rate–distortion optimization

Building on our introduction to rate and distortion in Chapter 4, we will consider this important area of compression in more detail here. As outlined previously, the bit rate for any compressed sequence will depend on a number of factors. These include:

- The video content (high spatio-temporal activity will in general require more bits to code).
- The encoding algorithm used (e.g. wavelet or DCT based, intra-only or motion compensated).
- The encoding parameters selected. At the coarsest level this includes things such as spatial resolution and frame rate. At a finer granularity, issues such as quantizer control, intra vs inter modes, and block size choices will be key. The difference in performance between an encoding based on good parameter choices and one based on poor choices can be very significant.

The rate vs distortion (or rate vs quality) characteristic for a given codec–sequence combination provides a convenient means of comparing codecs and assessing how parameter selections can influence performance. A plot of operating points might, for a range of different parameter values, look like Figure 10.9. The aim is to ensure that the parameters selected at any time are those that yield the lowest possible distortion for any given coding rate. To achieve this we must perform Rate–Distortion (RDO) or Rate–Quality Optimization (RQO).

For further reading on this topic, excellent reviews of rate–distortion theory and practice are provided by Ortega and Ramchandran [43] and Sullivan and Wiegand [44]. Wang et al. [45] also provide a good overview of R-D bounds. For more recent work on RDO related to current standards, the reader is referred to Refs. [46,47].

10.5.1 Classical rate–distortion theory

Rate–distortion theory lies at the heart of source coding. It refers to the trade-off between source fidelity and coding rate and can be stated simply as either:

1. representing a source with the minimum number of bits given a quality constraint, or,
2. achieving the optimum quality for a given bit rate.

Shannon's *separation principle* states that, for the case of transmission of data over a noisy channel (under certain assumptions), the optimal use of bandwidth results if

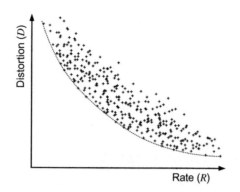

FIGURE 10.9

Rate–distortion plot for various coding parameter choices. The associated convex hull provides the operational R-D characteristic for the given codec–source combination.

source and channel coding are treated independently. In other words, we should first obtain the most efficient data representation for the source in an error-free environment and then add an appropriate level of protection to ensure error-free delivery over a lossy channel. As we have already discussed in Chapter 9, the underlying assumptions do not always apply but in general it is not a bad starting point.

Rate–distortion theory has conventionally been used to establish performance bounds (in terms of bit rate and quality) for given types of source, usually characterized in terms of their signal statistics. One of the problems faced in this respect is that simple statistical characterizations do not usually fit well with complex non-stationary sources like video.

Even in the simple case of optimizing a scalar quantizer, we still need to select an appropriate statistical model for the source. Only then can we investigate practical algorithms and assess how closely they perform to the bounds for that distribution. The problem with this approach is twofold:

1. If the model is poor then the bounds are meaningless and practical or heuristic designs may well outperform them. Ortega and Ramchandran [43] provide an example where a low complexity wavelet image encoder outperforms an encoder based on an i.i.d. Gaussian model by over 3 dB.
2. Even if the model is good, a practical solution may be unrealistically complex or require huge latency. For the example given in Ref. [43], the wavelet coder is relatively low complexity whereas the "R-D optimum" encoder is of infinite complexity.

Distortion measures

Let the distortion between a symbol of a coded source $\tilde{s}$ and its original version s be denoted as $d(s, \tilde{s})$. In the case where $s = \tilde{s}$ then $d(s, \tilde{s}) = 0$. The average distortion across the source is given by:

$$D = d\left(\mathbf{s}, \tilde{\mathbf{s}}\right) = E\left\{d\left(s, \tilde{s}\right)\right\} = \sum_{s}\sum_{\tilde{s}} P\left(s, \tilde{s}\right) d\left(s, \tilde{s}\right) \tag{10.14}$$

Normally the distortion measure used is either squared error, $d(s, \tilde{s}) = (s - \tilde{s})^2$, or absolute error, $d(s, \tilde{s}) = |s - \tilde{s}|$. The average distortion for a source over N samples, for the case of a squared error distortion, is the MSE:

$$d\left(\mathbf{s}, \tilde{\mathbf{s}}\right) = \frac{1}{N}\sum_{n=1}^{N} d_n\left(s, \tilde{s}\right) = \frac{1}{N}\sum_{n=1}^{N}\left(s[n] - \tilde{s}[n]\right)^2 \tag{10.15}$$

More often, in practical RDO applications the Sum of Squared Differences (SSD) is used, which is computed as in equation (10.15), but omitting the $1/N$ term.

The memoryless Gaussian source

As we have discussed above, closed form solutions for R-D bounds are in general difficult to obtain. However, for the case of a Gaussian i.i.d. source with variance σ^2, the bound can be computed (assuming high bit rates) and is given by:

$$\bar{R}(D) = \begin{cases} \frac{1}{2}\log_2\left(\frac{\sigma^2}{D}\right) & 0 \le D < \sigma^2 \\ 0 & D \ge \sigma^2 \end{cases} \tag{10.16}$$

where $R(D)$ denotes the rate–distortion function (which gives the bit rate R needed to deliver a distortion D for a given source) and $\bar{R}(D)$ is the rate–distortion bound (i.e. the minimum rate among all possible coders for an infinite length vector).

In practice, for more general and complex sources such as natural video, theoretical R-D bounds serve little purpose and more pragmatic approaches must be adopted. These are considered next.

10.5.2 Practical rate–distortion optimization

Operational control of a source encoder is a major issue in video compression and the aim of practical RDO is to select the best coding modes and parameter settings for the prevailing operating conditions. The available bit budget must thus be allocated in a manner such that the overall distortion is kept as low as possible in accordance with a given rate constraint. Parameter settings for modern codecs will cover a wide range of encoder attributes, such as block size, motion estimation references, inter- or intra-mode selection, and quantization step size. The optimization problem is made particularly difficult due to the non-stationary nature of video content, where the best encoder settings can vary significantly for different spatio-temporal regions. Furthermore the rate–distortion costs are not generally independent for all coding units (for example, due to the effects of prediction).

It should be noted that, especially for lower bit rates, it is not practical to consider quantization and entropy coding independently. Unlike higher bit rates and quality settings, where the number of alphabet symbols is large, for the case of lower bit rates it may for example be advantageous to choose a quantization point that delivers slightly higher distortion because it leads to significantly lower bit rate.

From source statistics to a parameterizable codec

Practical rate–distortion optimization methods do not normally start with a source model. Instead they are based on the assumption of a given parameterizable encode–decode combination which is known to perform well if the right parameters are chosen. In such cases, parameter choices are not based on source statistics but rather on some transformation of the source data (e.g. DCT coefficients or wavelet subband coefficients). This has the benefit of making the modeling more robust as the statistics of transformed data tend to be easier to model. For example, subband coefficients can be roughly approximated to a memoryless source modeled with an i.i.d. Laplacian process.

However, if the subband coefficients were indeed memoryless, we could entropy encode them based on this assumption using first order entropy. In practice the assumption is not sufficiently good as the coefficients, when scanned after quantization, typically present long chains of zeros. Hence the need for tricks such as zig-zag run-value coding that we explored in Chapter 7.

RDO methods must find the best operating points based on the optimum parameter vector, $\mathbf{p}_{opt}$. If we evaluate all parameter combinations for a given realization of a source (e.g. an image) over a range of rates then we can obtain the best rate–distortion curve for that coder–sequence combination. The convex hull, as shown in Figure 10.9, represents the operational bound against which any practical RDO heuristic algorithm can be measured.

RDO complexity

It is worth quickly mentioning the complexity of the RDO process. In off-line cases, where a single encode will be re-used multiple times, the value proposition will justify more complex or multipass encoding (a good example is encoding for a Blu-ray disk), while in other cases it may not.

Lagrangian optimization

Lagrangian methods have become widely accepted as the preferred approach for RDO in recent standards, primarily due to their effectiveness and simplicity [46,48]. Initially we will consider the case where each coding unit, i (the basic optimization building block), can be optimized independently of all others and where the only optimization parameter is a quantization index, j. While this assumption clearly breaks down in cases where context dependence or prediction are invoked, it can provide a useful and tractable solution.

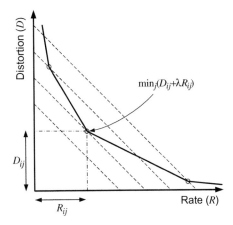

FIGURE 10.10

Lagrangian optimization for a coding unit, i.

Normally a discrete Lagrangian approach is adopted for RDO, as introduced by Shoham and Gersho [43,49], in order to find the maxima or minima of a function, subject to external constraints, where a closed form solution does not exist. The Lagrangian cost function for a coding unit i is given by equation (10.17):

$$J_{ij}(\lambda) = D_{ij} + \lambda R_{ij} \qquad (10.17)$$

where the quantization index j dictates the trade-off between rate and distortion and the Lagrange multiplier λ controls the slope of lines in the R-D plane that intersects the R-D characteristic to select specific operating points. This is shown graphically in Figure 10.10.

If we minimize equation (10.17) then we obtain:

$$\frac{\partial(J_{ij}(\lambda))}{\partial R_{ij}} = \frac{\partial(D_{ij} + \lambda R_{ij})}{\partial R_{ij}} = \frac{\partial D_{ij}}{\partial R_{ij}} + \lambda_{\text{opt}} = 0 \qquad (10.18)$$

This implies that the optimum value of λ is given by the negative slope of the distortion function:

$$\lambda_{\text{opt}} = -\frac{\partial D_{ij}}{\partial R_{ij}} \qquad (10.19)$$

Clearly if we minimize J for the case of $\lambda = 0$ we are minimizing the distortion and if we do the same when $\lambda = \infty$ we minimize the rate. However, in general, the slope is not usually known so we must estimate it using operational techniques as described below. It should be noted that Lagrangian approaches are constrained to select operating points on the convex hull of the R-D characteristic. In some cases, preferable non-optimal points may exist. This can be addressed using alternative but potentially more complex techniques such as dynamic programming [43].

10.5.3 **The influence of additional coding modes and parameters**

Modern coding standards offer high coding efficiency, not just because of their underlying architecture, but also because of their adaptability to different content types and channel constraints. We will examine this in more detail in Chapter 12, but the adaptation is typically based on some or all of the following decisions:

- Quantizer control.
- When to use block or picture skip modes.
- Determination of motion vectors and multiple reference frame indices.
- The use of integer-, half- or quarter-pixel motion estimation.
- The determination of intra or inter modes and the associated block partitions. For example, in H.264/AVC the standard uses a 16×16 luma macroblock with the following options:

  - Intra modes (for all frames): nine 4×4 modes and four 16×16 modes.
  - Inter modes (only for P- and B-frames): macroblock partitions of 16×16, 16×8, 8×16, 8×8, 8×4, 4×8, 4×4.

Each mode has an associated rate–distortion cost and the encoder must select the mode having the lowest cost. This is of course complicated by the fact that the optional modes will have different efficiencies for different content types and at different bit rates. The optimization problem is complicated even further because the rate–distortion costs are not generally independent for all coding units; for example, because of the spatial and/or temporal dependencies introduced through intra- and inter-prediction. This means that the Lagrangian optimization of equation (10.17) should theoretically be performed jointly over all coding units in the video sequence, something that in most cases would be prohibitively complex. A number of simplification techniques have therefore been proposed.

Perhaps the most obvious method of determining parameter choices or mode decisions is to use an exhaustive search. By comparing all possible options and parameter combinations, the optimum can be selected. This can be prohibitively expensive, so alternatives are generally used. Much simpler strategies for decision making can be made based on indicative thresholds. For example, the decision as to whether an intra or inter mode should be selected could be based on a comparison between the SAD of a 16×16 macroblock with respect to its mean, and the minimum SAD after integer motion search. Similar strategies have been reported for other decisions. However, the primary method of choice for RDO is based on the use of Lagrange multiplier methods and these are normally applied separately to the problems of mode selection and motion estimation.

Lagrangian multipliers revisited

Following the approach above and as described in Ref. [46], the aim is to solve the constrained problem:

$$\min_{\mathbf{p}} D(\mathbf{p}) \text{ s.t. } R(\mathbf{p}) \le R_T \tag{10.20}$$

where R_T is the target bit rate and $\mathbf{p}$ is the vector of coding parameters. As before, this can be represented as an unconstrained Lagrangian formulation based on the following cost function:

$$\mathbf{p}_{\text{opt}} = \arg \min_{\mathbf{p}} \{D(\mathbf{p}) + \lambda R(\mathbf{p})\} \qquad (10.21)$$

Intra modes: Let us first consider the problem of mode selection for *intra* coding modes. If we assume that the quantizer value Q is known and that the Lagrange parameter λ_{MODE} is given, then the parameter (or coding mode) selection is given by:

$$J_{\text{MODE}}(\mathbf{S}_k, \mathbf{p}_k \mid Q, \lambda_{\text{MODE}})$$
$$= D(\mathbf{S}_k, \mathbf{p}_k \mid Q, \lambda_{\text{MODE}}) + \lambda_{\text{MODE}} \cdot R(\mathbf{S}_k, \mathbf{p}_k \mid Q, \lambda_{\text{MODE}}) \quad (10.22)$$

where the parameter vector $\mathbf{p}$ is varied over all possible coding modes for the given coding unit (or sub-image), $\mathbf{S}_k$. As discussed previously, the distortion measure D is normally based on sum-of-squared differences between the original block and its reconstructed version and the rate R is normally measured after entropy coding.

Skip modes: For the case of computing *skip* modes, the distortion and rate calculations do not depend on the current quantizer value, but simply on the SSD between the current coding unit and that produced by prediction using the inferred motion vector. The rate for a skipped block is approximately 1 bit per coding unit in H.264.

Inter modes: For the case of *inter* modes, let us assume we know the Lagrange parameter λ_{MOTION} and the decoded reference picture $\tilde{\mathbf{S}}$. Rate constrained motion estimation can be performed by minimizing the cost function as follows:

$$\mathbf{p}_{\text{opt}} = \arg \min_{\mathbf{p}} \{D_{\text{DFD}}(\mathbf{S}_k, \mathbf{p}_k) + \lambda_{\text{MOTION}} \cdot R_{\text{MOTION}}(\mathbf{S}_k, \mathbf{p}_k)\} \qquad (10.23)$$

The distortion measure used in this case is normally either the SAD or the SSD, as defined by equation (10.24) with $l = 1$ or $l = 2$ respectively:

$$D_{\text{DFD}}(\mathbf{S}_k, \mathbf{p}_k) = \sum \left| s\left[x, y, z\right] - \tilde{s}[x - d_x, y - d_y, z - d_z] \right|^l \qquad (10.24)$$

where $\mathbf{d}$ is the motion vector for the current parameter set. The rate calculation for motion must include all the bits required to code the motion information, normally based on entropy coded predictions.

Wiegand et al. [44,46] demonstrated the power of the Lagrangian multiplier approach in the context of block size selection, motion estimation, and quantizer control. In the context of quantizer control, they demonstrated the relationship between quantizer step size and λ. They showed that, to a first order approximation, this relationship is content independent and (in the context of H.263) given by:

$$\lambda_{\text{MODE}} = 0.85 Q^2_{\text{H.263}} \qquad (10.25)$$

The Lagrange approach has been used widely since its introduction in H.263 and has been adopted in the test models for H.264/AVC and HEVC. For the case of

H.264/AVC, a different expression has been determined empirically to give for I and P frames:

$$\lambda_{MODE(I,P)} = 0.85 \cdot 2^{(QP_{H.264}-12)/3} \tag{10.26}$$

and for B frames:

$$\lambda_{MODE(B)} = \max\left(2, \min\left(4, \frac{QP - 12}{6}\right)\right) \cdot \lambda_{MODE(I,P)} \tag{10.27}$$

The relationship between λ_{MODE} and λ_{MOTION} is given by:

$$\begin{cases} \lambda_{MOTION} = \lambda_{MODE} : & \text{for SSD} \\ \\ \lambda_{MOTION} = \sqrt{\lambda_{MODE}} : & \text{for SAD} \end{cases} \tag{10.28}$$

10.5.4 From Rate–Distortion to Rate–Quality Optimization

The techniques described throughout this section have been based on two assumptions: (i) that the aim of the encoding process is to represent the picture sample values as closely as possible, and (ii) that MSE (or other L_p norms) accurately reflects the perceptual quality and importance of the spatio-temporal region being coded. In practice we have seen that, although the second assumption is clearly violated, MSE can provide monotonic and consistent indicators for parameter variations within a given coding strategy for given content [20].

In the context of the first assumption, new perceptually inspired coding strategies are emerging based on texture analysis and synthesis models (see Chapter 13). The aim of these is not to minimize the distance between the original and coded version, but instead to obtain a perceptually plausible solution. In this case, MSE is unlikely to provide any meaningful indication of perceived quality and will no longer be a valid objective function. Emphasis must therefore shift from Rate–Distortion Optimization to Rate–Quality Optimization, demanding new embedded perceptually driven yet low complexity quality metrics [23].

10.6 Rate control

Rate control is an important component in a practical video codec since it ensures that the coded video is delivered to the channel at a rate compatible with the prevailing content complexity and channel capacity. Putting to one side for the moment the influences of specific coding modes, the primary issues that influence the bit rate of an encoded video are:

1. The different requirements for I-pictures and P- or B-pictures (I-pictures can consume 5–10 times the number of bits of a P- or B-picture).
2. The influence of source complexity. Content with high spatio-temporal activity will normally require many more bits than for the case of low activity content. This is depicted in Figure 10.11.

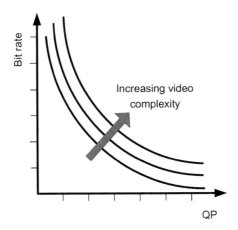

FIGURE 10.11

Managing bit rate in the context of changing source complexity.

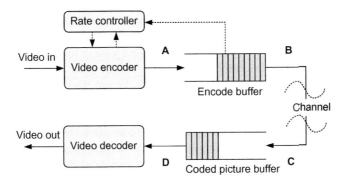

FIGURE 10.12

Transmission buffering at encoder and decoder.

Hence if the quantizer is kept fixed, then the peak-to-mean ratio of the output bitstream could be excessively high, translating to an efficient use of bandwidth. The peak output bit rate could also exceed the maximum channel capacity.

Even when rate control is used, bit rate fluctuations are inevitable and buffering is invariably employed to smooth out these variations. A simple depiction of the interface between an encoder and the channel and the channel and a decoder via buffers is shown in Figure 10.12. This figure explicitly includes a rate controller that is used to moderate the flow of coded data into the encode buffer. This is normally achieved by means of a rate-quantization model that delivers a quantizer parameter (QP) to the encoder based on dynamic measures of buffer occupancy and content complexity. If we consider the delays due to the buffering operations and the channel, we can represent the playout schedule for this type of configuration in Figure 10.13.

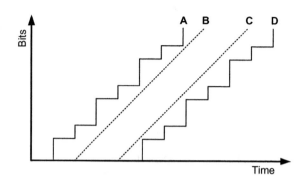

FIGURE 10.13

Streaming playout schedule on a picture by picture basis, corresponding to the points labeled in Figure 10.12.

The buffers described above and the associated rate control mechanisms are non-normative in most standards. However, to ensure that rate control is performed correctly at the encoder, recent standards have incorporated the concept of a *Hypothetical Reference Decoder* (HRD). This is used to emulate idealized decoder behavior in terms of buffering and bit consumption. Without rate control, decoder buffers would regularly underflow or overflow, resulting in playout jitter or loss.

Rate control and RDO jointly present a dilemma: in order to perform RDO for a coding unit, the quantization parameter must be known, and this is usually based on SAD or a variance measure. However, the actual SAD of the coding unit is not available until after the RDO process is complete; it is thus necessary to estimate the SAD of the current coding unit. Alongside this we also need to compute target bits for the current frame, and this is complicated by the fact that header information such as coding unit modes and MV information are not available before the RDO process completes.

10.6.1 Buffering and HRD

Bit rate allocation is normally associated with a buffer model that is specified as part of a Hypothetical Reference Decoder (HRD) in the video coding standard. The HRD is usually a normative component of a standard since it is important in dictating the requirements on compliant bitstreams.

The HRD is an idealized representation of a decoder that is conceptually connected to the output of the encoder. It comprises a decoder buffer as well as the decoder itself. The buffer is normally modeled using a leaky bucket approach which simply provides a constant rate flow of bits to and from the channel. Bits are assumed to flow into the decoder buffer (sometimes called the Coded Picture Buffer (CPB)) at a constant rate and are assumed to be extracted from it by the decoder instantaneously in picture-sized blocks. This is the scenario depicted in Figures 10.12 and 10.13 where, in the absence of loss in the channel, the content of the buffer at the receiver simply mirrors

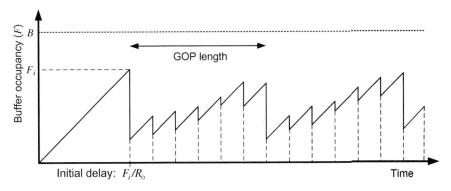

FIGURE 10.14
Example decoder buffer timing, indicating HRD parameters.

that at the encoder. An HRD-compliant bit stream must thus be processed by the CPB without overflow and underflow. This requirement must be satisfied by the rate control implemented in the encoder.

The HRD buffer model is normally described by three parameters, (i) the output rate R_o (assumed constant), the initial occupancy, F_i (which is the occupancy level at which decoding is assumed to start), and the buffer size, B. The rate controller must track the occupancy F of the hypothetical buffer and adjust the encoder parameters within its scope, in order to avoid buffer underflow and overflow (see Figure 10.14).

Example 10.2 (Buffering and HRD parameters)
Consider the operation of a constant bit rate video transmission system at 24 fps with the following parameters:

$$R_o = 500 \text{ kbps}; \quad B = 128 \text{ kb}; \quad F_i = 100 \text{ kb}; \quad GOP = 6 \text{ frames}$$

If the initial pictures transmitted have the following sizes, compute the occupancy of the decoder buffer and determine whether underflow or overflow occurs.

Frame no.	Picture size (kbits)
1	60
2–6	20
7	30
8–12	10
13	30
14–18	10
19	50
20–24	30

Solution. The ramp-up time for the buffer is $F_i / R_o = 0.2$ s. Plotting the buffer occupancy over the 1 s period of these initial pictures, we can observe that the decoder buffer overflows when pictures 18 and 19 are transmitted:

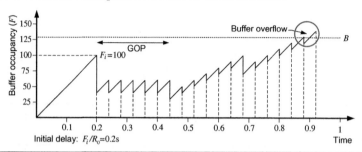

10.6.2 **Rate control in practice**

Figure 10.15 shows a generic rate control architecture. While simplified, this has many similarities to those used in MPEG-2, H.264/AVC, and HEVC. The basic elements are described below.

Buffer model

The buffer model represents the hypothetical reference decoder as described in Section 10.6.1. It is parameterized by its size and initial fullness, and takes a further input according to the bit rate of the channel. It provides an output of its current occupancy, F, to the bit allocation blocks described below.

Complexity estimation

In order to control the bit allocation process it is essential to estimate the complexity of the coded residual. The estimation of source complexity is normally based on the residual data (since this is what the quantizer influences). A measure such as Mean Absolute Difference (MAD), Sum of Absolute Values (SAD), Sum Squared Difference (SSD), or Sum of Absolute Transform Differences (SATD) [50] of the residual is often used for this purpose.

Rate-quantization model

Reflecting the variability indicated in Figure 10.11, the rate-quantization model describes how the available bits are related to quantizer values when content of varying complexity is encoded.

ΔQP limiter

In order to avoid excessive changes in QP values that might introduce noticeable artifacts or cause oscillatory behavior in the rate controller, the change in QP value is normally limited to a small increment, typically ± 2.

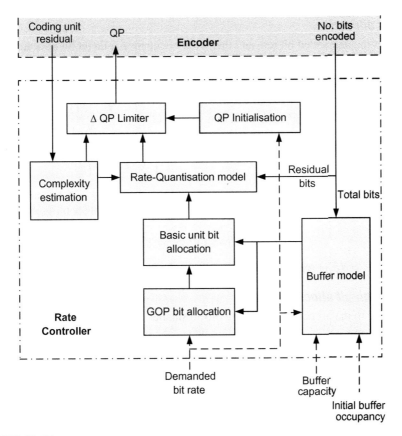

FIGURE 10.15

Generic rate controller.

QP initialization

The QP value must be set to some value prior to coding. An average value for the bit rate per pixel can be easily computed, based on the bit rate demanded and the video format, thus:

$$\text{bpp} = \frac{R_1(1)}{f \cdot X \cdot Y} \tag{10.29}$$

where $R_i(k)$ is the instantaneous available bit rate at the time of coding frame k of GOP i, f is the frame rate, and X and Y are the number of horizontal and vertical pixels in the picture. This can then be related to QP values by way of a look-up table such as that recommended for H.264/AVC in Ref. [51].

GOP bit allocation

Based on the demanded bit rate and the buffer occupancy, a target bit allocation for the GOP is computed. GOP level bit allocation normally initializes the IDR-picture and the first P-picture. A scheme used for this in H.264/AVC is as follows [52]:

$$
QP_i(1) = \max \left\{ QP_{i-1}(1) - 2, \min \left\{ QP_{i-1}(1) + 2, \frac{\sum_{k=2}^{N_{i-1}} QP_{i-1}(k)}{N_{i-1} - 1}, \right. \right.
$$
$$
\left. \left. - \min \left\{ 2, \frac{N_{i-1}}{15} \right\} \right\} \right\} \tag{10.30}
$$

where $QP_i(k)$ is the QP value assigned to picture k of GOP i and N_i is the total number of pictures in GOP i. If $QP_i(1) > QP_{i-1}(N_{i-1}) - 2$, this is further adjusted as follows:

$$
QP_i(1) \leftarrow QP_i(1) - 1 \tag{10.31}
$$

where $QP_{i-1}(N_{i-1})$ represents the QP value of the previous P-picture.

Coding unit bit allocation

This refers to the process of allocating bits to smaller coding elements, such as individual pictures, slices, or smaller coding units. The bit allocation to each picture is based on the target buffer level, actual buffer occupancy, the bit rate demanded, and of course the frame rate. QP values for each of the remaining P- and B-frames are thus assigned based on the number of remaining bits available for the GOP. The allocation process should also take account of whether the current picture is of P or B type since these may be provided with different bit allocations. The scheme used for this in H.264/AVC, based on a quadratic model derived assuming a Laplacian source distribution, is as follows [52]:

Firstly, the MAD of the current stored picture is estimated, $\tilde{\xi}_i(k)$, based on the corresponding value for the previous picture, as given in equation (10.32):

$$
\tilde{\xi}_i(k) = a_1 \cdot \xi_i(k-1) + a_2 \tag{10.32}
$$

where a_1 and a_2 are the coefficients of the prediction model, initialized to 1 and 0 and updated after each unit is coded. Next, the quantizer step size is computed for each of the remaining P-pictures, using equation (10.33).

$$
T_i(k) = c_1 \cdot \frac{\tilde{\xi}_i(k)}{Q_{\text{step},i}(k)} + c_2 \cdot \frac{\tilde{\xi}_i(k)}{Q_{\text{step},i}^2(k)} + m_{h,i}(k) \tag{10.33}
$$

where $m_{h,i}(k)$ is the total number of header bits and motion vector bits, c_1 and c_2 are the coefficients of the quadratic model, updated after each picture is processed, and $Q_{\text{step},i}(k)$ is the quantizer step size for picture k, corresponding to the target number of bits. In order to compute the quantizer step size in equation (10.33) we first need to calculate a value for the target number of bits for frame k, $T_i(k)$. This is usually computed based on a weighted combination of the buffer occupancy and the number of bits remaining for the rest of the GOP. Further details on this scheme can be found in Ref. [52].

Several other schemes have been proposed for rate control that improve prediction accuracy for both texture and header bits; for example, the contribution of Kwon et al. [50]. In the context of HEVC, several modifications to the above approach have been proposed and the reader is referred to the contributions by Sanz-Rodriguez and Schierl [53] and by Li et al. [54].

10.6.3 Regions of interest and rate control

In certain applications, and for certain types of content, a priori information about the scene, or areas within it, can be exploited to improve coding efficiency or better manage bit rate. For example, in many surveillance videos, there may be one or more *regions of interest* (ROI) that would benefit from prioritized bit allocation. A good discussion of the trade-offs involved in this is provided by Sadka [55].

Agrafiotis et al. [56] presented a modified rate control method that allows the definition of multiple priority regions, the quality of which varies, based on the ROI characteristics and the target bit rate. The results presented demonstrate increased coding flexibility which can lead to significant quality improvements for the ROI and a perceptually pleasing variation in quality for the rest of the frame, without violating the target bit rate. This work was presented in the context of low bit rate sign language coding for mobile terminals, but it can also be of benefit in many other applications including surveillance and healthcare.

10.7 Summary

This chapter has covered a range of methods for measuring and controlling visual quality. We first described the most common subjective testing methods and showed how these can provide a consistent means of comparing codec performance. We then went on to discuss a number of objective measures that can be used as an efficient substitute for subjective evaluation in certain cases. MSE-based methods were reviewed briefly and their strengths and weaknesses highlighted. We demonstrated cases where the perceptual distortion experienced by the human viewer cannot however be characterized using such simple mathematical differences. Perception-based video metrics have therefore emerged and a number of these were reviewed and compared, showing improved correlation with subjective scores.

The second part of this chapter addressed the issue of rate–distortion optimization, describing some of the most common techniques that enable us to select the optimum coding parameters for each spatio-temporal region of a video. After reviewing conventional rate–distortion theory, we focused on practical solutions such as those employing Lagrange multipliers, showing that these can provide a tractable solution for modern codecs. Related to this, the final part of the chapter studied the requirements of, and methods for, rate control as a means of adapting the output bit rate of an encoder according to the capacity of a given channel.

References

[1] S. Winkler, Digital Video Quality: Vision Models and Metrics, Wiley, 2005.

[2] F. Kozamernik, P. Sunna, E. Wycken, D. Pettersen, Subjective quality assessment of internet video codecs — phase 2 evaluations using SAMVIQ, EBU Technical Review (2005) 1–22.

[3] S. Winkler, Analysis of public image and video database for quality assessment, IEEE Journal of Selected Topics in Signal Processing 6 (6) (2012) 1–10.

[4] Recommendation ITU-R BT.500-13, Methodology for the Subjective Assessment of the Quality of Television Pictures, ITU-R, 2012.

[5] Recommendation ITU-T P.910, Subjective Video Quality Assessment Methods for Multimedia Applications, ITU-T, 1999.

[6] F. De Simone, L. Goldmann, J.-S. Lee, T. Ebrahomi, Towards high efficiency video coding: subjective evaluation of potential coding methodologies, Journal of Visual Communication and Image Representation 22 (2011) 734–748.

[7] P. Hanhart, M. Rerabek, F. De Simone, T. Ebrahimi, Subjective quality evaluation of the upcoming HEVC video compression standard, in: Applications of Digital Image Processing XXXV, vol. 8499, 2012.

[8] H. Hoffmann, T. Itagaki, D. Wood, T. Hinz, T. Wiegand, A novel method for subjective picture quality assessment and further studies of HDTV formats, IEEE Transactions on Broadcasting 54 (1) (2008) 1–13.

[9] D. Lane, Introduction to Statistics. <http://onlinestatbook.com/>, Rice University.

[10] Video Quality Experts Group, Final Report from the Video Quality Experts Group on the Validation of Objective Quality Metrics for Video Quality Assessment, VQEG, Technical Report, 2000. <http://www.its.bldrdoc.gov/vqeg/projects/frtv-phase-i/frtv-phase-i.aspx> phase I.

[11] Video Quality Experts Group, Final VQEG Report on the Validation of Objective Models of Video Quality Assessment, VQEG, Technical Report, 2003. <http://www.its.bldrdoc.gov/vqeg/projects/frtv> phase II.

[12] H. Sheikh, Z. Wang, L. Cormack, A. Bovik, LIVE Image Quality Assessment Database Release 2. <http://live.ece.utexas.edu/research/quality>.

[13] K. Seshadrinathan, A. Bovik, Motion tuned spatio-temporal quality assessment of natural videos, IEEE Transactions on Image Processing 19 (2010) 335–350.

[14] K. Seshadrinathan, R. Soundararajan, A.C. Bovik, L.K. Cormack, Study of subjective and objective quality assessment of video, IEEE Transactions on Image Processing 19 (2010) 335–350.

[15] P. Le Callet, F. Autrusseau, Subjective quality assessment IRCCyN/IVC database, 2005. <http://www.irccyn.ec-nantes.fr/ivcdb/>.

[16] Video Quality Experts Group, Report on the Validation of Video Quality Models for High Definition Video Content, VQEG, Technical Report, 2010. <http://www.its.bldrdoc.gov/vqeg/projects/hdtv/hdtv.aspx>.

[17] Video Quality Experts Group, Final Report from the Video Quality Experts Group on the Validation of Objective Models of Multimedia Quality Assessment, VQEG,

Technical Report, 2008. <http://www.its.bldrdoc.gov/vqeg/projects/multimedia-phase-i/multimedia-phase-i.aspx>.

[18] P. Marziliano, F. Dufaux, S. Winkler, T. Ebrahimi, A no-reference perceptual blur metric, in: Proceedings of the IEEE International Conference on Image Processing, vol. 3, 2002, pp. 57–60.

[19] R. Soundararajan, A. Bovic, Video quality assessment by reduced reference spatio-temporal entropic differencing, IEEE Transactions on Circuits and Systems for Video Technology 23 (4) (2013) 684–694.

[20] D. Huynh-Thu, M. Ghanbari, Scope of validity of PSNR in image/video quality assessment, Electronics Letters 44 (13) (2008) 800–801.

[21] Z. Wang, A. Bovik, Mean squared error: love it or leave it? IEEE Signal Processing Magazine 26 (1) (2009) 98–117.

[22] B. Girod, What's wrong with mean squared error? in: A. Watson (Ed.), Digital Images and Human Vision, MIT Press, 1998.

[23] F. Zhang, D. Bull, A parametric framework for video compression using region-based texture models, IEEE Journal of Selected Topics in Signal Processing 5 (7) (2011) 1378–1392.

[24] T. Pappas, T. Michel, R. Hinds, Supra-threshold perceptual image coding, in: Proceedings of the IEEE International Conference on Image Processing, 1996, pp. 234–240.

[25] V. Kayargadde, J. Martens, Perceptual characterization of images degraded by blur and noise: experiments, Journal of the Optical Society of America 13 (1996) 1166–1177.

[26] S. Karunasekera, N. Kingsbury, A distortion measure for blocking artifacts in images based on human visual sensitivity, IEEE Transactions on Image Processing 4 (6) (1995) 713–724.

[27] M. Carnec, P. LeCallet, D. Barba, An image quality assessment method based on perception of structural information, in: Proceedings of the IEEE International Conference on Image Processing, vol. 2, 2003, pp. 185–188.

[28] D. Chandler, S. Hemami, VSNR: a wavelet-based visual signal to noise ratio for natural images, IEEE Transactions on Image Processing 16 (9) (2007) 2284–2298.

[29] E. Larson, D. Chandler, Most apparent distortion: full reference image quality assessment and the role of strategy, Journal of Electronic Imaging 19 (1) (2010), pp. 011006-1–011006-21.

[30] S. Chikkerur, V. Sundaram, M. Reisslein, L. Karam, Objective video quality assessment methods: a classification, review and performance comparison, IEEE Transactions on Broadcasting 30 (2005) 17–26.

[31] W. Lin, C. Kuo, Perceptual visual quality metrics: a survey, Journal of Visual Communication and Image Representation 22 (2011) 297–312.

[32] S. Sheikh, A. Bovik, Image information and visual quality, IEEE Transactions on Image Processing 15 (2006) 430–444.

[33] S. Winkler, P. Mohandas, The evolution of video quality measurement: from PSNR to hybrid metrics, IEEE Transactions on Broadcasting 54 (3) (2008) 660–668.

[34] M. Pinson, S. Wolf, A new standardized method for objectively measuring video quality, IEEE Transactions on Broadcasting 50 (2004) 312–322.

[35] Z. Wang, A. Bovik, H. Sheikh, E. Simoncelli, Image quality assessment: from error visibility to structural similarity, IEEE Transactions on Image Processing 13 (2004) 600–612.

[36] Z. Wang, E. Simoncelli, Translation insensitive image similarity in complex wavelet domain, in: IEEE International Conference on Acoustics, Speech and Signal Processing, 2005, pp. 573–576.

[37] Z. Wang, E. Simoncelli, A. Bovik, Multi-scale structural similarity for image quality assessment, in: Proceedings of the IEEE Asilomar Conference on Signals, Systems and Computers, 2003, pp. 1398–1402.

[38] Z. Wang, L. Lu, A. Bovik, Video quality assessment based on structural distortion measurement, Signal Processing: Image Communication 19 (2) (2004) 121–132.

[39] P. Vu, C. Vu, D. Chandler, A spatiotemporal most apparent distortion model for video quality assessment, in: Proceedings of the IEEE International Conference on Image Processing, 2011, pp. 2505–2508.

[40] R. Mantiuk, K. Kim, A. Rempel, W. Heidrich, HDR-VDP-2: a calibrated visual metric for visibility and quality predictions in all luminance conditions, ACM Transactions on Graphics (Proceedings of SIGGRAPH'11) 30 (4) (2011) (article no. 40).

[41] A. Zhang, D. Bull, Quality assessment method for perceptual video compression, in: Proceedings of the IEEE International Conference on Image Processing, 2013, pp. 39–43.

[42] A. Zhang, D. Bull, A perception-based hybrid model for video quality assessment, IEEE Transactions on Circuits and Systems for Video Technology, submitted for publication.

[43] A. Ortega, K. Ramchandran, Rate-distortion methods for image and video compression, IEEE Signal Processing Magazine 15 (6) (1998) 23–50.

[44] G. Sullivan, T. Wiegand, Rate-distortion optimization for video compression, IEEE Signal Processing Magazine 15 (6) (1998) 74–90.

[45] Y. Wang, J. Ostermann, Y. Q Zhang, Video Processing and Communications, Prentice Hall (2002).

[46] T. Wiegand, H. Schwarz, A. Joch, F. Kossentini, G. Sullivan, Rate-constrained coder control and comparison of video coding standards, IEEE Transactions on Circuits and Systems for Video Technology 13 (7) (2003) 688–703.

[47] E.H. Yang, X. Yu, Rate distortion optimization for H.264 interframe coding: a general framework, IEEE Transactions on Image Processing 16 (7) (2007) 1774–1784.

[48] H. Everett, Generalised Lagrange multiplier method for solving problems of optimum allocation of resources, Operations Research 11 (1963) 399–417.

[49] Y. Shoham, A. Gersho, Efficient bit allocation for an arbitrary set of quantizers, IEEE Transactions on Acoustics, Speech, and Signal Processing 36 (1988) 1445–1453.

[50] D.-K. Kwon, M.-Y. Shen, C. Kuo, Rate control for H.264 video with enhanced rate and distortion models, IEEE Transactions on Circuits and Systems for Video Technology 17 (5) (2007) 517–529.

[51] G. Sullivan, T. Wiegand, K. Lim, Joint Model Reference Encoding Methods and Decoding Concealment Methods, JVT-I049, San Diego, 2003.

[52] K. Keng-Pang Lim, G. Sullivan, T. Wiegand, Joint Video Team (JVT) of ISO/IEC MPEG and ITU-T VCEG Document JVT-O079, Text Description of Joint Model Reference Encoding Methods and Decoding Concealment Methods, 2005.

[53] S.S. Sanz-Rodrıguez, T. Schierl, A rate control algorithm for HEVC with hierarchical GOP structures, in: Proceedings of the IEEE International Conference on Acoustics Speech and Signal Processing, 2013, pp. 1719–1723.

[54] B. Li, H. Li, L. Li, J. Zhang, Rate control by R-lambda model for HEVC, JCTVC-K0103, in: 11th JCTVC Meeting, China, October 2012.

[55] A. Sadka, Compressed Video Communications, Wiley, 2002.

[56] D. Agrafiotis, D. Bull, C. Canagarajah, N. Kamnoonwatana, Multiple priority region of interest coding with H.264, in: Proceedings of the IEEE International Conference on Image Processing, 2006, pp. 53–56.

CHAPTER OUTLINE

11.1 **The operating environment** **362**
    11.1.1 Characteristics of modern networks 362
    11.1.2 Transmission types 364
    11.1.3 Operating constraints 365
    11.1.4 Error characteristics 365
    11.1.5 The challenges and a solution framework 367
11.2 **The effects of loss** **368**
    11.2.1 Synchronization failure 368
    11.2.2 Header loss 370
    11.2.3 Spatial error propagation 370
    11.2.4 Temporal error propagation 372
11.3 **Mitigating the effect of bitstream errors** **373**
    11.3.1 Video is not the same as data! 373
    11.3.2 Error-resilient solutions 374
11.4 **Transport layer solutions** **375**
    11.4.1 Automatic Repeat request (ARQ) 375
    11.4.2 FEC channel coding 376
    11.4.3 Hybrid ARQ (HARQ) 379
    11.4.4 Packetization strategies 379
11.5 **Application layer solutions** **379**
    11.5.1 Network abstraction 379
    11.5.2 The influence of frame type 380
    11.5.3 Synchronization codewords 382
    11.5.4 Reversible VLC 382
    11.5.5 Slice structuring 384
    11.5.6 Error tracking 385
11.6 **Cross layer solutions** **386**
    11.6.1 Link adaptation 386
11.7 **Inherently robust coding strategies** **389**
    11.7.1 Error-Resilient Entropy Coding (EREC) 389
    11.7.2 Pyramid Vector Quantization (PVQ) 394

Communicating Pictures. http://dx.doi.org/10.1016/B978-0-12-405906-1.00011-8

11.8 **Error concealment** **397**
    11.8.1 Detecting missing information 397
    11.8.2 Spatial error concealment (SEC) 398
    11.8.3 Temporal error concealment 399
    11.8.4 Hybrid methods with mode selection 403
11.9 **Congestion management and scalable video coding** **405**
11.10 Summary **408**
References **408**

Video compression algorithms rely on spatio-temporal prediction combined with variable-length entropy encoding to achieve high compression ratios but, as a consequence, they produce an encoded bitstream that is inherently sensitive to channel errors. This becomes a major problem when video information is transmitted over unreliable networks, since any errors introduced into the bitstream during transmission will rapidly propagate to other regions in the image sequence.

In order to promote reliable delivery over lossy channels, it is usual to invoke various error detection and correction methods. This chapter firstly introduces the requirements for an effective error-resilient video encoding system and then goes on to explain how errors arise and how they propagate spatially and temporally. We then examine a range of techniques that can be employed to mitigate the effects of errors and error propagation. We initially consider methods that rely on the manipulation of network parameters or the exploitation of network features to achieve this; we then we go on to consider two methods, EREC and PVQ, where the bitstream generated by the codec is inherently robust to errors. Finally we present decoder-only methods that conceal rather than correct bitstream errors. These deliver improved subjective quality without adding transmission overhead.

11.1 The operating environment

11.1.1 Characteristics of modern networks

The fixed IP network

The *internet* is essentially a collection of independent packet-switched networks operating under the same protocol and connected via routers. Each packet has a header that identifies its source and destination IP addresses, where packet delivery to the destination is conducted using TCP/IP protocols. The *Internet Protocol* (IP) provides the basis for packet delivery and the *Transmission Control Protocol* (TCP) provides a best-effort delivery mechanism using *Automatic Repeat reQuest* (ARQ) techniques, but with no bounds on delivery time. It is thus an unreliable, best-effort, connectionless packet delivery service with no guarantee of the actual delivery of packets.

For less reliable networks and real-time delay-constrained applications, the *User Datagram Protocol* (UDP) is normally preferred. The *Real-time Transport Protocol* (RTP) operates on top of UDP and contains essential timing and sequence information.

For colour and a higher quality version of Figure 11.8 please refer to the electronic version or the website.

Alongside RTP, the *Real-Time Control Protocol* (RTCP) enables quality of service (QoS) feedback from the receiver to the sender to enable it to adjust its transmission characteristics to better suit the prevailing network conditions. This is often used in conjunction with the *Real-Time Streaming Protocol* (RTSP) for control of streaming media servers and for establishing and controlling media sessions.

The wireless network

Most modern wireless networks also operate under IP protocols to support seamless integration with the internet backbone. The band-limited and error-prone nature of the wireless environment, however, needs additional mechanisms for adaption. These are normally based around a suite of operating modes combining different error control and modulation schemes (MCS modes) that can be selected according to prevailing channel conditions. They are implemented in the *Physical Layer* (PHY) and *Medium Access Control* (MAC) layers of the modem.

There are two main classes of wireless network and these are converging rapidly under new 4G standards. *Wireless Local Area Networks* (WLANs) are invariably based on IEEE 802.11 standards and offer local high speed data transfer via hotspots. The *digital cellular network* was originally designed to support voice and data traffic but now, under the 3G and 4G standards, provides the primary means of video downloading and conversational video services.

We can see the effect of a typical wireless channel on system performance in Figure 11.1. This shows data logged on a wireless video research platform at the University of Bristol. The data logging traces show how an 802.11 WiFi link adapts as a user moves away from the access point (time: 1–70 s) and then back toward it (time: 70–140 s). This illustrates how the radio environment changes and the effect that this has on packet delays, packet delay variability, linkspeed (modulation and coding modes), and the bit rate at the application layer. Clearly, our source coding

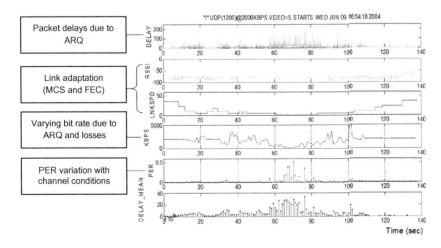

FIGURE 11.1

Logged radio link parameter variations with channel conditions for a wireless video trial.

methods need to adapt to these variable bandwidth conditions, while the addition of channel coding or other means of error resilience is needed to cope with the errors introduced during transmission.

11.1.2 Transmission types

Downloads and streaming

Access to a video file or stream can be achieved in many different ways. In the simplest case this can be in the form of a file transfer where data is stored on the local machine and then opened in an appropriate player. *Progressive downloads* are also file transfers, typically using the HTTP protocol, and the files are also stored locally, but they are more intimately linked to the player. The player can start playback before the download is complete but there is generally no interactive support. As the file is downloaded from the server, it will be buffered and when sufficient data is available for continuous display, it will begin to play.

In the case of streaming, the video is not usually stored on the user's device but is played directly from a buffer. A streaming server will interact with the user's player to enable interactive navigation through the video (e.g. FFWD, RWD, etc. via RTSP).

Interactive communication

With streamed video, playout will commence when the receiver buffer is sufficiently full to ensure sustained playout. Clearly, if conditions dictate that the network cannot support the encoded video bit rate, then the buffer will progressively empty and eventually the video will stutter and freeze. In cases where interaction is required, the demands on delay are much more stringent; typically 200–300 ms is cited as the maximum tolerable for conversational services to avoid unnatural hesitations.

Unicast transmission

In the case of unicast transmission, one sender is connected to one receiver. This is mandatory for conversational services, but can be inefficient for other services as every client demands its own bandwidth. For the case of a single transmitter unicasting a video at B bps to N receivers, the total bandwidth required is NB bps. The advantage of unicast transmission is that there is an intimate link between the sender and receiver and hence a feedback channel can be established to signal errors and thus support retransmission and error control. Unicast streaming is used by operators such as YouTube and Skype.

Multicast transmission

In the case of multicast transmission, many receivers join a single stream. This is much more efficient than the unicast case, as our N receivers now only consume B bps. However, all receivers will experience different channel conditions and, if the number of receivers is large, then feedback mechanisms become prohibitive and other error control mechanisms must be invoked. Multicasting thus offers much less service flexibility than unicasting, but provides much better bandwidth flexibility. Multicast transmission is used by operators such as Ustream.

11.1.3 **Operating constraints**

As discussed above, IP streaming, conferencing, and entertainment video have been primary drivers for the growth in multimedia communication services over fixed internet and mobile networks. The channels associated with these networked services can be characterized by the following three attributes:

1. **The bandwidth of the wireless channel is limited:** This necessitates the use of sophisticated source coding techniques to match the bit rate for the signal to the capacity of the channel.
2. **The channel quality for wireless networks can be highly variable:** This is due to signal fading and interference giving rise to erroneous transmission and the need for error protection, error correction, and retransmissions.
3. **Congestion can occur in packet switched networks:** This can lead to dropped packets and variable delays. These demand intelligent buffering and rate control strategies and impose strict bounds on retransmission.

We have seen how attribute (1) can be addressed by exploiting data correlations in a video sequence—using inter-frame prediction (motion estimation), intra-frame prediction, and variable length entropy encoding. However, as a consequence these methods produce an encoded bitstream that is inherently sensitive to channel errors, and this presents a particular problem in the context of attributes (2) and (3).

Clearly we want our encoder to exhibit excellent rate–distortion (or rate–quality) performance, but what the user is really interested in is the quality at the decoder, as this is what he or she experiences. In fact, the rate–distortion performance of the encoder is irrelevant if all packets are lost during transmission. In order to promote reliable delivery over lossy channels, it is usual to invoke various additional error detection and correction methods. In summary, the encoded video bitstream needs to be error resilient as well as efficient in a rate–distortion sense.

11.1.4 **Error characteristics**

Types of errors

Errors that occur during transmission can be broadly divided into two categories: *random bit errors* and *erasure errors*. The former are normally modeled with a uniform pdf whereas the latter are bursty in nature. Bit errors are normally characterized by their *bit error rate* or BER and can be caused by thermal noise. Erasure errors affect longer segments of data and can be caused by congestion in a packet network or fading, shadowing and drop-out in a wireless network. They will often exceed the correcting capability of any channel coding system and will demand retransmission or lead to the loss of a whole packet (itself an erasure at the *application layer*).

Test data

In order to model the effects of errors on a video signal, we need to ensure that the encoding method used is evaluated with representative test data. This can be obtained through data logging over prolonged periods using a real system, such as is illustrated

FIGURE 11.2

Radiowave propagation modeling using ray tracing. (Courtesy of A. Nix.)

in Figure 11.1, or by emulating system performance using wireless propagation modeling or by statistical channel modeling. We will not discuss Figure 11.1 in detail here, but will return to some of the things it depicts later in the chapter.

In practice, while it is useful to characterize the performance of a system using more generic models, for real-world usefulness it is preferable to use propagation models combined with appropriate network simulation tools to generate realistic bit error patterns at the application layer. Such methods are described in Ref. [1]. An example output from a ray tracing analyzer, that provides accurate representations of multipath effects in a given environment for a specific wireless configuration, is illustrated in Figure 11.2 and is described in Ref. [2].

Types of encoding

The encoding mode will have a significant influence on the way that errors manifest themselves and propagate spatially and temporally. For example, transform coding in combination with variable length coding can cause spatial error propagation within a block if symbol synchronization is lost. Propagation across multiple blocks will occur if an EOB codeword is incorrectly decoded; in this case any errors within a transform block will propagate spatially to adjacent regions of pixels. The situation is similar for wavelet subbands.

Spatial or temporal propagation of errors can also be caused by predictive coding. Errors within any pixel or transform coefficient used as the basis for prediction at the decoder will lead to incorrect reconstructions and these will propagate around the prediction loop, leading to errors in subsequent pixels or transform coefficients. For example, errors in a reference frame used for motion estimation will be propagated temporally by the motion compensation process.

11.1.5 The challenges and a solution framework

Up until this point we have attempted to achieve the best rate–distortion performance from our compression system. We have removed as much redundancy as possible from the video signal, without compromising quality. However, we are now faced with the dilemma that, in order to make the compressed bitstream robust in the presence of errors, we need to introduce redundancy back into the bitstream in the form of channel coding, thus offsetting some or possibly all of the gains made during compression. Clearly this is not a good situation!

The challenge of error-resilient coding is therefore to maintain the coding gain obtained through source compression while ensuring that the decoded bitstream retains integrity during transmission, so that it can be faithfully decoded. In order to achieve this we need to:

1. Understand the causes of errors and their effects (Section 11.2).
2. Characterize these from a psychovisual point of view to understand the impact of loss artifacts on the human visual system (Chapter 2 and Section 11.3).
3. Exploit appropriate channel coding and retransmission solutions but use these, where possible, in a manner suited to video content (Sections 11.3 and 11.4).
4. Exploit the adaptive mechanisms that exist in many wireless or internet protocols to work jointly across the application, access control and physical layers to produce joint source–channel coding or cross layer solutions (Sections 11.4 and 11.6).
5. Exploit the structure of the video coding process to make the encoded bitstream inherently more robust to errors (Sections 11.5 and 11.7).
6. Understand where the errors occur in the reconstructed video and take steps to conceal rather than correct them (Section 11.8).

Figure 11.3 shows a generic architecture that captures the essence of this approach. In practice, an error-resilient solution combines aspects of transport layer solutions and

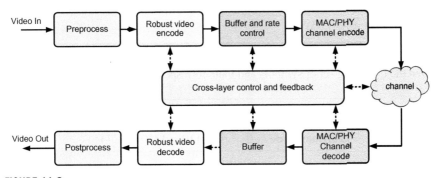

FIGURE 11.3

Generic error-resilient video transmission architecture.

application layer solutions in a sympathetic manner. That is, where possible, network parameters should be adapted to suit video requirements and video parameters should be managed to match the network constraints. These issues are discussed in more detail in the following sections.

11.2 The effects of loss

11.2.1 Synchronization failure

As we have already seen in Chapter 7, a coded video bitstream normally comprises a sequence of variable-length codewords. VLC methods such as Huffman or arithmetic coding are employed in all practical image and video coding standards. Fixed-length codewords (FLC) are rarely used in practice except when symbol probabilities do not justify the use of variable-length coding (e.g. in some header information). A further exception to this is the class of error-resilient methods including Pyramid Vector Quantization (PVQ) which we will study in Section 11.7.2.

As we will see in the following subsections, the extent of the problem caused will be related to the location where the error occurs and the type of encoding employed.

The effect of a single bit error

With FLC, any codeword affected by a single bit error would be decoded incorrectly, whereas all other codewords remain unaffected. This means that there is no loss of bitstream or symbol synchronization. In contrast with VLC, single bit errors can cause the decoder to decode a longer or shorter codeword, thus meaning that subsequent symbols, although delivered correctly, will most likely be decoded incorrectly due to the loss of synchronization at the decoder. This loss of synchronization can continue until the next explicit resynchronization point and will often result in the wrong number of symbols being decoded. This might mean that important implicit synchronization symbols such as EOB will be missed. In practice, most Huffman codes will resynchronize themselves, whereas arithmetic codes rarely do. The two scenarios in Example 11.1 illustrate the problems of synchronization loss in VLC, demonstrating the subtle differences between synchronization at bitstream, symbol and coding unit (e.g. transform block) levels.

Example 11.1 (Loss of VLC synchronization due to a single bit error)
Given the following set of symbols and their corresponding Huffman codewords,

Alphabet	A	B	C	D	E
Huffman code	0	10	110	1110	1111

with the following transmitted message:

Message	B	A	D	E	C	B	A
Encoded bitstream	10	0	1110	1111	110	10	0

(a) Compute and comment on the decoded sequence of symbols if there is a single bit error in the sixth bit.

(b) Repeat (a) for the case of a single bit error at the 10th bit position.

Solution.

(a) The table of decoded bits and symbols is given below with the errored bit in bold font:

Encoded message	B	A	D	E	C	B	A	–
Received bitstream	10	0	1**1**00	1111	110	10	0	
Parsed bitstream	10	0	110	0	1111	110	10	0
Decoded message	B	A	C	A	E	C	B	A

In this case the symbol D is incorrectly decoded as a C followed by an A and then bitstream resynchronization occurs. However, the fact that an extra symbol is decoded means that the subsequent symbols are displaced and hence symbol synchronization is lost.

(b) The table of decoded bits and symbols is given below with the errored bit in bold font:

Encoded message	B	A	D	E	C	B	A
Received bitstream	10	0	1110	11**0**1	110	10	0
Parsed bitstream	10	0	1110	110	1110	10	0
Decoded message	B	A	D	C	D	B	A

In this case, because of the lengths and encodings of the adjacent symbols, the error is constrained to only two symbols. After this, both bitstream and symbol resynchronization are achieved. Superficially this appears a lot better than the first case. However, consider the situation where the symbols represent {run/size} encodings of transform coefficients. The fact that there are errors in the fourth and fifth symbols means that, after the third symbol, the whole block will be decoded incorrectly because the runs of zeros corresponding to symbols 4 and 5 are incorrect. However, because symbol resynchronization is regained, any subsequent EOB symbol would be correctly detected and block level synchronization will be maintained. See Example 11.2 for more details.

11.2.2 Header loss

Errors in header information can cause catastrophic errors or error propagation. If any bit within a header is corrupted, then the result is often a complete failure of the decoder since the basic parameters needed to decode the bitstream are incorrect. There is therefore some justification for header information to be protected more strongly than other data as, without it, little else matters.

11.2.3 Spatial error propagation

Errors can propagate spatially during decoding for a number of reasons:

1. **DPCM prediction:** For example as used in the encoding of DC transform coefficients in JPEG. This will normally result in a luminance or chrominance shift in the image blocks decoded after the error.
2. **Intra-prediction:** Modern compression standards employ sophisticated intra-prediction modes that can result in the propagation of errors.
3. **Loss of VLC symbol synchronization within a block:** For example in a DCT-based system, this will result in the appearance of artifacts relating to superpositions of incorrect DCT basis functions.
4. **Loss of block synchronization:** Errors will propagate across blocks if EOB symbols are incorrectly decoded. This normally leads to a loss of positional synchronization.

The effect of a single error on a simple JPEG DCT/Huffman encoded image is illustrated in Figure 11.4. This shows the effects described in Examples 11.1 and 11.2. We can clearly see the block where the error has occurred as the corrupted DCT basis

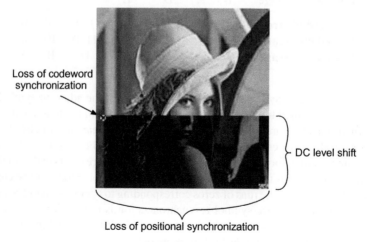

Loss of codeword synchronization

DC level shift

Loss of positional synchronization

FIGURE 11.4

The effect of a single bit error on a DCT/Huffman encoded Lena image.

FIGURE 11.5
Impact of increased error rates on a DCT image codec. Left: 0.01% BER. Middle: 0.1% BER. Right: 1% BER.

functions are evident. The figure also illustrates the effects of corrupted DC levels propagated by the DPCM method used. Also shown is the effect of a spatial shift in the reconstructed image due to EOB symbols being missed in blocks adjacent to the errored block. Eventually symbol synchronization is regained but the effects of the spatial displacement remain. The effect of higher error rates and the impact these have on subjective image quality are shown in Figure 11.5.

Example 11.2 (Spatial error propagation due to VLC errors)
Given the set of symbols and their corresponding Huffman codewords from Example 11.1, let us now assume that the symbols correspond to {run/size} values for an intra-coded DCT block in an image. If in this case, symbol B corresponds to EOB, then comment on the spatial error propagation associated with the following transmitted sequence, when a single error is introduced at bit position 12 in the first block of the image:

Message	A	D	E	C	B	D	A	E	C	B
Encoded bitstream	0	1110	1111	110	10	1110	0	1111	110	10

Solution. The table of decoded bits and symbols is given below with the errored bit in bold font:

Message	A	D	E	C	EOB	D	A	E	C	EOB
Received bits	0	1110	1111	11**1**	10	1110	0	1111	110	10
Parsed bits	0	1110	1111	1111	0	1110	0	1111	110	10
Decoded	A	D	E	E	A	D	A	E	C	EOB

In this case, symbols C and B are incorrectly decoded as E and A, hence the first EOB symbol is missed. So although bitstream and symbol resynchronization occurs, block synchronization is lost. All subsequently decoded blocks will thus be displaced by one block position to the left. So unless some explicit form of synchronization is introduced, the whole image after the second block will be corrupted.

11.2.4 Temporal error propagation

Temporal error propagation occurs when corrupted spatial regions are used as the basis for prediction of future frames. As shown in Figure 11.6, a single corrupted block in the reference frame can propagate to multiple (4 in this case) blocks in the predicted current frame. The spatial spread of the error is dependent on the local distribution of motion vectors and more active regions will tend to spread errors more widely than regions of low motion. Figure 11.7 shows two examples of temporal propagation in practice. Here, the error introduced locally due to spatial data loss still persists and has spread, several frames later.

It should be noted that corrupted motion vectors can also provide a source of errors due to incorrect motion compensation at the decoder.

Example 11.3 (Temporal propagation of errors due to prediction)
Consider the same symbol encoding and transmitted sequence as given in Example 11.2. Assume that the symbols represent {run/size} encodings related to the transform coefficients in two DFD blocks and that these are used as the basis for future motion compensated predictions. Comment on the temporal error propagation that would result from this single bit error in bit position 12.

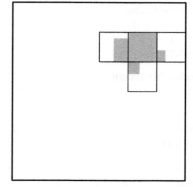

Reference frame k–1   Current frame k

FIGURE 11.6

Temporal error propagation from a single corrupted block in frame $k - 1$ to multiple corrupted blocks in frame k.

FIGURE 11.7

Example of temporal propagation; Top: across four frames of the table tennis sequence. Bottom: across 190 frames of Foreman.

Solution. As with Example 11.2, because the EOB symbol is missed, block synchronization is lost, and all blocks decoded after the errored block will be displaced.

In this case, however, because we are dealing with inter-coded DFD blocks, there is a second issue. The DFD blocks will be used as the basis of temporal prediction at the decoder and, since the decoded version will contain spatial errors, these will propagate temporally to subsequent frames because of motion prediction. Because all blocks after the errored block are displaced and because the errored block occurs early in the frame, if there is no explicit resynchronization point, almost all the data in future predicted frames are likely to be corrupted.

11.3 Mitigating the effect of bitstream errors

11.3.1 Video is not the same as data!

This might seem like a strange subsection heading but, to a certain extent, it is true. If we consider, for example, the case of a numerical database, then in order to guarantee its effectiveness we would expect it to be transmitted perfectly. If we were told it contained errors (especially if we did not know where they were), we probably would not trust it and avoid using it. However, in the case of an image or video sequence, the situation is, in general, quite different.

On one hand we have seen that a single bit error can be highly destructive, due to error propagation. On the other hand we know that the human visual system is tolerant to certain types of artifact and can accept loss if it is handled correctly. In fact

FIGURE 11.8

Subjective appearance of loss. PVQ encoded Riverbed sequence at 30 Mb/s. Left: encoded at 30 Mb/s (intra). Right: reconstructed after 25% PER. Bottom: magnified artifact from box shown in top right image.

the compression process itself is nothing but a managed process of introducing noise into an image and we are normally quite happy with the resulting approximations of the original content. Consider, for example, the video frames shown in Figure 11.8. On the left is the original compressed image (at 30 Mb/s coded using Pyramid Vector Quantization (PVQ)—see Section 11.7.2) while the one on the right has been reconstructed in the presence of 25% *packet error rate* (PER) (i.e. 1/4 of all packets contain uncorrectable errors). Artifacts do exist, as shown in the bottom sub-figure, but are in this case well masked and tolerable.

The process of managing bitstream errors is not straightforward since they are unpredictable. We can however model the impact of loss on the bitstream and take steps to avoid it, correct it, or conceal it and these approaches are all addressed in the following sections.

11.3.2 Error-resilient solutions

We have seen that the more redundancy we remove from the source, the more fragile our resulting bitstream becomes. We thus need to trade some rate–distortion performance in a clean channel for improved error resilience in the presence of errors.

Let us first consider the case of transmission of an encoded image over a noisy channel with varying degrees of loss. Table 11.1 compares three cases: (i) a DCT-only

Table 11.1 FLC vs VLC encoding for various channel conditions. 256×256 Lena image at 2 bpp.

PSNR (dB)	DCT only	DCT + DC pred.	DCT + DC pred. + VLC
Error-free	33.7	33.9	38.5
0.1% BER	24.6	21.0	6.4

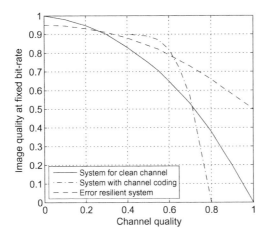

FIGURE 11.9
Idealized comparison of performance.

transform with fixed length encoding, (ii) a DCT transform-based system with DC prediction, and (iii) a DCT-based system with DC prediction and Huffman-based VLC. As can be seen, the VLC-based system provides excellent coding performance in a clean channel, yet it falls off badly as channel conditions deteriorate. On the other hand, the FLC scheme provides poor performance in a clean channel yet significantly outperforms the VLC case in lossy channels.

This trade-off between clean channel performance and error resilience is one of the challenges we face in designing our system. Ideally we want the best error resilience without compromising clean channel quality. In practice, however, it is necessary to strike a compromise and sacrifice some clean channel performance for improved error resilience in the presence of loss. These characteristics are illustrated in Figure 11.9 and methods for achieving this error resilience with graceful degradation are described in the following sections.

11.4 Transport layer solutions

In this section we use the term transport layer (a little loosely) to describe all aspects of the video transmission process that are not intimately associated with the application layer (i.e. the video codec). In many cases, the error control mechanisms will be implemented at the transport layer in the ISO 7 layer model. However, they may also be implemented in MAC and PHY layers. Many excellent references exist on this well-researched topic, for example Refs. [3–6].

11.4.1 Automatic Repeat request (ARQ)

ARQ is an interactive technique that is widely used in many communication systems. It relies on feedback from the decoder to determine whether a message (packet) has been correctly received. The detection of an error is normally performed through the addition of some form of *Cyclic Redundancy Check* (CRC) or through *Forward Error Correction* (FEC) coding. When an error is detected, the decoder returns a NACK (Negative ACKnowledge) signal which instructs the encoder to retransmit the data. There are a number of obvious advantages and problems with this method:

Advantages

- ARQ is simple and efficient. The redundancy introduced through retransmission is efficient as it is targeted specifically at lost data.
- It can be effective for coping with packet loss or large burst errors.

Problems

- A delay is introduced while the decoder waits for data to be retransmitted and, in real-time video applications, this may imply that frames have to be dropped (video frames have to be rendered at specific time intervals).
- ARQ requires a feedback channel which is not always available, for example when multicasting or broadcasting.

Delay-constrained retransmission

In practice, ARQ is a simple yet effective technique for ensuring reliable transmission of data but, in the context of real-time video, it has a usefulness lifetime, limited by the maximum tolerable latency between encoding and decoding. To address this problem, *delay-constrained retransmission* is normally used and this can be implemented at both receiver and transmitter. In the former case, the decoder will only request a retransmission of packet n if the following condition is satisfied:

$$T_c + T_{RT} + T_x < T_d(n) \tag{11.1}$$

where T_c is the current time, T_{RT} is the round trip delay for the network, $T_d(n)$ is the playout time for packet n, and T_x is a variable that accounts for other factors such as the tolerance on delay estimates and other uncertainties. Similarly, this decision could be made at the encoder if the following condition is satisfied:

$$T_c + T_{RT}/2 + T_x < T_d'(n) \tag{11.2}$$

where $T_d'(n)$ is in this case the estimated playout time for the packet.

11.4.2 FEC channel coding

In FEC methods, additional parity data is appended to the compressed signal, which allows the decoder to correct a certain number of errors. FEC can be combined with layered coding (or data partitioning) to provide unequal error protection, where different parts of the compressed bitstream are protected by different strength codes, according to importance.

The use of FEC increases the total data rate required for transmission and clearly offsets some of the benefits gained from source compression—redundancy can reach

100% for a 1/2 rate code. Furthermore, in general, FEC must be chosen with a particular worst case channel scenario in mind. For channels that have a highly variable quality, this worst case situation may imply the need for a very powerful FEC code which can severely reduce the compression performance. Furthermore, such a system will fail catastrophically whenever this worst case threshold is exceeded. One method that is frequently used to help mitigate the effects of burst errors is to use interleaving. However, this is sub-optimal and can introduce large delays.

Erasure codes

FEC can be very effective in cases of packet erasures as usually the positions of the lost packets are known. Efficient erasure codes include the Reed Solomon Erasure (RSE) correcting code. With an RSE(n, k) code, k data packets are used to construct r parity packets, where $r = n - k$, resulting in a total of n packets to be transmitted. The k source packets can be reconstructed from any k packets out of the n transmitted packets. This provides for error-free decoding for up to r lost packets out of n. The main considerations in choosing values of r and k are:

- **Encoding/decoding delay:** In the event of a packet loss, the decoder has to wait until at least k packets have been received before decoding can be performed. So, in order to minimize decoding delay, k must not be too large.
- **Robustness to burst losses:** A higher value of k means that, for the same amount of redundancy, the FEC will be able to correct a larger number of consecutive lost packets, or bursts. For example, an RSE$(4, 2)$ code and an RSE$(6, 4)$ code can both protect against a burst error length of 2, but the RSE$(6, 4)$ has only 50% overhead whereas RSE$(4, 2)$ has 100% overhead.

Cross packet FEC

If FEC is applied within a packet or appended to individual packets, in cases where packets are not just erroneous but are lost completely during transmission (e.g. over a congested internet connection), the correction capability of the code is lost. Instead, it is beneficial to apply FEC across a number of packets as shown in Figure 11.10. One

FIGURE 11.10

Cross packet FEC.

FIGURE 11.11

Performance of cross packet FEC for a coding depth of 8 (left) and 32 (right) for various coding rates.

problem with using FEC is that all the k data packets need to be of the same length, which can be an issue if GOB fragmentation is not allowed. The performance of cross packet FEC for the case of different coding depths (8 and 32) is shown in Figure 11.11. This clearly demonstrates the compromise between clean channel performance and error resilience for different coding rates.

Unequal error protection and data partitioning

We saw in Section 11.2 that not all types of video data are equally important. For example, header information is normally most important, as without it little else is possible. Motion information plays a particularly important role, especially because of the relatively small fraction of the total bit rate it occupies (approximately 10%). Faithful reconstruction of the encoded sequence depends heavily on the accuracy of the motion vectors. In order to reflect these priorities, the video data can be partitioned into separate data segments, each being allotted a protection level appropriate to its importance.

Prioritization of data may also be implemented in conjunction with layered coding as described in Section 11.9.

Rateless codes

Raptor codes, proposed by Shokrollahi [7], have gained significant traction as an effective means of channel coding for multimedia information. Raptor codes are a type of fountain code (first introduced by Luby [8] as LT codes) that combine LDPC and LT codes and offer the property of linear, $O(k)$, time encoding and decoding. Fountain codes encode k symbols into a (potentially) limitless sequence of encoding symbols and the probability that the message can be recovered increases with the number of symbols received. For example, in the case of Raptor codes, the chance of decoding success is 99% once k symbols have been received.

Raptor codes form part of the 3rd Generation Partnership Project (3GPP) for use in mobile cellular multicast. They offer a flexible and adaptive mechanism for dealing with variable wireless channel conditions.

11.4.3 Hybrid ARQ (HARQ)

HARQ [9] combines FEC coding with ARQ. It suffers from some of the basic limitations of ARQ, but offers some efficiency gains. In HARQ, the data is encoded with an appropriate FEC code, but the parity bits are not automatically sent with the data. Only when an error is detected at the decoder are these additional parity bits transmitted. If the strength of the FEC code is sufficient to correct the error, then no further action is taken. If however this is not the case, then the system reverts to a full ARQ retransmission. Typically a system will alternate between data and FEC packets during retransmission. HARQ is a compromise between conventional ARQ and conventional FEC. It operates as well as ARQ in a clean channel and as good as FEC in a lossy channel.

In practice, with multiple retransmissions of the same data, a receiver would store all transmissions and use these multiple copies in combination. This is often referred to as *Soft Combining*. This approach has been employed on up- and downlinks for high speed data in the UMTS 3G standard, and in the WIMAX (broadband wireless access—IEEE 802.16) and in 4G LTE.

11.4.4 Packetization strategies

The packetization strategy [10] can have an influence on performance in the presence of loss. The basic principles can be summarized as:

- Longer packets lead to improved throughput for clean channels (lower overhead).
- Shorter packets facilitate better error resilience as their loss will have less impact on the video sequence and error concealment will be easier.
- To support error resilience, video packet fragmentation should be avoided. Every NAL Unit (see Section 11.5.1) should thus be transmitted in one and only one UDP frame.

To illustrate the third bullet point, if a video packet containing encoded VLC symbols is split across two UDP packets, then the loss of one of these would render the other useless because of VLC dependencies.

11.5 Application layer solutions

11.5.1 Network abstraction

Standards since H.264/AVC have adopted the principle of a Network Abstraction Layer (NAL) that provides a common syntax, applicable to a wide range of networks.

A high level parameter set is also used to decouple the information relevant to more than one slice from the media bitstream. H.264/AVC, for example, describes both a Sequence Parameter Set (SPS) and a Picture Parameter Set (PPS). The SPS applies to a whole sequence and the PPS applies to a whole frame. These describe parameters such as frame size, coding modes, and slice structuring. Further details on the structure of standards such as H.264 are provided in Chapter 12.

11.5.2 The influence of frame type

I-frames, P-frames, and B-frames

The type of frame employed has a direct influence on the way errors propagate temporally. The main characteristics of I-frames, P-frames, and B-frames are listed below:

- **I-frames:** I-frames are not predicted with reference to any other frame. They therefore provide a barrier to temporal error propagation. They do however, in H.264/AVC and HEVC, exploit intra-prediction and hence can propagate errors spatially. Also I-frames form the basis for prediction P- and B- frames, so any spatial errors occurring in an I-frame can propagate temporally to subsequent frames.
- **P-frames:** P-frames are predicted from other P-frames or from I-frames. They will thus both propagate errors contained in their reference frames and introduce new spatial errors that can propagate spatially and temporally.
- **B-frames:** B-frames are normally not used as the basis for predicting other frames. They therefore do not propagate errors temporally and any spatial errors introduced in a B-frame are constrained to that frame.

Intra-refresh

We have seen that the Group of Pictures (GOP) structure for a video sequence defines the relationship between I-, P-, and B-frames. All encoded sequences will have at least one I-frame and most will have them inserted periodically, for example every 12 or 15 frames. Many codecs support intra-coding of macroblocks in P-frames. This provides a more flexible approach to rate–distortion optimization and helps to limit temporal error propagation. Macroblocks that are intra-coded in this way can be selected according to rate–distortion criteria, or according to their concealment difficulty, or randomly.

Some codecs such as X.264 take this further and support *Periodic Intra-Refresh* instead of key-frames. This effectively spreads the I-frame across several frames using a column of intra-coded blocks that progressively steps across the video frame. In this case, motion vectors are restricted in the same way as in slice structuring, so that blocks on one side of the refresh column do not reference blocks on the other side. This approach has the additional benefit that the peak-to-mean bit rate ratio is constrained (unlike the situation where large I-frames are used in isolation), offering better streaming and lower latency.

Reference picture selection

In simple motion-compensated hybrid codecs, the most recent previously coded frame is used as a reference for prediction of the current frame. More recent standards (from H.263+ onwards) support the use of multiple reference frames and reference picture selection (RPS) modes. These allow, with some constraints, the choice of any previously coded frame as a reference. This can be beneficial in limiting error propagation in error-prone environments.

In reference picture selection (RPS), both the encoder and the decoder store multiple reference frames. If, through feedback from the decoder, the encoder is instructed that the most recent reference frame has been received erroneously, then it will switch to another (error-free) reference picture for future predictions. The RPS selection mode was originally adopted in H.263+ as Annex N. This is shown in Figure 11.12 where an NACK(4) signal is fed back from the decoder and received during the encoding of frame 5. For frame 6, the encoder switches to use frame 3, which was received error-free, as the new reference frame.

Periodic reference frames

In terms of coding efficiency, it is normally more efficient to predict a picture from a frame several frames displaced from it, than to use intra-frame coding. Techniques such as Periodic RPS [11] exploit this by encoding every Nth frame in a sequence using the Nth previous frame as reference. All the other frames are coded as usual. This scheme is illustrated in Figure 11.13. Its advantage is that if any errors in a PR frame can be corrected through the use of ARQ or FEC before it is referenced by the next PR frame, then this will effectively limit the maximum temporal error propagation to the number of frames between PR frames.

Since PR frames require far fewer bits to code than intra-frames at the same quality, extra bits can be used to protect the PR frames, thus providing better performance under packet losses than intra-frame coding. Because this technique does not

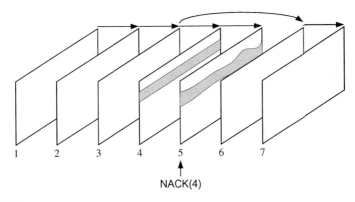

1   2   3   4   5   6   7

NACK(4)

FIGURE 11.12

Error resilience based on reference picture selection.

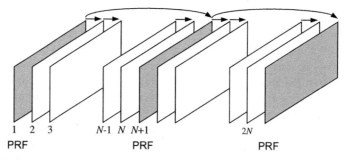

1 2 3 N-1 N N+1 $2N$
PRF   PRF   PRF

FIGURE 11.13

Periodic reference frames.

depend on feedback, it is applicable to broadcast and multicast scenarios. This type of approach is now common in coding standards, for example in H.264/AVC, referred to as *inter-prediction with subsequences*.

11.5.3 Synchronization codewords

Periodic resynchronization of a bitstream, and its associated symbols and data structures, can be achieved by the insertion of explicit synchronization codewords. For example, these might be inserted at the end of each row of blocks, slice, or frame. Resynchronization codewords are uniquely identifiable and are distinct from all other codes or concatenations of codes. They offer the following advantages and disadvantages:

Advantages

If an error is encountered, the maximum loss is limited by the distance between synchronization markers. Once a new synchronization marker is encountered, decoding can proceed again correctly.

Disadvantages

Resynchronization codewords incur a significant overhead, especially if used frequently. In the case of a JPEG encoded image, resynchronization markers at the end of each row typically incur a 1–2% overhead, whereas if they were located after each block, they would typically incur over 30% overhead.[1]

11.5.4 Reversible VLC

When an error is encountered in conventional VLC, there is a strong likelihood that the error will propagate spatially until the next resynchronization point. In the case of *Reversible VLC* (RVLC), backward decoding as well as forward decoding can be

---

[1]Overhead will varying according to compression ratio and content.

performed. Hence some of the data lost with conventional VLC can be faithfully decoded with RVLC. RVLC methods initially found favor in H.263+ and MPEG-4, and can be used effectively in combination with synchronization markers to provide enhanced error resilience. They were introduced by Takashima et al. [12], who showed how both symmetric and asymmetric codes can be generated.

RVLC codes [12, 13] should exhibit both prefix and suffix properties so that the codewords are decodable in both directions. The types of code that can offer this so-called biprefix property are limited and few are optimal. Codes such as Golomb–Rice codes or Exp-Golomb codes can be mapped into an equally efficient biprefix code, but in most cases the resulting biprefix code is less efficient. Girod [14] introduced an alternative method for generating bidirectionally decodable VLC bitstreams, based on modifications to prefix codewords. This elegant approach results in a coding efficiency equivalent to that of Huffman coding.

It is known that VLC data comprising DCT coefficients or differential motion vectors exhibit a *Generalized Gaussian Distribution*. As discussed in Chapter 7, these can be efficiently entropy coded using Exp-Golomb codes. These can be implemented without look-up tables and reversible versions are easy to design. Exp-Golomb codes form an important part of the H.264 and HEVC standards.

A simple example of RVLC is provided in Example 11.4.

Example 11.4 (Reversible VLC)

Consider the symbol–codeword mappings:

$$a \leftrightarrow 0; \quad b \leftrightarrow 11; \quad c \leftrightarrow 101 = \text{EOB}$$

The sequence $\{a, b, c, b, a, a, b, c, b, a\}$ is transmitted followed by a SYNC code-word. If the 9th and 11th bits are corrupted, show how this encoding can minimize errors through bidirectional decoding.

Solution.

Message	a	b	c	b	a	a	b	c	b	a	SYNC
Rx. bitstream	0	11	101	11	**1**	0	**01**	101	11	0	SYNC
Fwd. decode	a	b	c	b	–	–	X	X	X	X	SYNC
Rev. decode	X	X	X	X	X	X	–	c	b	a	SYNC

Note that the codewords are symmetrical, giving us a clue that they should be bidirectionally decodable. The table shows the forward and reverse decoding operations. Through bidirectional decoding we have successfully recovered most of the symbols, having lost only three symbols where the errors occur. If we assume that codeword c represents an EOB then we can see that the RVLC method recovers two of the three blocks of data transmitted.

11.5.5 Slice structuring

Most recent standards support the idea of decomposing a frame into slices prior to coding. A slice comprises a slice header followed by an integer number of macroblocks or coding units. Since all forms of prediction are constrained to within the slice boundary, slice structuring is effective in limiting error propagation. It therefore provides a simple basis for error-resilient coding. Some typical generic slice structures are shown in Figure 11.14.

Flexible macroblock ordering (FMO)

Flexible macroblock ordering (FMO) is one slice structuring approach that is particularly attractive for error resilience and for supporting concealment. It forms part of the H.264/AVC standard. The video frame is divided into several independently decodable slice groups (Figure 11.14) and these can assume checkerboard, scattered and interleaved patterns as shown. For example, in the checkerboard pattern, each frame can be represented as two slice groups with macroblocks interleaved as shown in the figure. It is easy to see that if a slice in one group is lost then the remaining one can be employed to provide a basis for the interpolation of the missing slice. Providing that only one of the two slices is lost, then FMO distributes errors across a larger region of the frame.

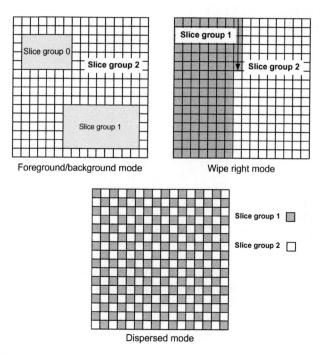

Foreground/background mode

Wipe right mode

Dispersed mode

FIGURE 11.14

Example slice structures.

FIGURE 11.15
Use of dispersed mode slice groups as a basis for error concealment.

A disadvantage of slice structuring is that coding gain is reduced. This is especially true in FMO, since neighboring macroblocks cannot be used as a basis for prediction. It has been reported that the efficiency drop or bit rate increase is approximately 5% at QP = 16 and up to 20% for QP = 28 [15].

An example of using dispersed mode slice groups is shown in Figure 11.15. Here, each slice corresponds to two rows of macroblocks and the figure shows the case where slices 2 and 3 are lost from one of the slice groups. It can be clearly seen that data from the remaining slice group is distributed in such a way as to facilitate interpolation of the lost data. This is the basis of the error concealment approach presented in Section 11.8.4.

Redundant slices

Redundant slices (RS) are a feature of standards such as H.264/AVC that supports the inclusion of redundant information for error resilience purposes. The redundant slice can be a simple duplicate of the original or it could be a lower fidelity copy. This copy can then be used in place of the original if it is lost. The question is—which slices should be included as redundancy? This has been addressed by a number of authors. Ferre et al. [16], for example, present a method whereby macroblocks are selected for inclusion on the basis of an end-to-end distortion model aimed at maximizing the reduction in distortion per redundant bit. This shows advantages over alternative methods such as cross packet FEC and Loss-adaptive Rate–Distortion Optimization.

11.5.6 Error tracking

This technique also relies on the existence of a feedback channel, not only to indicate that an error has occurred, but also to provide information on the location of the error (e.g. which slices or blocks have been corrupted). If an encoder knows which blocks have been received in error, it can predict how the errors will propagate to the currently encoded frame; it can then take one of two actions:

1. Encode the corresponding blocks in intra-mode.
2. Encode the corresponding blocks based on predictions only from blocks that have been correctly received.

This so-called *selective recovery* technique was proposed by Wada [17]. Steinbach et al. propose an efficient error tracking algorithm in Ref. [18].

11.6 Cross layer solutions

We have seen that compressed video (in the context of the human visual system) is tolerant to a certain level of errors, provided the associated artifacts are managed effectively. However, the aim of almost all networks is to deliver perfect data; anything else is deemed useless and is normally discarded by lower network layers without being sent to the application. Resource management and protection mechanisms that are traditionally implemented in the lower layers of the OSI stack (Physical (PHY), Medium Access Control (MAC), and transport) are, by intent, isolated from the application layer. We have seen however how the error characteristics of these lower layers are intimately coupled with their manifestations at the application layer.

There has thus recently been significant interest in performing optimization across the communication layers in order that the demands of the application can be reflected in the parameterization of the network layers and vice versa. This is referred to as *cross layer optimization*. Cross layer optimization differs from *joint source–channel coding* in that the aim of the latter is to optimize channel coding, dependent on the rate constraints of the channel. The former however takes a much more holistic approach, taking account of packetization strategies, retransmission strategies, delay constraints, and the loss tolerance of the data being encoded.

An excellent overview of the various aspects of cross layer optimization is given by van der Scharr et al. [4, 19].

11.6.1 Link adaptation

One form of cross layer optimization that is realizable in practice is link adaptation, based on end-to-end distortion metrics and modeling. Wireless local area networks (WLANs) such as IEEE 802.11a/g/n support numerous modulation and coding schemes (MCS), each providing different throughputs and reliability levels. Conventional algorithms for adapting MCS attempt to maximize the error-free throughput, based on RF signal measures such as residual signal strength indication (RSSI) and bit or packet error rates (BER or PER). They do not take account of the content of the data stream and they rely heavily on the use of ARQ and FEC to correct erroneous transmissions. In contrast, authors such as Ferre et al. [20] have presented ways of achieving this by minimizing the video distortion of the received sequence.

Ferre et al. use simple, local rate–distortion measures and end-to-end distortion models at the encoder to estimate the received video distortion at the current transmission rate, as well as at the adjacent lower and higher link speeds (MCS modes). This allows the system to select the mode which offers the lowest distortion, adapting to the channel conditions to provide the best video quality. Simulation results, based on H.264/AVC over IEEE 802.11g, show that the proposed system closely follows the optimum theoretic solution. Example results are provided in Figures 11.16–11.19. The first three figures illustrate the significant difference in switching characteristics

FIGURE 11.16

802.11g MCS switching characteristics based on throughput. (Reproduced with permission from Ref. [20]).

FIGURE 11.17

802.11g MCS switching characteristics based on video distortion. (Reproduced with permission from Ref. [20]).

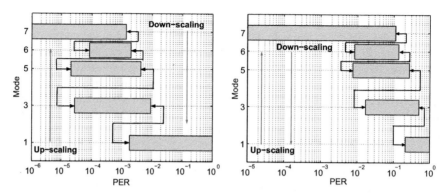

FIGURE 11.18

MCS switching characteristic comparison: video quality based (left) vs throughput based (right). (Reproduced with permission from Ref. [20]).

- —*— 250 kbps, BPSK 1/2 rate
- —⊖— 500 kbps, QPSK 1/2 rate
- —▽— 1000 kbps, 16QAM 1/2 rate
- —△— 1500 kbps, 16QAM 3/4 rate
- —◇— 2250 kbps, 64QAM 3/4 rate
- —□— Estimated
- —✳— Received

FIGURE 11.19

MCS link adaptation results. (Reproduced with permission from Ref. [20]).

between the cases of a throughput-based metric and a video quality-based metric. Figure 11.19 shows the result of the metric applied to the *Foreman* sequence plotted against channel carrier to noise ratio. It can be clearly seen that the switching characteristic closely tracks the optimum trajectory between MCS modes.

11.7 Inherently robust coding strategies

We have seen above that one of the contributing causes of error propagation in compressed video is variable length entropy coding. In this section we consider two methods that offer the benefits of fixed length coding (FLC) while preserving the compression performance offered by VLC. The first, *Error-Resilient Entropy Coding* (EREC), is a pseudo fixed length scheme where inherent synchronization is introduced through bitstream restructuring. The second, *Pyramid Vector Quantization* (PVQ), is a true fixed length scheme which exploits subband statistics to create a model that supports FLC.

11.7.1 Error-Resilient Entropy Coding (EREC)

Principle of operation

The Error-Resilient Entropy Code (EREC) [21,22] is a simple yet elegant approach to the coding of variable length blocks of data. It was introduced by Redmill and Kingsbury [21]. Based on a form of bin-packing, using a fixed length slotted structure, it enables the introduction of implicit synchronization points without the overhead of explicit synchronization codewords. The algorithm rearranges the variable length encoded blocks of data (for example, those resulting from block-based transformation and entropy coding) into a structure with dimensions that are predefined to ensure that it has sufficient capacity.

Assume that there are N variable length blocks, each of length $b_i : i = 1 \cdots N$ and we wish to pack these into M slots of length $s_j : j = 1 \cdots M$. The slotted EREC data structure should therefore satisfy the following conditions:

$$\begin{cases} N \geq M \\ T = \sum_{j=1}^{M} s_i \geq \sum_{i=1}^{N} b_i \end{cases} \tag{11.3}$$

where T is the total data size of the slotted structure. The first condition ensures that there are enough slots while the second ensures that the EREC structure is large enough to code all the variable length blocks of data. In most circumstances, for image or video coding, it is appropriate to set $N = M$ and we shall assume this from now on. We can thus compute the slot length for the structure as given in equation (11.4):

$$s_i = \left\lceil \frac{1}{N} \sum_{i=1}^{N} b_i \right\rceil \tag{11.4}$$

An N stage algorithm is then used to pack the blocks into the slotted structure. To enable a systematic encoding process that can be reversed at the decoder, an offset sequence is used. This defines the offset used at each stage in the algorithm. At stage n, the algorithm will search slot $k = i + \phi_n(\text{mod}(N))$, where ϕ is a pseudo-random offset sequence. In our examples we will simply use the sequence: $\phi = \{0, 1, 2, 3, 4, \ldots, N\}$. At each stage in the process, the EREC algorithm searches the slot at the appropriate offset to see if there is space to pack some or all of the excess

---

Algorithm 11.1 EREC encoding.

1. Define offset sequence ϕ_n;
2. INPUT N blocks of data;
3. Compute slot length, s_i, using equation (11.4);
4. Assign block b_i to slot s_i;
5. FOR $n = 1 \cdots N$
  6. Compute residual data for each slot: $r_i = b_i - s_i$;
  7. FOR $i = 1 \cdots N$
    8. Compute offset: $k = i + \phi_n (\mathrm{mod}(N))$;
    9. IF $r_i > 0$ AND $r_k < 0$ THEN pack slot k with residual from block i;
    IF $|r_k| \geq |r_i|$ THEN $r_k \leftarrow r_k + r_i$; $r_i \leftarrow 0$;
    IF $|r_k| < |r_i|$ THEN $r_i \leftarrow r_k + r_i$; $r_k \leftarrow 0$;
  10. END FOR;
  11. IF $\{r_i = 0 : i = 1 \cdots N\}$ THEN END;
12. END FOR.

---

data from overlong blocks—i.e. where $b_i > s_i$. This process is repeated until all excess data from overlong blocks is packed into the structure. The EREC encoding procedure is defined in Algorithm 11.1.

The decoding process is simply the inverse operation, progressively reconstructing, over N decoding stages, each original block until the block is fully decoded (e.g. an end-of-block code is detected). Thus, in the absence of channel errors, the decoder will correctly decode all the data.

A simple example of EREC encoding is shown in Figure 11.20 for six blocks of data. Referring to Figure 11.20: at stage 0, blocks 3, 4, and 6 are completely coded in slots 3, 4, and 6, with space left over. Blocks 1, 2, and 5 however are only partially coded and have data left to be placed in the space left in slots 3, 4, and 6. At stage 1, for an offset of $\phi = 1$, the remaining data from block 2 are coded in slot 3, and some data from block 5 are coded in slot 6. This process continues until, by the end of the sixth stage, all the data are coded. Further worked examples, illustrating both the encoding and decoding processes for a specific set of blocks, are given in Examples 11.5 and 11.6.

FIGURE 11.20

EREC operation.

Example 11.5 (EREC encoding)

A DCT/VLC-based encoder has generated a slice of data comprising eight blocks with the following lengths:

Block i	1	2	3	4	5	6	7	8
Length b_i	12	5	6	9	4	2	7	3

Assuming an offset sequence {0, 1, 2, 3, 4, 5, 6, 7}, show how the EREC algorithm would encode this slice of data.

Solution. The total length of the slice is 48 bits, hence the slot length is 6 bits. The encoding process proceeds as follows:

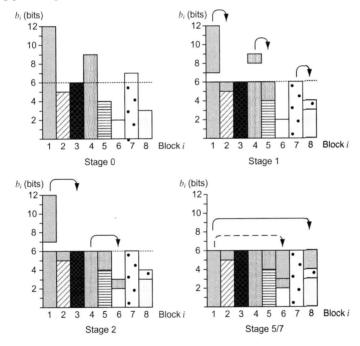

Stages 1 and 2 of the encoding process are shown above. Then, for stages 3 and 4, there is no spare space to fill in slots 4 and 5, hence no action is taken. At stage 5, 3 bits from block 1 are packed into slot 6 and, at stage 7, the remaining 2 bits from this block are packed into slot 8. These two operations are shown combined for convenience above.

Example 11.6 (EREC decoding)

Perform EREC decoding on the EREC-encoded blocks of image data from Example 11.5 above. Assume that each block is terminated with an EOB symbol.

Solution. For convenience, here we use a more compact, tabular structure to illustrate each stage of the decoding process, where **X** indicates an incomplete block and **C** indicates a completely decoded block.

Stage	Offset	bk1	bk2	bk3	bk4	bk5	bk6	bk7	bk8
0	0	X	C	C	X	C	C	X	C
1	1	X			X			C	
2	2	X			C				
3	3	X							
4	4	X							
5	5	X							
6	6	C							

At stage 0, we start decoding from the base of the first block and proceed until the end of the slot is reached or an EOB symbol is decoded. In the case of block 1, we reach the end of the slot with no EOB detected. We do the same for all other slots. Hence blocks 2, 3, 5, 6, and 8 are all fully decoded at stage 0. At stage 1 we search (for all incomplete blocks) for additional data at an offset of 1. For example, in the case of block 1 we search slot 2: we detect additional data beyond the EOB for block 2, but still no EOB is detected for block 1 so it remains incomplete. This proceeds until stage 6 in this case, when all blocks are fully decoded.

EREC performance

In a clean channel, the EREC decoder will correctly decode all the data with negligible overhead (typically only the value of T needs to be sent as side information). When channel errors occur, their effect depends on the particular distribution of codewords in the block and how rapidly symbol resynchronization is achieved. If symbol resynchronization is not regained before the end of the block, then this can cause the EOB symbols to be missed or falsely detected. This means that the decoder will incorrectly decode following information which was packed at later stages of the algorithm. This, in turn, implies that data placed in later stages (that toward the ends of long blocks) are more susceptible to error propagation than data placed in earlier stages. For typical compression methods, these propagation effects occur in high frequency data from active regions of the image. The important property of EREC is that, because the decoding of each block commences from a known location, block synchronization is preserved.

Examples of reconstructed images after EREC encoding for 0.1% BER and 1% BER are shown in Figure 11.21. Comparing these with their equivalents in Figure 11.5, the robust properties of EREC encoding can clearly be seen. Further improvements in subjective quality can be obtained if concealment is applied (see Section 11.8). An example of the relative performance of EREC vs a conventional sequential encoding is shown in Figure 11.22 showing up to 10 dB improvement in noisy channels.

Redmill and Bull [22] have shown that EREC can perform extremely well with arithmetic coding systems as well as with conventional VLC methods. Interestingly,

FIGURE 11.21

EREC performance examples of reconstructed DCT/Huffman encoded images after erroneous transmission. Left: 0.1% BER. Right: 1% BER. (Courtesy of D. Redmill.)

FIGURE 11.22

Graph of EREC performance for increasing BER compared to conventional DCT/VLC encoding. (Courtesy of D. Redmill.)

the paper demonstrates that, without EREC, Huffman coding is significantly more robust to errors than arithmetic coding. However, with EREC, significant improvement gains are manifest for both approaches and their performances are similar, with up to 15 dB improvement in PSNR.

EREC has been considered for adoption in standards such as MPEG-4, but has never been included in the final draft. Possible reasons for this are that it requires a number of blocks to be processed before they can be packed, although these delays can be managed in a slice structured environment. More likely, it is because the method does not fit well with conventional processing pipelines. Also, in the case of video, the use of predictive coding to remove inter-frame redundancy implies that an alternative approach to coding motion vectors must be employed. Swann and Kingsbury demonstrated the benefits of EREC for video coding in Ref. [23] showing significant benefits over conventional approaches in the context of MPEG-2.

11.7.2 Pyramid Vector Quantization (PVQ)

Principle of operation

Techniques such as EREC provide enhanced resilience through restructuring a bit-stream in such a way that pseudo fixed length coding is achieved. In contrast, Pyramid Vector Quantization (PVQ) prevents error propagation through the use of actual fixed length codewords. PVQ, introduced by Fischer [24], exploits the Laplacian-like probability distributions of transform coefficients to provide an efficient means of compressing them.

PVQ is a form of vector quantization optimized for i.i.d. Laplacian random variables. A vector $\mathbf{x}$ is formed by grouping L such random variables (e.g. transform coefficients). Based on the law of large numbers and the asymptotic equipartition property, it can be shown that almost all vectors will lie on a contour or region of constant probability. This region is a hyperpyramid in L-dimensional space with $2L$ vertices and $2L$ faces. Thus, for sufficiently large vector dimensions, there is negligible distortion if $\mathbf{x}$ is approximated by its nearest point on the hyperpyramid. This is the basis of the PVQ encoding algorithm.

The PVQ encoding algorithm consists of two steps. The first finds the closest vector $\hat{\mathbf{x}}$ to the input vector $\mathbf{x}$. The second step generates a bit code for $\hat{\mathbf{x}}$. Since all points on the hyperpyramid are equally probable, FLC is the optimum method for assigning bit codes. The PVQ-encoded bitstream thus comprises a sequence of codebook vector indices.

Chung-How and Bull [25] showed that PVQ is much more resilient to bit errors than either JPEG or H.263 and that wavelet-based PVQ outperformed DCT-based PVQ in terms of compression performance. In Ref. [26], Bokhari et al. showed that intra-mode PVQ is an effective codec for indoor wireless transmission of HD video. Intra-coding was employed since inter-frame encoding can lead to temporal error propagation. Although there is a loss in compression performance due to using an intra-only mode, this was justified by the increased error resilience.

Performance

Bokhari et al. [26] introduced an efficient rate–distortion (RD) algorithm for PVQ which yields impressive compression performance, comparable to intra-H.264/AVC and Motion JPEG 2000. They also evaluated the error resilience of the codec in the context of HD video using a realistic lossy wireless channel. Up to 11 dB PSNR improvement over H.264/AVC was demonstrated for lossy conditions. The basic compression performance of their approach is shown in Figure 11.23 for the Tennis sequence. It can be seen that intra-PVQ provides compression performance and video quality (43 dB PSNR @25 Mbps) comparable to other codecs in intra-mode. Importantly, it also obviates the need for sophisticated tools such as intra-prediction, variable size transforms, deblocking filters, or entropy coding (CABAC).

The key aspect of PVQ however is its error resilience. This was characterized and compared to other standardized codecs in Ref. [26] and the results of this are also shown in the right-hand graph of Figure 11.23. PVQ can be seen to offer up to 13.3 dB

FIGURE 11.23
RD optimized PVQ performance for Tennis. Left: RD performance in a clean channel.
Right: performance in a lossy 802.11n channel with correlated errors (Reproduced with
permission Ref. [26]).

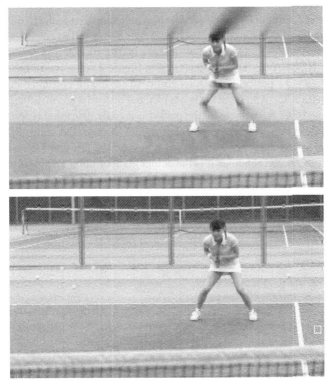

FIGURE 11.24
Frame shots for Tennis at 25 Mb/s and 25% BER. Top: H.264/AVC (HPS). Bottom: RD
optimized PVQ (Reproduced with permission from Ref. [26]).

FIGURE 11.25

Variability in frame distortions for PVQ and H.264/AVC (Reproduced with permission from Ref. [26]).

PSNR performance gain over the best performing H.264/AVC counterpart in terms of error resilience. Example PVQ encoded frames are compared with H.264/AVC (high profile) in Figure 11.24.

A major attribute of PVQ is the low variance in distortion across corrupted frames compared to H.264/AVC. This is illustrated in Figure 11.25. One of the factors that contributes to PVQ's performance is its ability to cope with corrupted data. In most wireless scenarios, the whole packet is not lost, but instead it has insufficient integrity to benefit from FEC. In such cases, the packet is usually discarded before being sent to the application layer. In PVQ and other robust formats, the codec benefits from the use of corrupted data. For VLC-based schemes, the limiting factor is the PER, whereas for FLC-based schemes it is the actual number of bit errors (in the received packet). This is defined as the residual bit error rate (RBER). This is demonstrated in Figure 11.26, which shows that, although the PER might be high, the corresponding

FIGURE 11.26

Packet error rate vs residual bit error rate for an 802.11n channel (Reproduced with permission from Ref. [26]).

RBER can be much lower. The figure shows sample packet error traces and their corresponding RBER traces. In bursty or correlated channels, the instantaneous RBER varies considerably. This leads to a temporally localized psychovisual impact. In an uncorrelated channel, the instantaneous RBER varies much less, giving a more consistent loss in quality throughout the sequence.

11.8 Error concealment

Error concealment is a decoder-oriented post-processing technique which attempts to conceal the effects of errors by providing a subjectively acceptable approximation of the original data. This exploits the spatial and temporal correlations present in a video sequence to estimate the lost information from previously received data. Lost data is reconstructed using spatial or temporal interpolation in a way that reduces the perception of spatio-temporal error propagation. Error concealment methods can benefit from complimentary error-resilient encoding so that errors are, as far as possible, localized and have local good data upon which estimates can be based.

Concealment methods can offer solutions where other techniques, particularly interactive methods which require a feedback channel, are not possible—for example, in broadcast or multicast systems. Concealment is a post-processing operation and is not mandated in any coding standard. Many algorithms exist and some of these are described below, classified as spatial, temporal, or hybrid methods. The reader is also referred to many excellent papers and reviews on the topic, including Refs. [3,5,6,27]. Reference [28] also contains details of the performance of specific methods based on motion field interpolation and multihypothesis motion estimation. Alternative methods that show some potential are those based on texture synthesis or inpainting. These are not covered further here, but the reader is referred for example to Ref. [30].

11.8.1 Detecting missing information

Firstly, before we can apply concealment methods, we need to understand where exactly the errors are located. This can be done in a number of ways and will depend to a degree on the implementation details. Some examples are:

- **Using header information:** For example, the sequence number in a packet header can be used to detect packet loss.
- **Using Forward Error Correction (FEC):** FEC can be used at the application layer as well as at the link layer.
- **Detection of syntax and semantic violations:** For example, illegal decoded codewords, invalid numbers of decoded units, out of range decoded parameters etc.
- **Detection of violations to the general characteristics of video signals:** For example, blocks with highly saturated colors (pink or green), blocks where most decoded pixels need clipping, strong discontinuities at borders of blocks etc.

None of these methods guarantee finding all errors and some can be subject to false detections. So, in practice, combinations are often employed. Let us now consider methods for concealing the errors once they have been detected.

11.8.2 Spatial error concealment (SEC)

Spatial error concealment methods exploit the spatial correlations that exist in most video signals. They operate under the assumption of spatial continuity and smoothness. The most common method for spatial interpolation is based on the weighted averaging technique proposed in Ref. [31]. Concealment is ideally based on data that has been correctly received; however, in areas of more extensive loss, a progressive approach can be used where previously concealed blocks or pixels are used as a basis for estimation.

Spatial interpolation, based on weighted averaging, is illustrated in Figure 11.27. After the detection of a lost block (indicated by the thicker lined boundary), each pixel, $s(x, y)$, is interpolated using the following expression:

$$s(x, y) = \frac{d_L s_R(x, y) + d_R s_L(x, y) + d_T s_B(x, y) + d_B s_T(x, y)}{d_L + d_R + d_T + d_B} \qquad (11.5)$$

where $s_R(x, y)$ etc. are the border pixels from the spatially adjacent macroblocks located to the right, left, top, and bottom of the missing block. An example of spatial error concealment applied to the *Foreman* sequence is shown in Figure 11.28. The result in this case has an MSE of 46.6 with respect to the original error-free frame.

There are a number of extensions to the basic SEC method that can provide additional benefit, but usually at the cost of additional complexity. For example, *edge continuity* can be enforced. This means that if an edge is detected in a neighboring block or blocks, then this edge should be preserved during concealment. Agrafiotis et al. [32] use an edge preserving method that not only preserves existing edges but also avoids introducing new strong ones. The approach switches between a directional interpolation and a conventional bilinear interpolation, based on the directional entropy of neighboring edges. The proposed strategy exploits the strengths of both methods without compromising the performance of either and offers performance

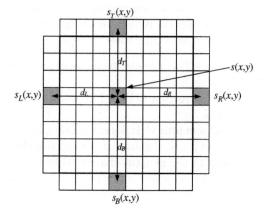

FIGURE 11.27

Spatial error concealment using nearest neighboring pixels.

FIGURE 11.28
Spatial error concealment applied to a lost row of macroblocks in the Foreman sequence.
Left: original with error shown. Middle: result of spatial error concealment. Right: ampli-
fied difference signal. MSE = 46.6. (Courtesy of D. Agrafiotis.)

improvements of over 1 dB for some cases. A similar approach was also presented
by Ma et al. [33].

Example 11.7 (Spatial error concealment)
Consider the missing block of data shown below with the correctly received boundary
pixels indicated. By performing SEC on this block, calculate the value of the missing
pixel shown encircled in the diagram.

$$
\begin{array}{ccccc}
 & 1 & 1 & 2 & 2 \\
0 & \text{X} & \text{X} & \text{X} & \text{X} & 2 \\
1 & \text{X} & \text{X} & \text{X} & \text{X} & 4 \\
3 & \text{X} & (\text{X}) & \text{X} & \text{X} & 7 \\
2 & \text{X} & \text{X} & \text{X} & \text{X} & 11 \\
 & 2 & 5 & 9 & 8 \\
\end{array}
$$

Solution. Using equation (11.5):
$$
s(x, y) = \frac{2 \times 7 + 3 \times 3 + 3 \times 5 + 2 \times 1}{10} = 4
$$
All other missing pixel values in the block can be interpolated in a similar manner.

11.8.3 Temporal error concealment

Temporal error concealment methods operate under the assumption of temporal continuity and smoothness of the motion field [28,29]. They exploit the highly correlated
nature of most natural sequences and conceal corrupted or lost pixels based on predictions from surrounding regions. They work well on inter-coded blocks, but are often
less successful on intra-coded blocks. Two basic techniques are considered below.

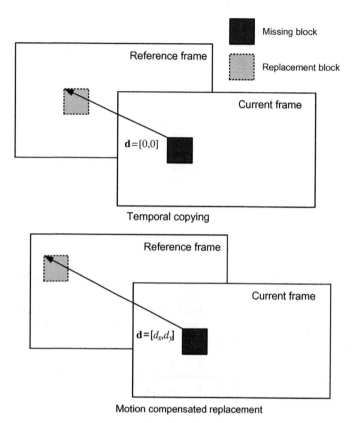

FIGURE 11.29

Temporal error concealment based on temporal copying (top) and motion compensated prediction (bottom).

Temporal copying (TEC_TC)

The simplest form of temporal error concealment is *temporal copying*. In this case, as shown in Figure 11.29, the underlying model is one of zero motion. This method is extremely easy to implement, but the underlying motion model is rarely sufficient in practice. Results for a single lost packet containing one slice of the Foreman sequence are shown in Figure 11.30. Although this result looks reasonable, the error is actually quite large due to the person's head moving. This displacement is particularly evident if you inspect the area on the left-hand edge of his face.

Motion compensated temporal replacement (TEC_MCTR)

As illustrated in Figure 11.29 (bottom), this method replaces the missing block with the best candidate region from the reference frame or frames. This is significantly more complex than the previous approach as it requires producing a list of candidate motion vectors that can be evaluated in order to choose the best replacement block. The method employs a two-stage process:

FIGURE 11.30

Temporal error concealment results for Foreman based on temporal copying. MSE = 19.86. (Courtesy of D. Agrafiotis).

1. Estimation of concealment displacement.
2. Displacement evaluation and replacement selection.

The first stage raises the question: How do we select the candidate motion vectors for TEC_MCTR? Assuming that the motion vector for the missing block is also lost, there are several likely candidates for replacement. These include:

- Motion vectors from spatially neighboring blocks.
- A rank-order statistic or average of the motion vectors from neighboring blocks.
- Motion vectors from temporally neighboring blocks.

We then need to evaluate these candidate motion vectors and select the one that produces the best match. So the second question is—What matching metric do we use to decide which of the candidate vectors is the best? This is normally done using a form of *Boundary Matching Error* (BME). Options for this metric are shown in Figure 11.31. The first method shown, referred to simply as BME, evaluates all candidate replacement blocks by computing the SAD value between the boundary pixels of the matched

FIGURE 11.31

Boundary Matching Error measures. Left: Boundary Matching Error (BME). Right: External Boundary Matching Error (EBME).

FIGURE 11.32

Temporal error concealment results for *Foreman* based on motion compensated replacement. MSE = 14.61. (Courtesy of D. Agrafiotis.)

block and the adjacent pixels from the surrounding blocks. The second method, referred to as an *External Boundary Matching Error* (EBME), again computes an SAD value but this time between the row(s) of pixels surrounding the missing block in the current frame and those surrounding the candidate block in the reference frame. EBME is a superior measure because it fits with the assumption that temporal concealment is preserving temporal smoothness. BME on the other hand is inconsistent with the model, as it uses a spatial continuity metric to assess temporal smoothness.

Temporal error concealment results for the *Foreman* sequence based on motion compensated replacement are shown in Figure 11.32. Because the underlying model is one of smooth translational motion, this technique generally performs significantly better than temporal copying. It can be observed from Figure 11.32 that the residual error is lower than in Figure 11.30 and that the artifacts at the side of the foreman's face are significantly reduced.

Example 11.8 (Temporal error concealment)

Consider the region of a current frame and the co-located region of an adjacent reference frame, as shown below. Assuming that the 2×2 block shown in gray has been lost during transmission, but all other data shown has been received correctly, compute the TEC_MCTR block using a BME metric.

Reference frame

8	8	8	9	7	9
8	7	7	6	7	6
6	5	8	6	4	4
7	4	7	4	7	8
5	6	8	6	7	7
4	6	7	4	4	8

Current frame

7	9	7	5	8	9
7	7	6	4	7	8
6	5	5	7	7	8
7	4	3	6	7	7
5	5	5	7	7	8
3	4	4	8	4	7

The motion vector values for adjacent blocks in the current frame are given as:

$$\mathbf{d}_L = [0, 0]$$
$$\mathbf{d}_R = [-1, 1]$$
$$\mathbf{d}_T = [-2, 2]$$
$$\mathbf{d}_B = [2, 0]$$

Solution. We can use these four motion vectors to select candidate blocks for replacement. In each case we compute the BME value as follows:

$$d = [0, 0] : BME = 19$$
$$d = [-1, 1] : BME = 11$$
$$d = [-2, 2] : BME = 7$$
$$d = [2, 0] : BME = 13$$

The best candidate block according to the BME metric is that associated with the $[-2, 2]$ motion vector. Hence the replacement pixels in this case are:

$$\begin{bmatrix} 5 & 6 \\ 4 & 6 \end{bmatrix}$$

11.8.4 Hybrid methods with mode selection

The choice of using either spatial or temporal concealment is not always straightforward. In most cases temporal error concealment provides the best results, and many simple systems use only a temporal copying approach, which works well if there is no motion! In practice, there are obvious situations when one method will be preferable over the other. For example, in the case of a shot cut or fade, spatial concealment is normally superior. In the case of a static scene, temporal block copying will produce good results and in the case of camera or object motion, it is most likely that motion compensated temporal replacement will produce the best results. However, if the scene motion does not fit well with the motion model used to find the candidate vectors then reverting to spatial concealment may be better.

We saw in Section 11.5.5 that slice structured modes can be beneficial in terms of error resilience. In particular, techniques such as *flexible macroblock ordering* (FMO) can help to ensure that, if a slice is lost, some good data is retained in an adjacent slice upon which an interpolation of the lost data can be based. This type of error-resilient encoding can provide significant benefits when combined with concealment. The question that arises when implementing a hybrid system is: How do we select the mode of operation—temporal or spatial? This can be done using a mode selection algorithm.

For example, Figure 11.33 shows the architecture for the EECMS algorithm as proposed by Agrafiotis et al. [27]. This is more formally defined in Algorithm 11.2. EECMS performs mode selection based on a comparison of the motion compensated

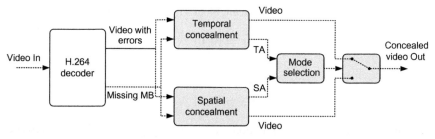

FIGURE 11.33

Enhanced error concealment with mode selection [27].

Algorithm 11.2 EECMS.

1. INPUT video with lost block;
2. REPEAT for each corrupted block:
   3. Perform TEC: Use EBME for matching;
   4. Compute spatial activity: $SA = E\left\{(s_c - \mu)^2\right\}$;
   5. Compute temporal activity: $TA = E\left\{(s_c - s_r)^2\right\}$;
   6. IF $(TA < SA)$ OR $(TA < \text{Threshold})$ THEN (use TEC, GOTO (8));
   7. Perform SEC; use directional interpolation;
8. UNTIL all missing blocks concealed;
9. END.

---

temporal activity and the *spatial activity* in the vicinity of the lost block. Spatial activity (SA) is computed as the variance of the surrounding MBs in the current frame whereas temporal activity (TA) measures motion uniformity and is based on the mean squared error between pixels (s_c) in the MBs surrounding the missing block in the current frame and those (s_r) surrounding the replacement MB in the reference frame. SEC is invoked only if SA is lower than SA, and SA is above a specified threshold. Selected results from the EECMS method are summarized in Table 11.2 and compared to the concealment algorithm used in the H.264/AVC joint model (JM). Across all sequences evaluated in Ref. [27], performance improvements up to 9 dB PSNR were reported.

Table 11.2 Selected performance results for EECMS (from Ref. [27]).

PSNR (dB)	0% PER	1% PER		20% PER	
		JM	EECMS	JM	EECMS
Foreman	40.14	38.11	39.44	27.33	31.10
Bus	37.20	35.08	36.33	23.79	26.99
Football	31.27	30.14	30.56	22.05	23.22

11.9 Congestion management and scalable video coding

In cases where streams are encoded on-line, it is possible to control the rate of the video to match that of the channel. However, in many cases, channel conditions can be highly variable and detailed knowledge (especially for multicast applications) may not be available. Also, with pre-encoded streams, it is not possible to adapt the rate unless multiple versions of the same content are encoded. A solution to this problem has been to represent the bitstream in multiple hierarchical layers or in terms of multiple independent descriptions. These approaches are briefly described below.

Scalable video encoding

It is common in video transmission for the signal to be encoded and transmitted without explicit knowledge of the downstream network conditions. In cases where network congestion exists in a packet switched network such as the internet, routers will discard packets that they are unable to forward due to congestion. In some cases it is possible for packets within a stream to be prioritized or embedded such that the least important information is discarded before the more important data. This mechanism lends itself to scalable or layered encoding, where a video signal is composed of a hierarchical layering in terms of spatial resolution, temporal resolution or SNR. This enables devices to transmit and receive multilayered video streams where a base level of quality can be improved through the use of optional additional layers that enhance resolution, frame rate, and/or quality.

The first standardized codec to include optional scalable modes was MPEG-2. However, these were rarely used in practice. More recently, H.264/AVC has introduced a comprehensive set of scalable modes in its Scalable Video Coding (SVC) Annex G extension of H.264/MPEG-4 AVC. SVC has many attractive features, not least that it is compatible with the conventional H.264/AVC bitstream. As we have already seen in Chapter 6, scalable methods have also found favor in still image coding standards such as JPEG2000. Scalable coding methods are now being adopted for some surveillance and video conferencing applications.

In scalable codecs, the aim is to create a hierarchical bitstream where a meaningful representation of the video can be reconstructed at the decoder even if some of the information is lost during transmission or if the terminal is incapable of decoding it. The basic architecture of a scalable video encoder and decoder pair is shown in Figure 11.34. The figure shows the case for spatial or temporal scalability. The other case—SNR scalability—is a subset of this where the up- and downsampling operators are omitted.[2]

Consider the case of spatial scalability as shown in Figures 11.34 and 11.35. The raw video input is first downsampled to the resolution of the base layer and is then transformed and quantized as usual. This layer is entropy coded and transmitted as the base layer. The base layer is then subject to inverse transformation and inverse

---

[2]For the case of SNR scalability it is normal for the encoder in Figure 11.34 to be modified so that only the quantization and inverse quantization operations are included in the upscaling path.

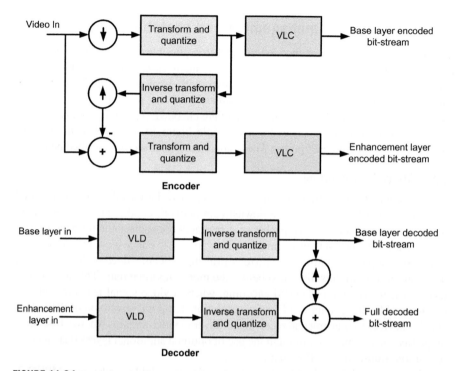

FIGURE 11.34

Scalable (spatial or temporal) video codec architecture.

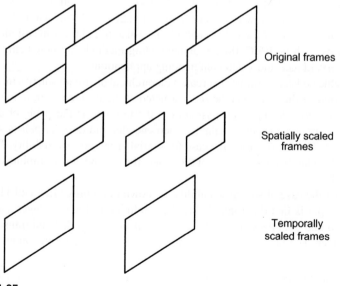

FIGURE 11.35

Frame types for scalable encoding.

quantization before being upsampled to its original resolution. The reconstructed base layer is then subtracted from the original video to form a residual signal which is encoded as the enhancement layer. Clearly the base layer must have higher priority than its enhancement layers, thus justifying the use of unequal error protection.

At the decoder, the base layer is decoded as usual but, if the enhancement layer is available, the base layer can be upsampled and added to the decoded enhancement layer to form a full quality enhanced output. This process is similar to that for temporal and SNR scalability. For further information the reader is referred to Refs. [4,34,35].

Multiple description coding (MDC)

Multiple description coding addresses the problem of congestion management in a different way to scalable encoding. Instead of creating a hierarchy of dependent information, where the decoding of each layer depends on the existence of the next most coarse layer, MDC creates a number of independent encodings (descriptions). These can be transmitted independently and may take different paths from source to destination. This path diversity is exploited and MDC ensures that, if any combination of descriptions are received correctly, then a meaningful reconstruction can be achieved at the decoder. If all descriptions are received then the full quality reconstruction is possible. MDC is finding application in methods such as live streaming and peer to peer distribution. This approach is shown diagrammatically in Figure 11.36.

Formation of efficient independent descriptions for video, where coding gain is predicated on the use of predictive encoding, is not straightforward. Many approaches have been proposed and a review of these and their performance is provided by Aposolopoulos et al. [36] and by Wang et al. [37]. For example, the use of FMO slice structuring methods, as described in Section 11.5.5, has been exploited by authors such as Wang et al. [38]. This approach modifies H.264/AVC to use three motion compensation loops and two slice groups to form two independent descriptions. The central encoder is identical to a single description H.264 encoder while the side encoders each process one of the slice groups.

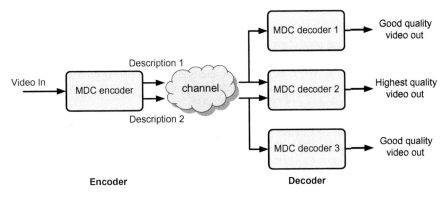

FIGURE 11.36

Multiple description codec architecture.

Tesanovic et. al. [39] extend the concept of path diversity to the case of multiple-input-multiple-output (MIMO) wireless technology based on spatial multiplexing. Their paper proposes the use of multiple description video coding (MDC) as a means of emulating the spatial diversity lacking in conventional spatially multiplexed (SM) systems. Singular value decomposition (SVD) is employed to create orthogonal subchannels which provide an efficient means of mapping video content to the wireless channels. Results indicate improvements in average PSNR of the decoded test sequences of around 5–7 dB, compared to standard, single-description video transmission. This is further enhanced by an additional 2–3 dB in the case of channels with low SNR values through the use of unequal power allocation.

11.10 Summary

We have examined in this chapter a number of methods that can be employed to improve the reliability of video delivery over lossy channels. We considered why errors propagate spatially and temporally under the influence of variable length and predictive coding regimes and saw examples which clearly demonstrate that the basic coding methods, considered earlier in this book, need to be adapted in a way that preserves rate–distortion performance while improving bitstream resilience to errors. We saw how network techniques such as FEC and ARQ can be exploited to provide better performance, but also how the source coding itself can be modified to deliver improved error resilience with less overhead. It is clear that methods such as PVQ provide greater tolerance to corrupted data while obviating the need for complex encoding tools. Finally we saw how error concealment, possibly in combination with other error resilience methods, can deliver significant improvements in subjective quality with little or no transmission overhead.

References

[1] C. Chong, C.-M. Tan, D. Laurenson, S. McLaughlin, M. Beach, A. Nix, A new statistical wideband spatio-temporal channel model for 5-GHz band WLAN systems, IEEE Journal on Selected Areas in Communications 21 (2) (2003) 139–150.

[2] G. Athanasiadou, A. Nix, J. McGeehan, A microcellular ray-tracing propagation model and evaluation of its narrow-band and wide-band predictions, IEEE Journal on Selected Areas in Communications 18 (3) (2000) 322–335.

[3] T. Stockhammer, M. Hannuksela, T. Wiegand, H.264/AVC in wireless environments, IEEE Transactions on Circuits and Systems for Video Technology 13 (7) (2003) 657–673.

[4] M. van der Schaar, P. Chou (Eds.), Multimedia over IP and Wireless Networks: Compression, Networking, and Systems, Academic Press, 2011.

[5] Y. Wang, S. Wenger, J. Wen, A. Katsaggelos, Review of error resilient coding techniques for real-time video communications, IEEE Signal Processing Magazine 17 (4) (2000) 61–82.

[6] Y. Wang, Q. Zhu, Error control and concealment for video communications: a review, in: Proceedings of the IEEE, 1998, pp. 974–997 (Special issue on Multimedia Signal Processing).

[7] A. Shokrollahi, Raptor codes, IEEE Transactions on Information Theory 52 (6) (2006) 2551–2567.

[8] M. Luby, LT-codes, in: Proceedings of 43rd Annual IEEE Symposium Foundations of Computer Science (FOCS), 2002, pp. 271–280.

[9] S. Lin, P. Yu, A hybrid ARQ scheme with parity retransmission for error control of satellite channels, IEEE Transactions on Communications 30 (I) (1982) 1701–1719.

[10] P. Ferre, A. Doufexi, J. Chung-How, A. Nix, Robust video transmission over wireless LANs, IEEE Transactions on Vehicular Technology 57 (4) (2008) 2596–2602.

[11] J. Chung-How, D. Bull, Loss resilient H.263 video over the internet, Signal Processing: Image Communication 16 (2001) 891–908.

[12] Y. Takashima, M. Wada, H. Murakami, Reversible variable length codes, IEEE Transactions on Communications 43 (1995) 158–162.

[13] A. Fraenkel, S. Klein, Bidirectional Huffman coding, Computer Journal 33 (4) (1990) 297–307.

[14] B. Girod, Bidirectionally decodable streams of prefix code-words, IEEE Communications Letters 3 (8) (1999) 245–247.

[15] JVT-B027.doc, Scattered Slices: A New Error Resilience Tool for H.26L, Joint Video Team of ISO/IEC MPEG and ITU-T VCEG, February 2002.

[16] P. Ferre, D. Agrafiotis, D. Bull, A video error resilience redundant slices algorithm and its performance relative to other fixed redundancy schemes, Signal Processing: Image Communication 25 (2010) 163–178.

[17] M. Wada, Selective recovery of video packet loss using error concealment, IEEE Journal on Selected Areas in Communications 7 (5) (1989) 807–814.

[18] E. Steinbach, N. Farber, B. Girod, Standard compatible extension of H.263 for robust video transmission in mobile environments, IEEE Transactions on Circuits and Systems for Video Technology 7 (6) (1997) 872–881.

[19] M. van der Scharr, S. Krishnamachari, S. Choi, X. Xu, Adaptive cross layer protection strategies for robust scalable multimedia video transmission, IEEE Journal on Selected Topics in Communications 21 (10) (2003) 1752–1763.

[20] P. Ferre, J. Chung-How, A. Nix, D. Bull, Distortion-based link adaptation for wireless video transmission, EURASIP Journal on Advances in Signal Processing (2008) 17, (Article ID 253706).

[21] D. Redmill, N. Kingsbury, The EREC: an error resilient technique for coding variable length blocks of data, IEEE Transactions on Image Processing 5 (4) (1996) 565–574.

[22] D. Redmill, D. Bull, Error resilient arithmetic coding of still images, in: IEEE International Conference on Image Processing, 1996, pp. 109–112.

[23] R. Swann, N. Kingsbury, Transcoding of MPEG-2 for enhanced resilience to transmission errors, in: IEEE International Conference on Image Processing, 1996, pp. 813–816.

[24] T. Fischer, A pyramid vector quantizer, IEEE Transactions on Information Theory 32 (4) (1986) 568–583.

[25] J. Chung-How, D. Bull, Robust image and video coding with pyramid vector quantisation, in: IEEE International Symposium on Circuits and Systems, vol. 4, 1999, pp. 332–335.

[26] S. Bokhari, A. Nix, D. Bull, Rate–distortion-optimized video transmission using Pyramid Vector Quantization, IEEE Transactions on Image Processing 21 (8) (2012) 3560–3572.

[27] D. Agrafiotis, D. Bull, C. Canagarajah, Enhanced error concealment with mode selection, IEEE Transactions on Circuits and Systems for Video Technology 16 (8) (2006) 960–973.

[28] M. Al-Mualla, C. Canagarajah, D. Bull, Video Coding for Mobile Communications, Academic Press, 2002.

[29] M. Al-Mualla, C. Canagarajah, D. Bull, Temporal error concealment using motion field interpolation, Electronics Letters 35 (3) (1999) 215–217.

[30] H. Lakshman, M. Koppel, P. Ndjiki-Nya, T. Wiegand, Image recovery using sparse reconstruction based texture refinement, in: IEEE International Conference on Acoustics Speech and Signal Processing (ICASSP), 2010, pp. 786–789.

[31] P. Salama, E. Shroff, E. Coyle, E. Delp, Error concealment techniques for encoded video streams, in: IEEE International Conference on Image Processing, vol. 1, 1995, pp. 9–12.

[32] D. Agrafiotis, D. Bull, C. Canagarajah, Spatial error concealment with edge related perceptual considerations, Signal Processing: Image Communication 21 (2006) 130–142.

[33] M. Ma, O. Au, G. Chan, M.-T. Sun, Edge-directed error concealment, IEEE Transactions on Circuits and Systems for Video Technology 20 (3) (2010) 382–395.

[34] Y. Wang, J. Ostermann, Y.-Q. Zhang, Video Processing and Communications, Prentice Hall, 2002.

[35] H. Schwarz, D. Marpe, T. Wiegand, Overview of the scalable video coding extension of the H.264/AVC standard, IEEE Transactions on Circuits and Systems for Video Technology 17 (9) (2007) 1103–1120.

[36] J. Aposolopoulos, M. Trott, W.-T. Tan, Path diversity for multimedia streaming, in: M. van der Schaar, P. Chou (Eds.), Multimedia over IP and Wireless Networks: Compression, Networking, and Systems, Academic Press, 2011.

[37] Y. Wang, A. Reibmann, S. Lin, Multiple description coding for video communications, Proceedings of the IEEE 93 (1) (2005) 57–70.

[38] D. Wang, N. Canagarajah, D. Bull, Slice group based multiple description video coding with three motion compensation loops, in: IEEE International Symposium on Circuits and Systems, vol. 2, 2005, pp. 960–963.

[39] M. Tesanovic, D. Bull, A. Doufexi, A. Nix, Enhanced MIMO wireless video communication using multiple-description coding, Signal Processing: Image Communication 23 (2008) 325–336.

Video Coding Standards

CHAPTER OUTLINE

12.1 **The need for and role of standards** **412**
    12.1.1 The focus of video standardization 412
    12.1.2 The standardization process 413
    12.1.3 Intellectual property and licensing 414
12.2 **H.120** **414**
    12.2.1 Brief history 414
    12.2.2 Primary features 415
12.3 **H.261** **415**
    12.3.1 Brief history 415
    12.3.2 Picture types and primary features 415
12.4 **MPEG-2/DVB** **417**
    12.4.1 Brief history 417
    12.4.2 Picture types and primary features 417
    12.4.3 MPEG-2 profiles and levels 418
12.5 **H.263** **419**
    12.5.1 Brief history 419
    12.5.2 Picture types and primary features 419
12.6 **MPEG-4** **423**
    12.6.1 Brief history 423
    12.6.2 Picture types and primary features 424
12.7 **H.264/AVC** **426**
    12.7.1 Brief history 426
    12.7.2 Primary features 426
    12.7.3 Network abstraction and bitstream syntax 427
    12.7.4 Pictures and partitions 429
    12.7.5 The video coding layer 430
    12.7.6 Profiles and levels 434
    12.7.7 Performance 435
    12.7.8 Scalable extensions 435
    12.7.9 Multiview extensions 435

Communicating Pictures. http://dx.doi.org/10.1016/B978-0-12-405906-1.00012-X

12.8 **H.265/HEVC** **436**
    12.8.1 Brief background 436
    12.8.2 Primary features 436
    12.8.3 Network abstraction and high level syntax 437
    12.8.4 Pictures and partitions 438
    12.8.5 The video coding layer (VCL) 440
    12.8.6 Profiles and levels 446
    12.8.7 Extensions 446
    12.8.8 Performance gains for HEVC over recent standards 446
12.9 **Other de-facto standards and proprietary codecs** **447**
    12.9.1 VP9 447
    12.9.2 VC-1 447
    12.9.3 RealVideo 447
    12.9.4 Dirac 447
12.10 **Summary** **447**
References **448**

The process of standardizing video formats and compression methods has been a major influence on the universal adoption of video technology. Standards are essential for interoperability, enabling material from different sources to be processed and transmitted over a wide range of networks, or stored on a wide range of devices. This interoperability provides the widest possible range of services for users. It also reduces risk for manufacturers, stimulates investment in research and development, and has created an enormous market for video equipment, with the advantages of volume manufacturing.

Video coding standards have, from the introduction of H.120 in 1984 through to the most recent H.265/HEVC codec, delivered a halving of bit rate for the equivalent video quality every 10 years. This chapter overviews the features that have enabled this progress. It is not intended to be a definitive guide to the structure and implementation of modern coding standards, but rather a description that enables the reader to understand how the architectures, approaches, and algorithms described in previous chapters are employed in the codecs that are in common use today.

12.1 The need for and role of standards

A chronology of video coding standards is represented in Figure 12.1 (repeated for convenience from Chapter 1). Their primary features are described in the following sections. Let us first, however, revisit why we need standards and consider the process that has led to a succession of successful coding standards over the last 30 years or so.

12.1.1 The focus of video standardization

In order for a video coding standard (or any other standard for that matter) to be successful it must clearly satisfy a demand. In particular: it must be superior in

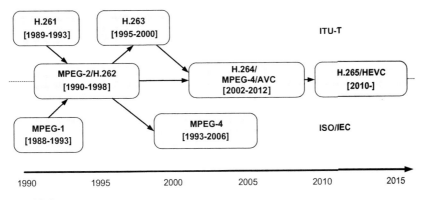

FIGURE 12.1

A chronology of video coding standards from 1990 to the present date.

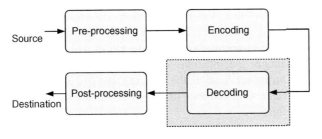

FIGURE 12.2

The scope of standardization.

performance to any previous standards, it must not stifle competition between producers of products (i.e. it must allow innovation) and, of course, it must provide interoperability through independence from specific communication networks or storage devices. In most cases it is advantageous if a standard is backward compatible—i.e. it should be able to decode bitstreams from prior standards. The standardization process must also be mindful that future standards will need to be backward compatible with it.

Video coding standards conventionally define the bitstream format and syntax and the decoding process, not (for the most part) the encoding process. This is illustrated in Figure 12.2 where the dashed box indicates the normative aspects of the standard. A standard-compliant encoder is thus one that produces a compliant bitstream and a standard-compliant decoder is one that can decode a compliant bitstream. In this context it is important to highlight the fact that the standard compliance of an encoder provides no guarantee of quality; the challenge remains for manufacturers to differentiate their products through innovative low complexity and efficient coding solutions.

12.1.2 The standardization process

Early video coding standards were produced independently either by the International Telecommunication Union (ITU) (previously CCITT) or by the International

Standards Organization (ISO), specifically the joint ISO/IEC (International Electrotechnical Commission) Technical Committee 1 (JTC 1) that addresses all computer-related activities including video compression. Most recent standards have however benefited from a joint approach between these two organizations.

The standardization process [1] follows a well defined path, with the following stages:

Requirements Definition: Here the scope of the standard is finalized and the requirements and goals of the standardization process defined.

Divergence: During this phase, competition is introduced, initially through the identification of Key Technical Areas (KTAs). A formal call for proposals is then normally issued, enabling experts from industry and academia to present and compare their methods and results, usually against a predefined set of test data.

Convergence: The aim of this phase is to create a solution for the standard by selecting the best algorithms contributed during the divergence phase. This is normally achieved through the use of an evolving test model such as the JM (Joint Model) in H.264/AVC or the HEVC Test Model (HM). This model evolves through a number of iterations through its use as a reference for subsequent proposals. These, if successful, are incorporated in a new improved version of the model. This process continues until the required performance specifications are met.

Verification: During verification, the resulting standard and its bitstream are validated for conformance. Conformance testing processes are then defined for compliance testing of products.

12.1.3 Intellectual property and licensing

When there is mass investment in producing products according to an international standard, the manufacturers and users rightly expect some degree of protection against patent infringement litigation. For most standardized codecs, vendors and users are required to pay royalties to the owners of associated intellectual property. This is, in the main, handled by a US organization (not affiliated with MPEG) called the MPEG Licensing Authority. MPEG LA administers patent licenses in connection with the patent pools for MPEG-2, MPEG-4, VC-1, and H.264/AVC. It is also pooling patents for licensing of HEVC.

Between 2005 and 2007, a dispute between Qualcomm and Broadcom, on the infringement of H.264 patents, came before the US Court. The associated patents were judged to be unenforceable as they had not been disclosed to MPEG JVT prior to the H.264 standardization in 2003. This type of ruling goes some way to provide confidence in the robustness of the licensing procedures adopted for video standardization.

12.2 H.120

12.2.1 Brief history

Study Group SG.XV of the CCITT commenced work on H.120 in 1980 and produced the first international digital video coding standard in 1984, followed by a second version in 1988 [2]. H.120 addressed videoconferencing applications at 2.048 Mb/s and 1.544 Mb/s for 625/50 and 525/60 TV systems respectively. This standard was never a commercial success, partially because it was based on different coding strategies for different international regions, but mainly because its picture quality (especially temporal quality) was inadequate.

12.2.2 Primary features

H.120 implemented a *conditional replenishment* strategy, whereby each frame was divided into changed and unchanged regions. The changed regions were coded with intra-field DPCM in parts 1 and 2 of the standard, although part 3 (for use in the USA) employed background prediction and motion-compensated inter-field prediction.

12.3 H.261

12.3.1 Brief history

H.261 [3] followed H.120 in 1989 and was the first video codec that achieved widespread product adoption. It was based on a $p \times 64$ kbps ($p = 1, \ldots, 30$) model, targeted at ISDN conferencing applications. H.261 was the first block-based hybrid compression algorithm to use a combination of transformation (the Discrete Cosine Transform (DCT)), temporal DPCM, and motion compensation. This architecture has stood the test of time as all major video coding standards since have been based on it.

12.3.2 Picture types and primary features

Macroblock, GOB, and frame format

H.261 introduced a basic hierarchy into the picture coding process that, with some modification, is still in use today. H.261 pictures, in CIF or QCIF format, are represented as a sequence of Groups of Blocks (GOBs), each comprising a number of macroblocks (MBs). A macroblock, as described in Chapter 4, comprises four spatial 8×8 DCT transformed luma blocks and 2 (sub-sampled) spatial 16×16 DCT-transformed chroma blocks. A diagram showing the picture hierarchy is given in Figure 12.3 for the case of a CIF frame. As can be seen, there are 33 MBs in a GOB, giving a GOB dimension of 176×48 pixels, and there are 12 GOBs in a CIF frame, giving a frame dimension of 352×288 pixels.

H.261 supports only I and P frames—B frames are not supported. Motion estimation and compensation are an integral part of the coding process with a fixed block size of 16×16 and a search window of $[-15 \cdots + 15]$ pixels.

Coder control

H.261 offered some flexibility in allowing switching between inter- and intra-frame modes and controlling quantizer step sizes at the macroblock level. Advanced features, such as a spatial loop filter, were incorporated for removing high frequency noise

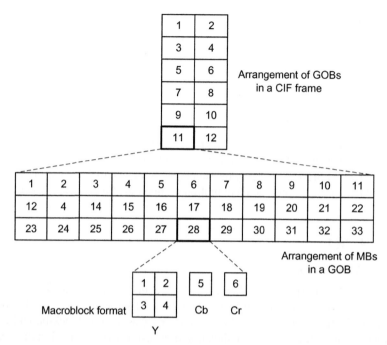

FIGURE 12.3

H.261 macroblock, GOB, and CIF frame format.

(a three-tap filter with coefficients of 0.25, 0.5, and 0.25). This allowed a reduction of prediction error by smoothing the pixels in the reference and output frames. For error resilience, an optional BCH error detection correction (511,493) scheme was also incorporated. The primary features of H.261 are summarized in Table 12.1.

Table 12.1 Primary features of H.261.	
Codec feature	**Approach**
Formats supported	CIF, QCIF
Picture format	Four layers (picture, GOB, MB, and block)
Color sub-sampling	4:2:0 YCbCr
Frame types	I, P
Intra-coding transform	Block DCT
Inter-coding transform	Block DCT
Entropy coding	VLC and Huffman coding
Quantizer	Uniform (DC) and deadzone (AC)
Motion compensation	Optional $[-15 \cdots +15]$
Coding control	Selection of inter/intra and quantizer step size
Loop filter	Three-tap spatial filter
Error protection	Optional BCH (511, 493) coding

12.4 **MPEG-2/DVB**

12.4.1 **Brief history**

In 1988, the Moving Picture Experts Group (MPEG) was founded, delivering, in 1992, a video coding algorithm (MPEG-1) intended for digital storage media at 1.5 Mbs/s. This was followed in 1994 by MPEG-2 [4], which specifically targeted the emerging digital video broadcasting market. MPEG-2 was instrumental, through its inclusion in all set-top boxes for more than a decade, in truly underpinning the digital broadcasting revolution.

Because of its focus on broadcasting, where there is an accepted imbalance between encoder and decoder complexity, this led to significant investment in high cost, high complexity encoding technology to produce high performance studio and head-end based encoders. An example of how the performance of MPEG-2 professional encoders improved over a decade of development is shown in Figure 12.4, together with the innovations that led to a series of step changes in that performance.

12.4.2 **Picture types and primary features**

MPEG-2 (ISO/IEC-13818:2000) [4], also known as H.262, is a generic audiovisual coding standard supporting a range of applications at bit rates from about 2 to 30 Mbps. The standard comprises four main parts: 13818-1: systems, -2: video, -3: audio, and -4: conformance. It employs a hybrid motion-compensated block-based DCT architecture (similar to H.261), using 8×8 blocks and 16×16 macroblocks

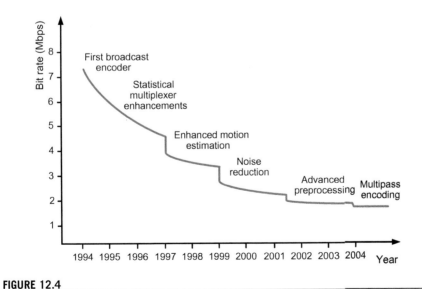

FIGURE 12.4

MPEG-2 encoder performance improvements between 1994 and 2004. (Adapted from an original presented by Tandberg.)

with translational block-based motion estimation to half-pixel accuracy. Because of its focus on digital TV broadcasting, MPEG-2 necessarily supports both progressive and interlaced picture formats. The coding standard was defined primarily for error-free environments and channel coding was added according to the application scenario (e.g. terrestrial, cable, or satellite broadcasting). MPEG-2 was defined to support a broad range of applications from studio processing (capture and editing) to distribution and broadcast delivery. It is intimately coupled with the DVB-T, DVB-S, and DVB-C broadcast content delivery standards.

MPEG-2 supports three picture types: Intra (I)—coded without reference to other frames (least efficient), Predicted (P)—coded based on prediction from previous I or P frame, and Bidirectionally predicted (B)—predicted from P and or I frames but not used as a basis for further predictions (most efficient). It utilizes a GOP structure as illustrated in Figure 12.5 and was the first standard to introduce bidirectionally predicted B frames.

12.4.3 **MPEG-2 profiles and levels**

To match performance against decoder capability or capacity, MPEG-2 introduced an extensive range of *profiles* and *levels*. A profile is a defined subset of the entire bitstream syntax and profiles are further partitioned into levels. Each level specifies a range of allowable values for the parameters in the bitstream.

MPEG-2 supports six profiles—Simple, Main, SNR, Spatial, High 4:2:2, and Multiview. Provisions for scalability are included in the SNR, Spatial, and High profiles whereas the Simple Profile and Main Profile allow only single-layer coding. It also offers four possible levels—Low, Main, High1440, and High—in each profile. The parameters for the Main Level approximately correspond to normal TV resolution, the Low Level corresponds to CIF resolution, and the values for

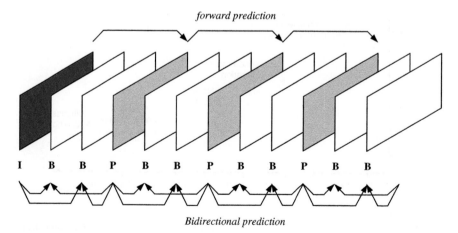

FIGURE 12.5

MPEG-2 GOP structure.

Table 12.2 MPEG-2 main profile@main level (MP@ML) bounds.

Parameter	Bound
Samples/line	720
Lines/frame	576
Frames/sec	30
Samples/sec	10,368,000
Bit rate	15 Mb/s
Buffer size	1,835,008 bits
Chroma format	4:2:0
Aspect ratio	4:3, 16:9 square pixels

High1440 and High correspond to HDTV resolution. Only two profiles were ever used in practice; the 4:2:2 Profile for studio work and post-production and the Main Profile for broadcast TV delivery. Table 12.2 provides the specification for the MPEG-2 Main Profile at Main Level (MP@ML) as used in most delivery applications.

12.5 H.263

12.5.1 Brief history

H.263 [5] was defined by ITU-T SG15, starting in 1993 with the goal of coding at bit rates below 64 kbps. The initial application focus was on PSTN and early mobile radio applications at bit rates between 10 and 24 kb/s. Despite mobile video being slower to take off than expected, H.263 nonetheless had a significant impact in conferencing and surveillance applications, as well as in early internet streaming. In particular, H.263 was used to encode Flash Video content for sites such as YouTube and MySpace. RealVideo was initially based on it, and it was specified in several ETSI 3GPP video services. The codec was first designed to be integrated in the H.324 framework for circuit-switched applications, but has been extensively used in H.323 (RTP-based video conferencing) and other IP streaming wrappers.

12.5.2 Picture types and primary features

Macroblock, GOB, and frame format

H.263 supports the following picture formats: sub-QCIF (88×72), QCIF (176×144), CIF (352×288), 4CIF (704×576), and 16CIF (1408×1152) all in YCbCr 4:2:0 format. As with H.261, each picture is divided into a number of groups of blocks (GOBs) defined as an integer number, k, of rows of macroblocks (MBs). $k = 1$ for SQCIF, QCIF, and CIF, $k = 2$ for 4CIF, and $k = 4$ for 16CIF. The structure of an H.263 QCIF picture is shown in Figure 12.6.

Primary features of H.263

Despite using a similar coding architecture to H.261 and MPEG-2, H.263 achieved significantly enhanced performance through the incorporation of a comprehensive

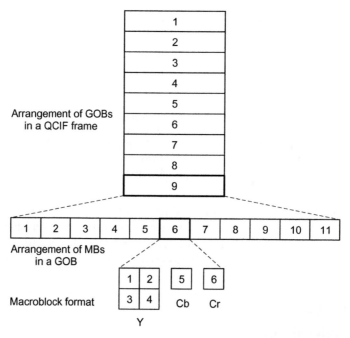

FIGURE 12.6

H.263 picture format (QCIF).

Table 12.3 Primary features of H.263.

Codec feature	Approach
Formats supported	SQCIF, QCIF, CIF, 4CIF, 16CIF
Picture format	Four layers (picture, GOB, MB, and block)
Color sub-sampling	4:2:0 YCbCr
Frame types	I, P, B, (PB)
Intra-coding transform	Block 8×8 DCT
Inter-coding transform	Block 8×8 DCT
Quantizer	Uniform (DC) and deadzone (AC) QP $= 1 \cdots 31$
Motion compensation	Half-pixel, unrestricted vectors, overlapped.
Entropy coding	3-D VLC, Huffman, and arithmetic coding
Loop filter	None
Coding control	Inter/intra, quantizer step size, and transmit/Skip at MB level
Error protection	Provided by transport layer or BCH-FEC

set of advanced features. These are summarized in Table 12.3. The standard is based on a hybrid motion-compensated DCT, using zig-zag scanning with quantization step sizes (QP $= 1 \cdots 31$) changeable at each MB. 3-D VLC tables are used in H.263, incorporating the events {RUN, LEVEL, LAST} where LAST signals the last

Normal mode   Advanced mode

— — — Picture or GOB border **MV** = Current motion vector **MVP** = Median (**MV1, MV2, MV3**)

FIGURE 12.7

Motion vector coding in H.263.

non-zero coefficient in the block. This avoids the need for an explicit EOB symbol and provides a compact representation of the 8×8 DCT block. H.263 uses motion compensation with half-pixel precision and motion vectors are coded predictively as shown in Figure 12.7.

In terms of PSNR, H.263 can provide up to 3–4 dB improvement over H.261. It does this because of its half-pixel motion prediction but also by incorporating four new optional coding modes:

- **Unrestricted motion vectors:** Motion vectors are allowed to point outside of the reference picture area and the search range is extended to $[-31.5, +31.5]$.
- **Syntax-based arithmetic coding:** The conventional Huffman entropy coder is replaced with an arithmetic coder, enabling fractional wordlengths per encoded symbol.
- **Advanced prediction mode:** This includes two submodes:
  - The use of four MVs per block, i.e. one MV for each 8×8 block rather than one per 16×16 block. This enables the encoder to better deal with multiple motions within a block.
  - Overlapped Motion Compensation (OMC): Here, each pixel in an 8×8 luma block is predicted as a weighted sum of three prediction values. These prediction values correspond to the vector of the current block and two of four predictions from the blocks to the top, bottom, left, and right of the current block. The two blocks selected are those closest to the pixel being computed— e.g. a pixel in the top left quadrant of the block uses the predictions from above and left. This is illustrated in Figure 12.7 (right).

- **PB-frames mode:** In a PB-frame, two pictures are coded as a single unit—a P-frame predicted from the previous P-frame and a B-frame, predicted from both adjacent P-frames using bidirectional prediction. A PB-frame macroblock comprises 12 blocks—six for the predicted P-block and six for the B-block. The benefit of a PB-frame is that motion vectors are not transmitted for the B-blocks, but are instead derived using a scaled version of that for the P-block based on the local temporal activity relative to the corresponding P-block in the previous P-frame. This process enables an effective increase in frame rate without a significant increase in bit rate.

H.263 extensions (H.263+ and H.263++)

H.263 Version 2, also known as H.263+, was standardized in 1998 and extends H.263 with many new modes and features that further improve compression efficiency. H.263++ (Version 3) provided still further enhancements in 2000. The 12 new modes in H.263+ improve coding gain, improve error resilience, enable scalable bit streams, introduce flexibility in picture size and clock frequency, and provide supplemental display capabilities. Many of these new modes were pulled through into the later H.264/AVC standard.

Firstly, several new modes to support error resilience were introduced and these are summarized below. We have covered many of these already in Chapter 11 and they are also reviewed by Wenger et al. [6].

- **Slice Structured Mode (Annex K):** In this mode, the GOB structure is replaced by a set of slices. All macroblocks in one slice can be decoded independently since prediction dependencies are not permitted across slice boundaries.
- **Independent Segment Decoding Mode (Annex R):** A segment boundary acts in the same way as a picture boundary. A segment can be a slice, a GOB, or a collection of GOBs and the shape of a segment must be identical from frame to frame. Independent segments support error resilience as error propagation is eliminated between defined parts of the picture. They also enable special effects in a similar manner to the object planes in MPEG-4.
- **Reference Picture Selection Mode (Annex N):** As described in Chapter 11, this allows flexibility in the choice of reference picture. It is also possible to apply the reference picture selection mode to individual segments rather than to full pictures. This mode supports error resilience if a feedback channel exists, but is also a precursor to the multiple reference frame methods used in H.264 and HEVC.
- **Temporal, SNR, and Spatial Scalability Mode (Annex O):** This mode introduced layering into the H.263 standard to support flexibility in delivery to terminals of different capabilities and in congestion management. As discussed in Chapter 11, a scalable bitstream comprises a base layer and enhancement layers, where the base layer provides an acceptable level of quality which can be further

enhanced by the other layers (in terms of improved signal to noise ratio, improved temporal resolution or improved spatial resolution) if they are available.

Other H.263 modes that support enhanced coding gain are:

- Advanced Intra-coding Mode (Annex I).
- Modified Unrestricted Motion Vectors Mode (Annex D).
- Improved PB-Frames Mode (Annex M).
- Deblocking Filter Mode (Annex J).
- Reference Picture Resampling Mode (Annex P).
- Reduced Resolution Update Mode (Annex Q).
- Alternative Inter VLC Mode (Annex S).
- Modified Quantization Mode (Annex T).

Finally one mode was introduced to provide additional bitstream information:

- **Supplemental Enhancement Information Mode (Annex L):** This enables the incorporation of additional information that may or may not be decodable by a specific decoder. It can contain information such as chroma keying, time segments, picture freezing, and picture resizing.

H.263++ provided further enhancements to H.263 in terms of introducing a range of nine profiles and seven levels similar to those in MPEG-2. The profiles enabled formalization of specific combinations of modes to support different application scenarios (e.g. Baseline and Wireless). The levels enabled matching of coder performance to environmental constraints (e.g. maximum bit rate, spatial resolution, etc.). It also introduced some new modes and revised others—a Data Partitioned Slice Mode (Annex V), Additional Supplemental Enhancement Information (Annex W), and an Enhanced Reference Picture Selection Mode (Annex U).

12.6 MPEG-4

12.6.1 Brief history

MPEG-4 [7] was a very ambitious project that sought to introduce new approaches using object-based as well as waveform-based methods. The MPEG-4 standard was ratified in 2000, as ISO/IEC 14496, potentially offering significant encoding improvements over and above MPEG-2. It was however, in general, found to be too complex. It is interesting to compare the standards documents for H.261 and MPEG-4. The former was 25 pages long while MPEG-4 comprised 500 pages for the visual part alone. Only the Advanced Simple Profile (ASP) was used in practice, and this gained some traction forming the basis for the emerging digital camera technology of the time. The early MPEG-4 part 2 offered little practical advantage over MPEG-

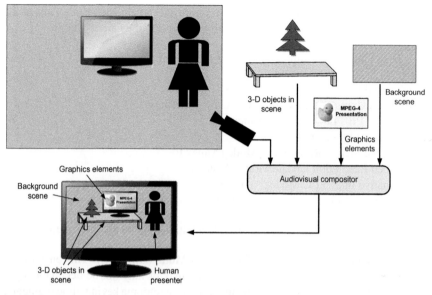

FIGURE 12.8

An example of an MPEG-4 audiovisual scene.

2 or H.263 and was rapidly overtaken by MPEG-4 part 10/AVC, now known as H.264/AVC.

12.6.2 Picture types and primary features

MPEG-4 was aimed at encoding video at lower bit rates and higher video qualities than MPEG-2, effectively creating a multimedia transmission and storage framework. MPEG-4 (part 2) was designed to be suitable for a wide range of video encoding scenarios ranging from the studio and movies to video applications on mobile phones. In this context it focused on object-based as well as waveform-based coding.

MPEG-4 provided a set of coding tools for audiovisual scenes, supporting the coding of arbitrarily shaped objects as shown in Figure 12.8. The scene in this figure represents a 2-D background, a video playing on the screen, a presenter, and 3-D objects such as the table and tree. MPEG-4 was aimed at the independent processing of such audiovisual objects as well as their compositing and allowed modification of both natural and synthetic (computer generated) content.

As with MPEG-2 and H.263++, MPEG-4 initially defined a number of conformance points using a Simple Profile, a Core Profile, and a Main Profile. The Simple and Core Profiles address scene sizes of QCIF and CIF at 64, 128, 384 kbps, and 2 Mbps. The Main Profile is targeted at CIF, ITU-R 601, and HD sizes, with bit rates at 2, 15, and 38.4 Mbps. The standard also incorporated support for sprites, scalability, and error resilience.

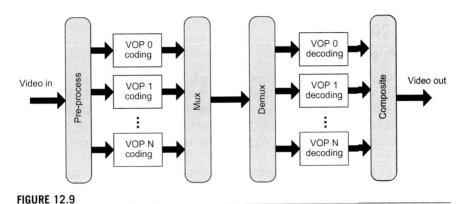

FIGURE 12.9

Generic representation of MPEG-4 Video Object Plane (VOP) coding.

MPEG-4 also provided, for the first time, support for integration with the MPEG-7 content description scheme to facilitate the use of metadata. An excellent overview of MPEG-4 natural video coding is provided by Ebrahimi and Horne [8].

Coding framework

Although we will not devote a lot of time to describing the details of MPEG-4, it is worth briefly putting its aims into context.

MPEG-4 supported the composition or decomposition of video content into visual objects, allowing these to be processed independently as a number of Video Object Planes (VOPs) (Figure 12.9). In order to describe these object planes it was necessary to code their textures, shapes, and motions. Shape coding is done using a binary mask, or a gray-scale alpha channel, allowing transparency. Both motion compensation and DCT-based texture coding have to take account of object boundaries. Error resilience is provided by resynchronization markers, data partitioning, header extension codes, and reversible variable length codes. Scalability is provided for both spatial and temporal resolution enhancement. MPEG-4 also provided scalability on an object basis, with the restriction that the object shape has to be rectangular.

MPEG-4 part 2 Advanced Simple Profile (ASP)

The Advanced Simple Profile was the most commonly used MPEG-4 profile, offering support for interlaced video, B-pictures, quarter-pixel motion compensation, and global motion compensation. The quarter-pixel accuracy was subsequently adopted in H.264/AVC whereas the global motion compensation feature was not generally supported in most implementations. As the reader will notice, ASP does not offer a lot that is not present in H.263; in fact the uptake of H.263 was far more widespread than that of ASP.

Soon after the introduction of ASP, MPEG-4 visual work refocused on part 10 and this became a joint activity with ITU-T under the H.264/AVC banner. This major step forward is the topic of the next section.

12.7 H.264/AVC

12.7.1 Brief history

ITU-T/SG16/Q6 (Video Coding Experts Group) commenced work in 1998 on a project called H.26L. The JVT (Joint Video Team) between VCEG and MPEG (Moving Picture Experts Group) was formed in 2001 to establish a joint standard project known as H.264/MPEG-4-AVC. The final draft for formal approval was produced in March 2003 [9], with the scalable video coding (SVC) extension finalized in 2007.

MPEG-4 (part 10), or H.264/AVC, is by far the most ubiquitous video coding standard to date. It will remain so, probably until 2015 when volume production and infrastructure changes enable a major shift to H.265/HEVC. In the same way that MPEG-2 underpinned the revolution in digital broadcasting, H.264/AVC has played a key role in enabling internet video, mobile services, OTT services, IPTV, and HDTV. H.264/AVC is a mandatory format for Blu-ray players and is used by most internet streaming sites including Vimeo, YouTube, and iTunes. It is used in Adobe Flash Player and Microsoft Silverlight and it has also been adopted for HDTV cable, satellite, and terrestrial broadcasting.

ITU-T, in partnership with ISO/IEC, delivered H.264/AVC in 2004. We have already examined many of the features that contribute to the success and performance of H.264 in previous chapters and we summarize these below.

12.7.2 Primary features

The aim of the H.264/AVC project was to provide:

- **Improved coding efficiency:** A targeted average bit rate reduction of 50% given fixed fidelity, compared to any other video standard.
- **Error robustness:** Tools to deal with packet loss and congestion management.
- **Network friendliness:** Major targets were mobile networks and the internet— separation of coding layer through network abstraction.
- **Adaptation to delay constraints:** To provide low delay modes.
- **Simple syntax specification:** Avoiding an excessive quantity of optional features or profile configurations.

These are all sensible objectives, but the last one is interesting in as much as it acknowledges the over-complex nature of its predecessor.

Let us now consider some of these attributes in more detail. A summary is provided in Table 12.4; although not all the details of a complex codec such as H.264 can be captured in a table of this type, it can be seen that many of the features are similar to those in the extensions to H.263. For further details, the reader is referred to the standards documents [9], or overviews such as that by Wiegand et al. [10]. A very accessible introduction to the standard is provided by Richardson [11].

Table 12.4 Primary features of H.264/AVC.

Codec feature	Approach
Formats supported	QCIF to UHDTV (4k)
Picture format	Four layers (picture, GOB, MB and block)
Color sub-sampling	4:0:0, 4:2:0, 4:2:2, 4:4:4 YCbCr
Partitions	Flexible slice structuring
Frame types	I, P, B, SP, SI
Intra-coding transform	Block 8×8 or 4×4 integer transform
Inter-coding transform	Block 8×8 or 4×4 integer transform
Intra-prediction	Multi-direction, multi-pattern
Quantizer	Scalar with logarithmic control, $QP = 0 \cdots 51$
Motion compensation	Quarter pixel, unrestricted vectors, multiple reference frames
ME block size	$16 \times 16, 16 \times 8, 8 \times 16, 8 \times 8, 8 \times 4, 4 \times 8, 4 \times 4$
Entropy coding	CAVLC or CABAC
Coding control	Many sophisticated coding control modes—RDO
Loop filter	In-loop non-linear deblocking filter
Error resilience	Slices, FMO, SI, SP frames, multiple reference frames

12.7.3 Network abstraction and bitstream syntax

Figure 12.10 shows the layered structure of H.264/AVC coding. This ensures that the coding layers are independent of the network, relying on protocols as shown to provide mapping to the appropriate video transport mechanism. VCL data can thus easily be mapped to transport layers such as RTP/IP for real-time wired or wireless

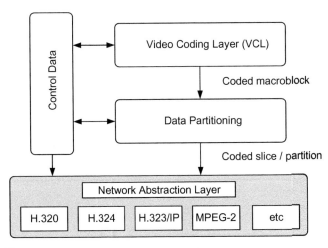

FIGURE 12.10

H.264/AVC layer structure.

internet, MP4 for storage and MMS, H.32X for wireline and wireless conversational services or MPEG-2 systems for broadcasting.

In H.264/AVC, the bitstream is organized hierarchically, comprising the Network Abstraction Layer (NAL) and the Video Coding Layer (VCL). When coded, H.264 information is represented as a series of NAL Units (NALUs). The NALU header indicates the type of NALU which may be a Sequence Parameter Set (SPS), a Picture Parameter Set (PPS), VCL coded slice data, or Supplemental Enhancement Information (SEI). The SPS contains information about the whole sequence such as Profile, Level, frame size, and other characteristics important to the decoder. A PPS contains more localized information, relevant to a subset of frames, such as the number of slice groups, entropy coding mode, whether weighted prediction is used, and other initialization parameters. SEI relates to information that is not essential for decoding but may assist in, for example, buffer management or other non-normative tasks. A diagram showing the hierarchical H.264 syntax is shown in Figure 12.11.

Each sequence starts with an instantaneous decoder refresh (IDR) access unit. This is an intra-coded picture which indicates to the decoder that no subsequent picture will require reference to pictures prior to it. The video slice data corresponds to a sequence of coded macroblocks, each of which contains information relevant to decoding plus the actual encoded residual data. The following describes the MB level information, with reference to Figure 12.11:

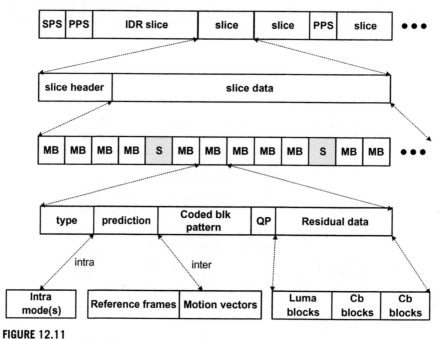

FIGURE 12.11

H.264 syntax.

- **Type:** I, P, B.
- **Prediction:** I-block mode ($16 \times 16, 8 \times 8$ or 4×4), P-block partition (16×16, 16×8, 8×16 or 8×8 (contains sub-block partitions)), B-block partition (as for P-block), choice of reference frames, and motion vector data. Note: P- and B-blocks may indicate a SKIP condition to signal that no residual data is sent.
- **Coded Block Pattern:** Indicates for which blocks coefficients are present.
- **Quantization Parameter:** QP value for the MB.
- **Residual data:** Coded residual data where indicated by the CBP parameter.

12.7.4 Pictures and partitions

Picture types

H.264/AVC supports the processing of both progressive and interlaced formats in a consistent manner, as a single unit. The fact that the frame is progressive or interlaced is indicated in the PPS and adaptive field-frame encoding can be used (either to encode a frame as a single unit or as two fields) but this has no impact on the way the picture is decoded via the decoder buffer.

H.264/AVC supports I-frames, P-frames, and B-frames. B-frames are referred to as generalized B-frames in H.264 in that they can be predicted from one or from two reference frames. H.264/AVC supports a very wide range of picture formats from QCIF to 4K UHDTV.

Slices and slice groups

As discussed in Chapter 11, slices can be arranged into slice groups (SG) where each SG can contain one or more slices. The number of slices is signaled by a PPS parameter. In cases where there is more than one slice group, the mapping arrangement of slices must be signaled using a value in the range of 0–6. As examples, SG type 0 is interleaved, SG type 1 refers to the dispersed allocation, and SG type 2 represents the foreground and leftover case. Depending on the slice group map type, additional syntax elements are required to enable the decoder to remap the received macroblocks to the slice groups used at the encoder. We introduced the various types of slice used in H.264/AVC in Section 12.7.3, and selected slice group patterns are illustrated in Figure 12.12.

Slice groups in H.264 support independent decoding, for example as required in multiple description coding and error concealment (both described in Chapter 11). Switching slices are also included in H.264. SI and SP slices direct a decoder to jump to a different point in the stream or to a different stream. These features are useful for video streaming where operations such as fast-forward are used.

Blocks

The macroblock in H.264 is based on a 16×16 sample arrangement, comprising luminance and chrominance residual data. The difference with H.264/AVC is that this structure can be subdivided in various ways according to the available prediction modes. This is discussed in more detail below.

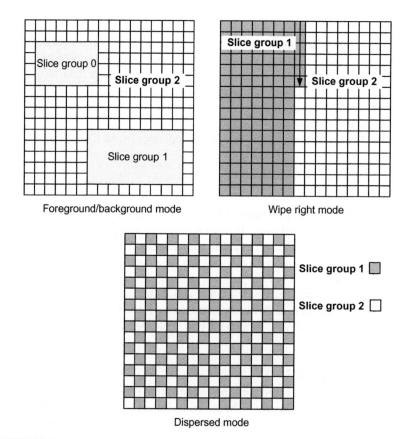

FIGURE 12.12

Example slice structures.

12.7.5 The video coding layer

Intra-coding

- **I_PCM mode:** This is the most basic H.264 mode and switches off all of the normal prediction and transform modes associated with the standard. This does not offer any compression but can be useful in high quality regimes where the PCM approach may offer better rate–distortion performance than a prediction–transform approach.
- **Intra-prediction:** H.264/AVC has benefited from the introduction of a range of intra-prediction modes. These offer significant benefits when there is a high level of local detail, particularly with oriented features such as edges. H.264 supports intra-prediction for 4×4, 8×8 (High profiles only), and 16×16 blocks. Coding of a given block is performed with reference to a subset of samples from previously coded blocks that lie to the left and above the current block. These modes were discussed in detail in Chapter 9.

FIGURE 12.13

Hadamard transformation of prediction residuals and coded block ordering.

- **Transforms:** In the case of intra_4 × 4 modes, the residual signal after prediction is transformed using the core 4 × 4 integer transform as described in Chapter 9. For the case of intra_16 × 16 modes, each of the 16 4 × 4 residual blocks is transformed as shown in Figure 12.13 and each of the 16 DC coefficients are further transformed using a 4 × 4 Hadamard transform. A similar approach is taken for the DC coefficients of the transformed chroma residuals, but this time a 2 × 2 Hadamard transform is applied.

Inter-coding

- **Variable block sizes:** We saw in Chapter 9 how the use of multiple block sizes for predictive coding of frames can provide significant benefits in dealing with complex textures and motions, offering up to 15% bit rate saving over the use of fixed block sizes. The range of block sizes available in H.264/AVC is shown again for convenience in Figure 12.14 and an example of a typical block decomposition is given in Figure 12.15. In order to make the most benefit of this flexibility, however, significant complexity must be added to the encoding process in terms of rate–distortion optimization (see Chapter 10).
- **Transforms:** H.264/AVC adopts the 4 × 4 DCT-like integer transform (described in Chapters 5 and 9) for residual coding in its baseline and main profiles. This is complemented by the use of a 2 × 2 Hadamard transform on the DC values for the case of chroma signals. For the High profiles, the standard also includes an 8 × 8 transform as this is often better suited to content with higher spatial resolutions.
- **Multiple reference frames:** Up to 16 frames or 32 fields (depending on profile) can be buffered in the H.264 reference buffer. The reference buffer does not have to contain the 16 most recent frames; frames can be updated according to their utility, provided that the encoder and decoder remain synchronized. For further details on MRF-ME, the reader is referred back to Chapter 9.

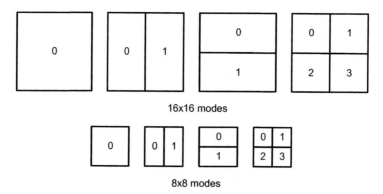

FIGURE 12.14

Variable block sizes supported by H.264/AVC. Top: 16 × 16 modes 1–4. Bottom: 8 × 8 modes 1–4.

FIGURE 12.15

Example H.264/AVC macroblock partition for inter-coding.

- **Prediction structures:** The H.264 coding regime enables a much more flexible approach to temporal prediction. For example, in the case of low delay, low memory constraints, an IPPP ⋯ I structure is most suitable, as offered by the baseline profile. If some delay is tolerable, along with the availability of B-frames, and multiple reference frames, then improved compression performance can be achieved. An example of this is shown in Figure 12.16, where a hierarchical prediction structure is imposed. Using this approach it has been demonstrated [12] that, if the quantization parameter is increased in a controlled fashion, with layer, then more efficient compression performance can be achieved.
- **Weighted prediction:** This is a useful tool, introduced in H.264, that enables the weighting/offsetting of prediction sample values in B-frames. This provides coding gains in the case of accelerating motion or during scene cuts.
- **Quarter-pixel prediction:** H.264 benefits from the availability of subpixel motion compensation to quarter-pixel accuracy. This has been estimated to provide up to 20% saving over the case of integer-pixel accuracy. A comprehensive description of the H.264/AVC sub-pixel motion estimation and compensation process was provided in Chapter 9. H.264/AVC uses a five-tap filter for half-pixel interpolation and a simple two-tap filter for local refinement to quarter-pixel accuracy.

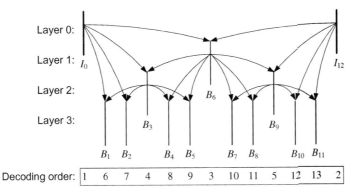

FIGURE 12.16

Hierarchical B-pictures in H.264.

Deblocking operations

The H.264 deblocking filter has been assessed to reduce the coded bit rate for the same subjective quality by up to 10%. It improves subjective visual and objective quality of the decoded picture. This highly content-adaptive non-linear filter removes blocking artifacts and does not unnecessarily blur the visual content. Its structure and operation were described in Chapter 9.

Variable length coding

- **Exp-Golomb coding:** This is a highly structured and simple variable length coding method, used almost universally in H.264 for all symbols apart from transform coefficients.
- **CAVLC:** Context-Adaptive VLC is a relatively low complexity but effective method, used as the Baseline and Extended Profiles for entropy coding of transform coefficients. Local contexts are used at the encoder to select between different VLC tables according to the local statistics, in this case the number of non-zero coefficients in neighboring blocks.
- **CABAC:** CABAC is a more complex, arithmetic coding-based method that is only supported in the H.264 Main and High Profiles. It uses local contexts to adjust its conditioning probabilities. CABAC has been found to reduce bit rate by between 10 and 20% compared to CAVLC, dependent on content type and quantization level.

Further details on all of the above entropy coding methods can be found in Chapter 7.

Coder control

Coder control is a non-normative part of H.264/AVC. It is however a key element in achieving the optimum performance from the standard, since without good RDO, many of the advanced features (e.g. block size selections) included will not deliver major savings. The goal of coder control is to select what parts of the video signal should be encoded using which methods and with what parameter settings. Often a Lagrangian optimization approach is adopted where the distortion measure, D, is based on sum of squared difference and the rate, R, includes all bits associated with the decision including header, motion, side information, and transform data. Further details on the RDO methods used in H.264/AVC can be found in Chapter 10.

12.7.6 Profiles and levels

H.264/AVC started with a relatively small number of profiles and levels. The number has however grown considerably, since its introduction, to reflect its popularity in a widening range of applications and the demand for higher quality profiles to deal with emerging formats. The standard now supports 21 profiles ranging from the simplest Constrained Baseline Profile (CBP) and Baseline Profile (BP) to the High 4:2:2 Profile (Hi422P—used for studio work) and the High 4:4:4 Predictive Profile (Hi444PP). The Main Profile (MP) is typically used for standard definition (4:2:0) DVB broadcasting. The Extended Profile (XP) is used in streaming and the High Profile (HiP) is used in Blu-ray players and for DVB HDTV delivery.

H.264 also supports a range of decoder constraints on memory and computational power in terms of its 17 levels, specifying parameters such as decoded bit rate, decoded picture buffer size, picture size, and processing speed in terms of numbers of macroblocks per second. A selection of the H.264 profiles are shown in Figure 12.17 and a good summary of all of profiles and levels is provided in Ref. [13].

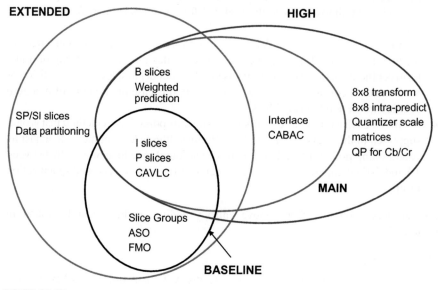

FIGURE 12.17

Selected H.264 profiles.

FIGURE 12.18

Typical H.264/AVC (MP) performance relative to MPEG-2 for standard definition entertainment content.

12.7.7 Performance

The performance of H.264/AVC relative to previous standards is summarized in Ref. [14]. An example RD profile is shown in Figure 12.18 for the case of an entertainment application. This shows approximately 50% saving in bit rate for the same PSNR value between H.264/AVC and MPEG-2. A more recent comparison including HEVC is provided by Ohm et al. [15] (see Table 12.7).

12.7.8 Scalable extensions

A scalable extension to H.264/AVC (SVC) was approved in 2007. As described in Chapter 11, bitstream scalability is useful for providing graceful degradation of performance in the context of dynamic channel conditions. SVC contains five extra profiles, offering a layered approach to scalability. This creates an embedded bitstream providing SNR, spatial and/or temporal scalability, where packets can be discarded in an orderly fashion to manage congestion or terminal limitations. As can be observed from Figure 12.16, the use of hierarchical B-frames is a very useful tool in providing temporal scalability. The reader is referred to Ref. [16] for more details on SVC.

12.7.9 Multiview extensions

Finally, to cater for the predicted boom in 3-D stereoscopic video content and broadcasting, a multiview extension (MVC) was approved in 2009 comprising three additional H.264 profiles. As its name suggests, MVC supports the coding of content acquired from multiple cameras. Its primary focus is stereoscopic (3-D) video, but it also supports free-viewpoint and multiview 3-D applications, television, and multiview

3-D television. The Stereo High profile is used in 3-D Blu-ray devices. For further information on MVC, the reader is referred to Refs. [17,18].

12.8 H.265/HEVC

12.8.1 Brief background

HEVC, formally known as ISO/IEC MPEG-H Part 2 and ITU-T H.265, is the most recent outcome from collaborative working between ISO/IEC MPEG and ITU-T VCEG. The standard was approved in April 2013 [19] and, with a patent licensing framework in place and the announcement of several HEVC-enabled products, it is poised to make a major impact on the video industry. HEVC offers the potential for up to 50% compression efficiency improvement over AVC and will go some way to providing a solution to the explosion in mobile video and the need to support broadcast and download services with ever-increasing quality and experience levels.

HEVC is targeted at the same application portfolio as H.264/AVC, but with a specific focus on bit rate reduction for increased video resolutions and on support for parallel processing as well as loss resilience and ease of integration with appropriate transport mechanisms. An excellent introduction to the range of HEVC features is provided by the papers in a special issue of the *IEEE Transactions on Circuits and Systems for Video Technology* published in December 2012, in particular the overview by Sullivan et al. [20] and the performance comparisons by Ohm et al. [15]. For full details, the reader should consult the standards documentation [19].

12.8.2 Primary features

As we will see later, HEVC reduces bit rate requirements roughly by half compared to existing codecs while retaining image quality. It does this using the proven hybrid approach that has served us well since H.261, but with the addition of several new and important features. While some of these add complexity to the codec, many actually offer simplicity compared to H.264/AVC and facilitate parallel processing, compatible with recent hardware developments. HEVC, like previous standards, can trade off computational complexity, compression rate, robustness to errors, and processing delay time, according to application requirements, and this range of operating points is defined by its profile and level structure. A major focus for HEVC is the next generation of HDTV displays and acquisition systems which feature progressively scanned frames, higher frame rates, and resolutions up to UHDTV.

In the following sections, we examine more closely the video coding and syntactic structure of HEVC. The reader will see that a large amount of HEVC technology has been pulled through from H.264/AVC, with some H.264 features omitted and other new and modified features added where appropriate to improve performance. The primary new features in HEVC relate to (i) its block structure, where super-macroblocks known as Coding Tree Units (CTUs) are used, where the size of a CTU can range up to 64 × 64 samples; (ii) the way in which these CTUs are partitioned into

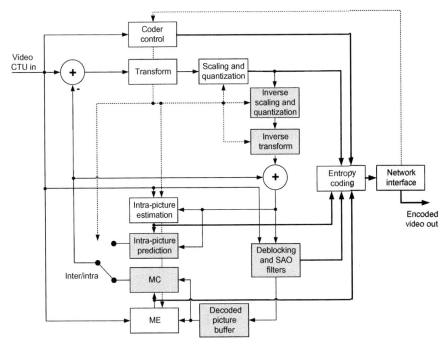

FIGURE 12.19

HEVC video encoder architecture.

variable sized coding, transform, and prediction blocks using a quadtree structure; (iii) significantly enhanced intra-prediction, supporting 35 modes; (iv) the use of Sample Adaptive Offset (SAO) in conjunction with the deblocking filter in the inter-prediction loop to reduce banding and ringing artifacts.

The architecture of the HEVC codec, illustrating many of these features, is shown in Figure 12.19. This will be described in more detail in the following sections.

12.8.3 Network abstraction and high level syntax

NAL units

The high level syntax of HEVC retains a number of the elements used in H.264/AVC to provide network abstraction and the basis for error resilience. NALUs are still employed as the basic interface to the transport mechanism and HEVC supports 63 NALU (VCL and non-VCL) types [19,20].

Slice structures

Slices enable independent decoding and assist with resynchronization and error robustness. The slice structure in HEVC has been largely pulled through from H.264/AVC and the reader is referred to Chapters 9 and 11 for a more detailed description.

Parameter sets

The same parameter sets as in H.264/AVC are employed, supplemented by a new Video Parameter Set (VPS). VPS can include metadata to describe the general characteristics of coded layers, such as could be useful in future scalable extensions. Information in the NAL Unit header also supports the identification of temporal layers.

Reference picture sets and reference picture lists

It is important to manage the content of the Decoded Picture Buffer (DPB) to ensure synchronization between encoder and decoder. The retained set of pictures is referred to as a Reference Picture Set (RPS). HEVC supports two reference picture lists, RPL0 and RPL1. In the case of unidirectional prediction, a reference picture can be selected from either list and in the case of bidirectional prediction, one picture must come from each list. Some improvements have been made in HEVC to make the identification of the RPS more robust than in H.264/AVC.

12.8.4 Pictures and partitions

HEVC supports a very wide range of picture sizes and frame rates as illustrated in the level examples in Table 12.5. The basic picture types supported in HEVC are identical to those in H.264/AVC, with the small exception of a modified version of the IDR frame known now as a Clean Random Access (CRA) picture.

An HEVC Coding Unit (CU) can contain up to three blocks (arrays of luma and chroma samples) and any information required to decode these blocks. Each CU is subdivided into Prediction Units (PUs) and these are coded using either intra- or inter-prediction. The appropriate prediction parameters (inter- or intra-) are signaled within each PU. A Residual Quadtree (RQT) is then used to divide the CU into transform units (TUs) within which the residual information is coded. These units all contain coding blocks (CBs), prediction blocks (PBs), and transform blocks (TBs) for the constituent luma and chroma components, as described above. This structure is described in more detail in Ref. [21]. The decomposition of the HEVC picture is described in more detail below.

Table 12.5 Example HEVC levels.

Level	Max. luma sample rate	Max. luma picture rate	Max bit rate (kbps)	Example resolution
1	552,960	36,864	128	176 × 144@15
3	16,588,800	552,960	6,000	720 × 576@30
5	267,386,880	8,912,896	160,000*	3,840 × 2160@60
6.2	4,278,190,080	35,651,584	800,000*	7680 × 4320@120

*High Tier specification, others are Main Tier.

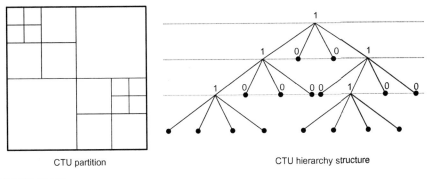

CTU partition   CTU hierarchy structure

FIGURE 12.20

CTU partitioning in HEVC—an example.

Coding tree units (CTUs) and coding tree blocks (CTBs)

HEVC replaces the macroblock of earlier standards with an $L \times L$ CTU where $L = 16, 32, 64$. This provides more efficient coding for higher resolution formats. The CTU is made up of luma and chroma CTBs.

Quadtree CTU structure

CTBs can be partitioned into Coding Units (CUs) comprising luma and chroma Coding Blocks (CBs) with smaller block sizes using a quadtree decomposition. The size of the largest coding unit (LCU) is 64×64. The manner in which the CTU is decomposed is not specified in the standard (it is non-normative) but nonetheless forms an important part of the encoder's RDO processing. An example of CTU partitioning is given in Figure 12.20. Each CU is further partitioned into Transform Units (TUs) and Prediction Units (PUs). This approach provides the encoder with much more flexibility to optimize block partitions according to the video content.

Prediction units (PUs)

A much larger variety of block decompositions for intra- and inter- coding is supported in HEVC, from 4×4 to 64×64.

Transform units (TUs)

Prediction residuals, whether from intra- or inter-coding, are transformed using integer transforms similar to the DCT, with sizes from 4×4, 8×8, 16×16, and 32×32. For the case of intra-prediction residuals, a 4×4 integer sine transform is also available as this provides better performance on less correlated residuals.

Random access points (RAPs) and clean random access (CRA) pictures

The concept of CRA pictures is the same as that for IDR frames in H.264/AVC. CRAs generalize this by embodying independently coded pictures, where following frames have no dependence on frames received before the CRA picture. CRA pictures can be placed at locations where random access is required, known as RAPs.

12.8.5 The video coding layer (VCL)

Similar to H.264, HEVC uses a hybrid coding structure combining motion compensation with block transform and entropy coding. Several innovations have been incorporated into HEVC that provide significant gains in performance and these are described below.

Intra-coding

In a similar manner to H.264/AVC, the boundary pixel values from adjacent decoded blocks can be used as a basis of spatial prediction. Whereas H.264/AVC supports only eight prediction modes, this has been significantly enhanced in HEVC with 35 modes including 33 directional modes (Figure 12.21) plus surface fitting and DC prediction. Further details on the computation of the predictions for these modes can be found in Ref. [19]. The amplitude of the planar surface in mode 0 is computed based on horizontal and vertical slopes derived from the reference pixel values in the surrounding blocks. The DC mode (1) is computed in the same way as for H.264/AVC—based on the average of the block boundary reference values.

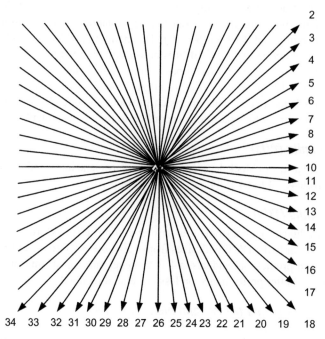

FIGURE 12.21

HEVC intra-prediction modes (mode 0 = planar and mode 1 = DC).

HEVC also extends the concept of reference and boundary sample smoothing. These operations help to reduce residual and reconstruction edges by filtering either the reference samples (from surrounding blocks) or the boundary samples (within the predicted block) prior to prediction.

When intra-modes are used the prediction block (PB) size is the same as that for the coding block (CB), except for the smallest CB size where a further partition into four PB quadrants of 4×4 is permitted. After prediction, residuals are coded using a forward integer transform, quantized, and entropy coded.

Inter-coding

The approach to inter-prediction and coding is similar to H.264/AVC. Multiple reference frames are stored in the Decoded Picture Buffer and are used in a similar way to H.264/AVC, supporting both uni predictive and bipredictive modes. As with H.264/AVC, weighted prediction can be employed. Some modifications included in HEVC are as follows:

- **Subpixel motion compensation:** Quarter-pixel precision is used for estimating motion during temporal prediction. Local luma interpolation to half-pixel resolution is provided by an eight-tap filter, rather than the six-tap filter in H.264/AVC. A seven-tap filter is then used for quarter-sample interpolation rather than the two-tap filter used in H.264/AVC. These are applied separably and (unlike H.264) without intermediate rounding. This improves precision and simplifies the prediction architecture. The filter coefficients for interpolation are given in equations (12.1) and (12.2). Further details on the use of these can be found in Refs. [19,20].

$$h_h[i] = \{-1, 4, -11, 40, 40, -11, 4, -1\}/64 \qquad (12.1)$$
$$h_q[i] = \{-1, 4, -10, 58, 17, -5, 1\}/64 \qquad (12.2)$$

- **Motion vector prediction:** An Advanced Motion Vector Prediction mode (AMVP) is used for initialization and derivation of likely motion vector candidates. Also, improved skip modes are included along with a capability of merging MVs between adjacent blocks.
- **Coding block partitioning into prediction blocks:** In the case of inter-prediction, luma and chroma CBs can be divided into one, two, or four PBs (again the latter only for the smallest CB size). These partitioning options are illustrated in Figure 12.22.

Transforms in HEVC

Residual data after prediction is coded with a separable integer transform whose structure depends on prediction mode and block size. The transform block sizes supported in HEVC are: $4 \times 4, 8 \times 8, 16 \times 16$, and 32×32. As for the case of H.264/AVC, the transform basis functions are derived from the DCT. The transform is defined as a 32×32 matrix of basis function and the smaller transform sizes are obtained by sub-sampling this larger matrix. For example, the 16×16 matrix is

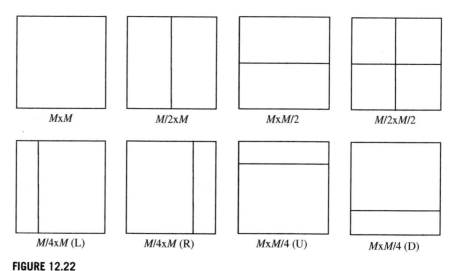

FIGURE 12.22

CB partitions for inter-prediction PBs.

derived from the 32×32 matrix by taking the first 16 values from rows 0, 2, 4, etc. A similar sub-sampling is used to produce 8×8 and 4×4 transform matrices. As an example, the 8×8 transform is given below.

$$
\mathbf{A}_{8\times8} =
\begin{bmatrix}
64 & 64 & 64 & 64 & 64 & 64 & 64 & 64 \\
89 & 75 & 50 & 18 & -18 & -50 & -75 & -89 \\
83 & 36 & -36 & -83 & -83 & -36 & 36 & 83 \\
75 & -18 & -89 & -50 & 50 & 89 & 18 & -75 \\
64 & -64 & -64 & 64 & 64 & -64 & -64 & 64 \\
50 & -89 & 18 & 75 & -75 & -18 & 89 & -50 \\
36 & -83 & 83 & -36 & -36 & 83 & -83 & 36 \\
18 & -50 & 75 & -89 & 89 & -75 & 50 & -18
\end{bmatrix}
\tag{12.3}
$$

In the case of intra-coded 4×4 residuals, an alternative integer transform is used—derived from the DST. It has been observed that the statistics of intra residuals better fit the basis functions of the DST since the error values increase with distance from the boundary samples used to predict them. The 4×4 transform matrix is thus:

$$
\mathbf{A}_{4\times4I} =
\begin{bmatrix}
29 & 55 & 74 & 84 \\
74 & 74 & 0 & -74 \\
84 & -29 & -74 & 55 \\
55 & -84 & 74 & -29
\end{bmatrix}
\tag{12.4}
$$

Quantization in HEVC

HEVC employs a similar quantization method for transform coefficients to that used in H.264. It uses uniform quantization with a range of scaling matrices available for each block size.

Coefficient scanning

The scanning of coefficients in an HEVC TB [22] is based on the use of multiples of 4×4 blocks for all TB sizes. The three coefficient scanning options available to represent each Coefficient Group (CG: a group of 16 consecutive coefficients) are shown in Figure 12.23. In the case of intra-blocks, a new method referred to as Mode-Dependent Coefficient Scanning (MDCS) is used, where the mode closest to the orientation of the intra-prediction mode is selected. For inter-coded blocks, only the diagonal mode is used. For larger inter-blocks, coefficients are scanned from the least significant to the most significant as shown in Figure 12.23 for the case of an 8×8 TB. In contrast to the zig-zag scanning used in H.264/AVC and previous standards, this simplified approach enables better matching of scanning to coefficient contexts. In terms of coding efficiency, improvements of around 3.5% compared to H.264/AVC have been reported [22].

A further modification in HEVC relative to AVC is related to the signaling of the last significant coefficient in a scan. As we have seen, signaling the last significant coefficient reduces the number of coded bins by eliminating the need to explicitly code runs of trailing zeros. In HEVC the position of the last significant coefficient is coded explicitly followed by the flags to indicate significant coefficients.

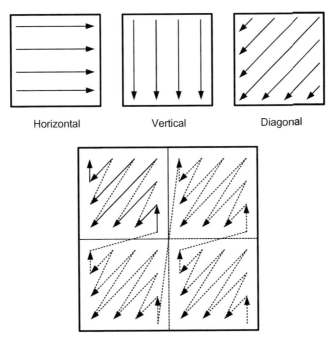

Diagonal scan for 8x8 TB

FIGURE 12.23

Coefficient scanning in HEVC.

FIGURE 12.24

Significance flag contexts for 4 × 4 and 8 × 8 HEVC transform blocks.

Contexts and significance

As with H.264/AVC, HEVC signals the significance of a TB using a Coded Block Flag (CBF) which indicates whether the TB contains any non-zero coefficients. The context of a CBF depends on the level of the block in the quadtree decomposition. The significance maps within HEVC are coded in a way that exploits the sparsity of the significance map, using two passes. Firstly the significance of coefficient groups is encoded and then the significance of the coefficients within them. The coefficient significance contexts in HEVC for 4 × 4 and 8 × 8 blocks are shown in Figure 12.24. The significance flags are coded for each coefficient between the last and the DC based on a context model that depends on its location and the significance of its neighbors. In the case of 4 × 4 TBs, coefficients are grouped according to frequency and the distribution, where coefficients with similar distributions are grouped together.

For the case of 8 × 8, 16 × 16 and 32 × 32 TBs in HEVC, context modeling is again based on position but according to the templates shown in Figure 12.24 (right). One of the four templates is selected according to the values of s_R and s_L where s_R and s_L correspond to the coded subblock flag for the right-hand and lower sub-blocks respectively.

Variable length coding with CABAC

CABAC, as described in Chapter 7, is a powerful arithmetic coding method that was supported in the H.264/AVC Main and High Profiles. Because of its performance advantages, HEVC has adopted CABAC for all entropy coding purposes. As with H.264/AVC, CABAC in HEVC uses local contexts to adjust its conditioning probabilities. CABAC has been improved in HEVC through:

1. Better context modeling: based on exploitation of HEVC's tree-structured approach.
2. Reducing the number of context coded bins: although the number of contexts is fewer than in H.264/AVC, its performance is better.
3. Introducing an enhanced bypass mode: more extensive use of bypass is made to reduce the amount of data that needs to be processed using CABAC contexts, resulting in a throughput increase.

In-loop filters

HEVC includes two stages of in-loop filtering prior to updating its decoded picture buffer: a Deblocking Filter (DBF) and a Sample Adaptive Offset (SAO) process.

- **Deblocking Filter:** The deblocking operation in HEVC is very similar to that for H.264/AVC which was described in Chapter 9. It is slightly simplified to make it more amenable to parallel processing. DBF is applied to all TU and PU boundaries, except when they coincide with picture or slice boundaries. In the case of 4×4 blocks the DBF is simplified by only using it when block boundaries coincide with 8×8 boundaries. This simplifies processing generally with no noticeable degradation in performance. In the case of HEVC, only three deblocking strengths are available rather than the five with H.264/AVC.
- **Sample Adaptive Offset:** SAO is a new feature in HEVC that is invoked after DBF, in order to increase picture quality through a reduction in banding and ringing artifacts. It is a non-linear adjustment that adds offsets to samples based on a look-up table created at the encoder (and transmitted to the decoder) using histogram analysis of signal amplitudes. Two SAO modes are available: *edge offset mode* and *band offset mode*. The first operates on a CTB basis using one of four gradient operators to compare a sample with adjacent values in a given orientation. Five conditions can result according to the relative value of the sample compared to its neighbors. Consider, for example, the case of a horizontal orientation, if a sample p has a left neighbor n_0 and a right neighbor n_1, then the regions are classified according to Table 12.6. In the second case, *band offsetting*, the offset is applied directly based on the sample amplitude. The full amplitude range is divided into 32 bands and the values of samples in four consecutive bands are modified using transmitted offsets. This can help to reduce banding artifacts in smooth areas.

Table 12.6 SAO edge classification.

EdgeIdx	Condition	Case
0	Exceptions to cases (1...4)	Monotonic
1	$p < n_0$ and $p < n_1$	Local min.
2	$p < n_0$ and $p = n_1$ or $p < n_1$ and $p = n_0$	Edge
3	$p > n_0$ and $p = n_1$ or $p < n_1$ and $p = n_0$	Edge
4	$p > n_0$ and $p > n_1$	Local max.

12.8.6 Profiles and levels

Version 1 of HEVC specified three profiles, Main, Main 10, and Main Still Picture.

Main profile

The Main Profile supports a bit depth of 8 bits per sample and 4:2:0 (and 4:0:0) chroma sampling, employing the features described in the previous subsections.

Main 10 profile

The Main 10 profile supports bit depths up to 10 bits and up to 4:2:2 chroma sampling. In recent tests, it has been demonstrated that the Main 10 profile outperforms the Main Profile with a bit rate reduction of approximately 5% for identical PSNR values.

Main Still Picture profile

HEVC also includes, like H.264, a still image coding profile, which is a subset of the Main Profile. Performance comparisons based on MOS and PSNR scores have provided indications that HEVC still image coding outperforms JPEG 2000 in terms of bit rate reduction by approximately 20% based on PSNR and 30% for MOS scores. Likewise it has been shown to outperform JPEG by 40% and 60%.

Levels

In the same way as previous standards, HEVC also supports a range of levels that impose parameter limits that constrain decoder performance. There are currently 13 levels associated with HEVC. Some examples are shown in Table 12.5.

12.8.7 Extensions

At the time of writing, a number of HEVC range extensions and five additional profiles are being worked on. These include a Main 12, Main 4:2:2 10, Main 4:2:2 12, Main 4:4:4 10, and Main 4:4:4 12. The extended ranges in terms of color sampling and bit depths are obvious from their titles.

12.8.8 Performance gains for HEVC over recent standards

An excellent comparison of coding standards is provided by Ohm et al. [15], where a comprehensive comparison between HEVC and previous standards is reported. We reproduce their results here in Table 12.7, noting that even greater improvements were

Table 12.7 Comparison of video coding standards for entertainment applications. Average bit rate savings are shown for equal objective quality values measured using PSNR.

Video standard	Relative bit rate savings (%)		
	H.264/AVC	MPEG-4	MPEG-2
HEVC MP	35.4	63.7	70.8
H.264/AVC HP	X	44.5	55.4
MPEG-4 ASP	X	X	19.7
H.263 HLP	X	X	16.2

reported for interactive applications. As a rule of thumb, we have stated previously that video coding standards have, since the introduction of H.120 in 1984, delivered a halving of bit rate for the equivalent video quality every 10 years. This is evidenced by the results in Table 12.7.

12.9 Other de-facto standards and proprietary codecs

A number of other video codecs exist. While these are worthy of mention, we will not cover them in detail here.

12.9.1 VP9

Perhaps foremost is Google's VP9. VP9 is a successor to VP8 and is open-source and royalty-free. VP8 was created by On2 Technologies, who were purchased by Google in 2010. YouTube, also owned by Google, uses Adobe Flash Player and H.264/AVC, but there is a reported move toward VP9. In 2013 a version of the Google Chrome web browser was released that supports VP9 decoding. VP9 is also likely to find favor in some cloud-based applications.

12.9.2 VC-1

The Microsoft WMV 9 codec was a proprietary codec but was in 2006 adopted as international standard SMPTE 421M or VC-1. VC-1 is a recognized format for Blu-ray devices.

12.9.3 RealVideo

RealVideo comprises a series of proprietary video compression formats developed by RealNetworks for their players. The current version, rv40, is thought to be based on H.264/AVC.

12.9.4 Dirac

Dirac was developed by BBC Research and Development as a royalty-free alternative to the mainstream standards. Dirac employs wavelet compression rather than block transforms such as the DCT, and offers good performance for HDTV formats and beyond. An I-frame-only version, Dirac Pro, has been used by the BBC internally for studio and outside broadcast applications and was standardized by SMPTE as VC-2 in 2010. A portable version, known as Schrödinger, also exists.

12.10 Summary

We have examined in this chapter the primary features of all major video compression standards from H.120 through H.261, H.263, and MPEG-2, to the current H.264/AVC standard and the new H.265/HEVC codec. Although these have been largely built on the same hybrid motion-compensated transform architecture, continued innovations over the past three decades or so have delivered, on average, a halving of bit rate for the equivalent video quality every 10 years. While this is impressive progress, the demand for high volumes of high quality video content keeps growing and new approaches will continue to be needed to fulfill future requirements. Some of the potential candidate solutions are considered in the next chapter.

References

[1] S. Okubu, Reference model methodology—A tool for the collaborative creation of video coding standards, Proceedings of the IEEE 83 (2) (1995) 139–150.

[2] CCITT/SG XV, Codecs for Videoconferencing Using Primary Group Transmission, Rec. H.120, CCITT (now ITU-T), 1989.

[3] International Telecommunication Union-Telecommunication (ITU-T), Recommendation H.261, Video Codec for Audiovisual Services at px64 kbit/s, Version 1, 1990; Version 2, 1993.

[4] ITU-T and ISO/IEC JTC 1, Generic Coding of Moving Pictures and Associated Audio Information—Part 2: Video, ITU-T Rec. H.262 and ISO/IEC 13818-2 (MPEG-2 Video), Version 1, 1994.

[5] ITU-T, Video Coding for Low Bitrate Communication, ITU-T Rec. H.263, Version 1, 1995, Version 2, 1998, Version 3, 2000.

[6] S. Wenger, G. Knorr, J. Ott, F. Kossentini, Error resilience support in H.263+, IEEE Transactions on Circuits and Systems for Video Technology 8 (7) (1998) 867–877.

[7] ISO/IEC JTC 1, Coding of Audio-Visual Objects—Part 2: Visual, ISO/IEC 14496-2 (MPEG-4 Visual), Version 1, 1999, Version 2, 2000, Version 3, 2004.

[8] T. Ebrahimi, C. Horne, MPEG-4 natural video coding: an overview, Signal Processing Image Communication 15 (2000) 365–385.

[9] ITU-T and ISO/IEC JTC 1, Advanced Video Coding for Generic Audiovisual Services, ITU-T Rec. H.264 and ISO/IEC 14496-10 (AVC), Version 1, 2003, Version 2, 2004, Versions 3, 4, 2005, Versions 5, 6, 2006, Versions 7, 8, 2007, Versions 9, 10, 11, 2009, Versions 12, 13, 2010, Versions 14, 15, 2011, Version 16, 2012.

[10] T. Wiegand, G. Sullivan, G. Bjøntegaard, A. Ajay Luthra, Overview of the H.264/AVC video coding standard, IEEE Transactions on Circuits and Systems for Video Technology 13 (7) (2003) 560–576.

[11] I. Richardson, The H.264 Advanced Video Compression Standard, second ed., Wiley, 2010.

[12] H. Schwartz, D. Marpe, T. Wiegand, Analysis of hierarchical B-pictures and MCTF, in: International Conference on Multimedia and Expo, 2006, pp. 1929–1932.

[13] <http://en.wikipedia.org/wiki/H.264/MPEG-4_AVC>, 2013 (accessed August 2013).

[14] J. Ostermann, J. Bormans, P. List, D. Marpe, M. Narroschke, F. Pereira, T. Stockhammer, T. Wedi, Video coding with H.264/AVC: tools, performance, and complexity, IEEE Circuits and Systems Magazine 4 (1) (2004) 7–28.

[15] J.-R. Ohm, G. Sullivan, H. Schwartz, T. Tan, T. Wiegand, Comparison of the coding efficiency of video coding standards—including High Efficiency Video Coding (HEVC), IEEE Transactions on Circuits and Systems for Video Technology 22 (12) (2012) 1669–1684.

[16] H. Schwarz, D. Marpe, T. Wiegand, Overview of the scalable video coding extension of the H.264/AVC standard, IEEE Transactions on Circuits and Systems for Video Technology 17 (9) (2007) 1103–1120.

[17] P. Merkle, A. Smolic, K. Müller, T. Wiegand, Efficient prediction structures for multiview video coding, IEEE Transactions on Circuits and Systems for Video Technology 17 (11) (2007) 1461–1473.

[18] A. Vetro, T. Wiegand, G.J. Sullivan, Overview of the stereo and multiview video coding extensions of the H.264/MPEG-4 AVC standard, Proceedings of the IEEE 99 (4) (2011) 626–642.

[19] Joint Collaborative Team on Video Coding (JCT-VC) of ITU-T SG 16 WP 3 and ISO/IEC JTC 1/SC 29/WG 11 ISO/IEC 23008-2 and ITU-T Recommendation H.265, High Efficiency Video Coding (HEVC), January 2013.

[20] G. Sullivan, J.-R. Ohm, W. Han, T. Wiegand, Overview of the High Efficiency Video Coding (HEVC) standard, IEEE Transactions on Circuits Systems, Video Technology 22 (12) (2012) 1648–1667.

[21] I. Kim, J. Min, T. Lee, W. Han, J. Park, Block partitioning in the HEVC standard, IEEE Transactions on Circuits and Systems for Video Technology 22 (12) (2012) 1697–1706.

[22] J. Sole, R. Joshi, N. Nguyen, T. Ji, M. Karczewicz, G. Clare, F. Henry, A. Duenas, Transform coefficient coding in HEVC, IEEE Transactions on Circuits and Systems for Video Technology 22 (12) (2012) 1765–1777.

Communicating Pictures—The Future

13

CHAPTER OUTLINE

13.1 The motivation: more immersive experiences 452
13.2 Emerging formats and extended video parameter spaces 452
    13.2.1 Influences 453
    13.2.2 Spatial resolution 453
    13.2.3 Temporal resolution 455
    13.2.4 Dynamic range 459
    13.2.5 Parameter interactions and the creative continuum 461
13.3 Challenges for compression 462
13.4 Parametric video compression 464
13.5 Context-based video compression 466
13.6 Summary 467
References 468

In this chapter we briefly review the demands of future content types in the context of an *extended video parameter space*, to examine how parameters such as spatial resolution, temporal resolution, dynamic range, and color gamut affect our viewing experience, both individually and collectively. The delivery requirements associated with this extended parameter space are considered, with a focus on how compression might affect and influence the video quality.

Emphasizing the impact of, and the interactions between, the video parameters and their content dependence, the justification is made for an increased use of perceptual models. It is argued that perception-based compression techniques can improve video quality and that a new *rate–quality optimization* approach is required if we are to efficiently and effectively code the immersive formats of the future. In this context, new approaches to compression are proposed based on texture analysis and synthesis, moderated by in-loop quality metrics.

For colour and higher quality versions of Figures 13.3, 13.6, 13.12 and 13.13 please refer to the electronic version or the website.

Communicating Pictures. http://dx.doi.org/10.1016/B978-0-12-405906-1.00013-1
© 2014 Elsevier Ltd. All rights reserved.

13.1 The motivation: more immersive experiences

The moving image industry is a trillion dollar business worldwide, across film, broadcast, and streaming. And, according to the USA Bureau of Labor Statistics [1] we spend one fifth of our waking lives watching movies, television, or some other form of edited moving image.

Visual experiences are thus important drivers for business and technology development; by 2020 it is predicted that the number of network-connected devices will reach 1000 times the world's population, i.e. there will be 7 trillion connected devices for 7 billion people [2]. YouTube video currently accounts for around 25% of all internet traffic and, by 2016, Cisco predicts that video will account for 54% of all traffic (86% including P2P distribution) with total annual IP traffic rising to 1.3 ZB (zettabyte $= 10^{21}$ bytes $= 1000$ EB) [3]. Some mobile network operators predict a doubling of wireless traffic every year for the next 10 years driven primarily by video.

Despite the impressive facts and predictions above, for high value content shown in cinemas, attendance and revenues across much of the developed world in 2012 were flat. Although there was an increase of 30% in the number of digital 3-D screens, according to statistics from the British Film Institute, the revenues from 3-D movies actually decreased. 3-D has been at least as much driven by technology capabilities and the need for businesses to add value, as it has been by user demands or indeed because it provides a more engaging experience. It is now clear that the industry emphasis is shifting from stereoscopic 3-D toward UHDTV formats with larger and brighter screens. Audiences *may* still want to watch 3-D, but it seems that many are not willing to pay the premium.

So there is a significant demand for new, more immersive content: from users who want to experience it, from producers who want to add value to their content, and from operators who want to charge you and I more to watch it. The questions we need to ask are: what is the format of such content, how do we assess its quality and how do we preserve its values during acquisition, delivery and display?

13.2 Emerging formats and extended video parameter spaces

The drive for increased immersion, coupled with greater mobility, means that compression efficiency is a priority for the media, communications, and entertainment industries. Two particular recent events—the explosion of video consumption on portable and personal devices, and the increased investment in UHDTV—both demand significant increases in compression efficiency.

However, to produce and deliver more immersive content, it is not sufficient just to increase spatial resolution or screen size. Although these parameters play an important role, other factors such as frame rate, dynamic range, and color gamut are also key to delivering high value experiences. We will examine some of these briefly below.

13.2.1   **Influences**

There are many things that can influence our perception of video quality and affect the immersiveness of the viewing experience, not least our interest in the subject matter. However, putting the narrative to one side, the main factors are:

- Resolution: temporal, spatial and SNR.
- Motion characteristics: blurring and masking effects.
- Colorimetry: gamut bit depth, tone mapping, and grading.
- Display: size, aspect ratio, dynamic range, and viewing environment.
- Views: monoscopic, stereoscopic, and multiview.
- Cinematography: including set design, lighting, shot length, scene geometry, motion, and color.
- Other psychovisual characteristics: such as peripheral vision, saliency, depth cues, and image noise.
- And of course audio: ambisonics and wavefields are both exciting developments that complement the visual to enhance the experience.

Complex interactions between these exist and they will, of course, be confounded by strong content dependence. Their impact is however highly moderated by the method and degree of compression employed.

13.2.2   **Spatial resolution**

Why spatial detail is important

Spatial resolution is important as it influences how sharply we see objects. The key parameter is not simply the number of pixels in each row or column of the display, but the angle subtended, θ, by each of these pixels on the viewer's retina. Thought of in this way, increased spatial resolution is particularly important as screen sizes become larger and viewing distances closer. Also, the *effective spatial resolution* will depend on how the compression system interacts with the displayed content; it is very easy to turn HD content into SD through compression. Contrary to popular advertising, there is no such thing as "HD quality." Let us look at this a little closer.

UHDTV and ITU-R REC.2020

The Ultra High Definition Television (UHDTV) standard, finalized recently by ITU-R under the title of Rec.2020 [4], will be a major influence on compression methods in the future. It is intended to deliver the highest quality immersive viewing experience to the home.

Table 13.1 shows the basic parameters for UHDTV and some interesting points can be observed from this table. Firstly, the assumed viewing distance places the viewer very close to the picture at 0.75 times the screen height, compared to $3H$ for HDTV (Rec.709). Obviously large screens are assumed as the viewing angle is stated as 100°, in contrast to that for conventional HDTV of around 33°. This will mean significant changes in the way we interact (psychovisually) with the content through

Table 13.1 Parameter set for UHDTV/ITU-R Rec.2020.

Parameter	Values
Picture aspect ratio	16 × 9
Pixel count ($H \times V$)	7680 × 4320, 3840 × 2160
Sampling lattice	Orthogonal
Pixel aspect ratio	1:1 (square pixels)
Frame frequency (Hz)	120, 60, 60/1.001, 50, 30, 30/1.001, 25, 24, 24/1.001
Color bit depth	10 or 12 bits per component
Scan mode	Progressive
Viewing distance	0.75H
Viewing angle	100°

increased head and eye movements. In practice this author suspects that the viewing distance will be greater than this for most people in many environments. Nonetheless, different cinematographic methods will be required. Secondly, the maximum frame rate is 120 Hz, rather than the 25, 30, 50, or 60 Hz we are familiar with. This is in response to pressure from organizations such as the BBC and NHK who have gained significant experience of spatio-temporal interactions in the Super HiVision project [5]. We will consider this aspect a little more in Section 13.2.3.

Compression performance

Firstly let us take a look at the general interactions between spatial resolution and compression performance and dispel the myth that HD is always better than SD. Considering Figure 13.1, this shows how the RD curves for different formats overlap as the bit rate increases, clearly indicating threshold bit rates where it is better to switch

FIGURE 13.1

Does increased resolution mean increased quality?

format in order to ensure the best quality. The problem with this is that it is highly content dependent and most codecs will not provide seamless switching. However, adaptive methods are under investigation in some laboratories and simple approaches have already appeared in some streaming products. The implication of this is quite simple—there is no point increasing spatial resolution unless the compression system can preserve the perceptual benefits associated with that resolution.

A primary and stated goal of HEVC is to provide efficient compression for resolutions beyond HDTV. However, it is interesting to note that the subjective evaluations conducted in HEVC were primarily at HDTV resolution and the metrics used were mainly based on PSNR and subjective tests.

Hanhart et al. [7] have however reported work on evaluation of HEVC at extended spatial resolutions, and the results are encouraging. Subjective tests were conducted at the EPFL laboratory on a limited dataset of 4k content at 24 and 30 fps (one of the issues currently is that data is limited) and the results clearly indicate that greater relative savings are possible for resolutions beyond HDTV. They reported a bit rate reduction of between 51% and 74% for natural content based on MOS scores compared to H.264/AVC. Interestingly, the corresponding PSNR scores were 28% and 38%, highlighting once more the shortcomings of PSNR measurements.

13.2.3 Temporal resolution

Why rendition of motion is important

We saw in Chapter 2 that motion is important. Our retina is highly tuned to motion, even in the periphery, and our visual system is designed to track motion through retinal and head movements. In terms of video content, motion is conveyed through a combination of frame rate and shutter angle. Frame rates for film were standardized at 24 fps in the 1920s and TV frame rates have been fixed at 25 or 30 fps (using 50 or 60 fields) since the BBC adopted Marconi's 405-line system in 1937. These rates were appropriate at the time, were compatible with mains frequencies and provided an excellent trade-off between bandwidth, perceived smoothness of motion, and the elimination of flicker. However, as screen sizes grow, along with spatial resolutions, there has developed a significant mismatch between frame rate and spatial resolution that few people appreciate [6]. In an attempt to mitigate this, TV manufacturers have introduced sophisticated up-sampling at the receiver to smooth motion at 300 or 600 Hz. While this approach can reduce peripheral flicker for larger screens, it does little or nothing to improve the motion blur.

The portrayal of motion for a given frame rate is a trade-off between long shutters that create motion blur and short shutters that, in cases of lower frame rates, can cause jerky motion and aliasing. We explore this further below.

Frame rates and shutter angles—static and dynamic resolutions

Considering Figure 13.2, this shows, for a fixed frame rate, the effect that shutter angle has on motion blur over a sequence of three frames. The difference between high frame rate capture and conventional rates is also demonstrated by Figure 13.3,

FIGURE 13.2

Frame rate and shutter angle. The figure illustrates the effect of a ball thrown from left to right across a static scene, captured at various shutter angles.

FIGURE 13.3

The influence of frame rate (*Outdoor* sequence). Left: 25 fps. Right: 600 fps.

where the effects of motion blur on the hands and spokes can clearly be seen. To reduce motion blur, we could shorten the shutter time. However, this leads to aliasing effects and jitter, especially in areas that are not being tracked by the viewer.

Let us define dynamic resolution as the effective spatial resolution of a format in the presence of motion (see Example 13.1). The relationship between resolution and frame rate for the case of TV systems is shown in Figure 13.4. This illustrates an important relationship between static and dynamic resolution. The top dashed line shows the frame rate needed to preserve dynamic resolution, extrapolated from a baseline of 50 Hz SDTV, as the spatial resolution is increased. It can be seen that, even at HDTV spatial resolutions, there is a significant reduction in dynamic resolution. Taking this further, to the 8k UHDTV image format, this extrapolation indicates that the corresponding frame rate should be something approaching 600 fps! In contrast,

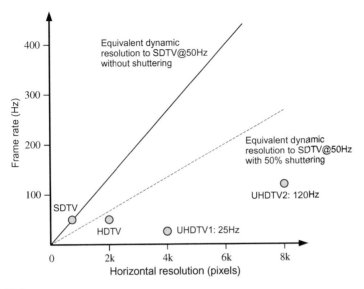

FIGURE 13.4

Static vs dynamic resolution—the relationship between frame rate and spatial resolution (adapted from an original by Richard Salmon, BBC).

UHDTV has been standardized with a maximum frame rate of 120 Hz, which is clearly well below this line.

Shorter shutter times and higher frame rates hold the potential to significantly improve the rendition of motion, especially for large screens where viewers will engage much more in wide angle tracking. They can reduce motion judder and the occurrence of background distractors caused by motion artifacts and flicker; they also provide an increased perception of depth in the image.

Compression methods and performance

The question to ask is how does frame rate interact with compression performance? A number of factors come into play here. Firstly, the temporal correlation between adjacent frames increases with frame rate. This produces smaller motion vector magnitudes and more correlated motion vectors. Secondly, with reduced frame periods, motion will more closely fit the translational model assumed in most codecs. Thus residual energy will be reduced and fewer bits will be required to code it. Higher frame rates, however, can introduce more high frequency spatial detail which can be harder to code in some cases. We can see these effects in Figures 13.5 and 13.6.

We would thus expect that the bit rate increase would be lower than the frame rate increase. Initial results in the Bristol-BBC Immersive Technology Laboratory indicate approximately a 2:1 ratio between the frame rate increase and the bit rate increase using H.264/AVC or HEVC. This is however highly content dependent.

FIGURE 13.5

Motion vector PDF comparisons for 25 fps and 600 fps.

FIGURE 13.6

Motion vectors and residual content for high frame rate content (*Mouse* sequence). Top: 400 fps. Bottom: 25 fps. Left: DFD signal. Right: spatial distribution of motion vectors.

Example 13.1 (Static and dynamic resolution)

Consider the case of an SD format with 720 horizontal pixels, compared with an 8k UHDTV system with 7680 horizontal pixels, both at 30 fps. What is the effective dynamic resolution of both formats for the case of an object traversing the width of the frame in 4 s?

Solution. Clearly the horizontal static resolution of the UHDTV format is over 10 times that of the SD format. However, when an object tracks horizontally across the image frame in 4 seconds, then this corresponds to a blurring across 6 pixels per frame for the SD image, compared with 64 pixels per frame for the UHD image. If the screens are the same size then the dynamic resolution of both formats will be the same (i.e. 120 pixels). However, if the UHD screen is larger and/or the viewer is closer to it, then its perceived dynamic resolution will actually be significantly worse than that for SD, due to the increased angle subtended at the retina by each pixel. Clearly this is not a great situation for our enhanced format.

Let us say, for example, that the viewing angle for the UHDTV screen is 100° and that for the SD screen is 20°. The dynamic resolution for each case is then given by:

$$\text{SD}: \ 120/20 \ = 6 \text{ pixels per degree}$$
$$\text{UHD}: 120/100 = 1.2 \text{ pixels per degree}$$

The expression for the viewer-screen-normalized dynamic resolution in pixels per degree is thus:

$$r_\theta = \frac{f}{2v_o \tan^{-1}\left(\frac{W}{2D}\right)}$$

where v_o is the object velocity in terms of frame widths per second, f is the frame rate, W is the width of the screen, and D is the viewing distance.

13.2.4 Dynamic range

Why dynamic range is important

It has been shown that increasing the dynamic range of content when displayed, both in terms of bit depth and screen bright to dark range, can make video appear more lifelike, increasing the perceived depth in the image and even imparting a sense of 3-D. The human visual system (HVS) can cover between 10 and 14 stops without adaptation and with adaptation it can accommodate 20 stops for photopic vision, and more for scotopic vision.

A typical modern flat panel TV, however, has a dynamic range capability of only around 8 stops. In order to fully exploit this limited dynamic range, methods such as black-stretch and pre-gamma mappings are used. The grading process in production also lifts detail in darker areas, and compresses highlights into the normal displayed range. New displays such as the SIM2 HDR47E offer the prospect of getting much closer to the capabilities of the HVS—delivering from approximately 0.4 to 4000 cd/m^2. This is achieved using a high dynamic range (HDR) (but low resolution) back-light LED array, in conjunction with a conventional LCD panel. Commercial cameras (e.g. the RED EPIC) are emerging with "HDR" capability but this is achieved using a complementary, delayed exposure for each frame.

The technological challenges of understanding and realizing HDR video are still significant. We must, for example, understand the masking effects of this type of content as well as the impact of the viewing environment. We also need to better understand the impact of compression on the immersive properties of such content.

Perceptual compression methods and performance

Several examples of approaches that exploit human perception during coding have been reported but, while these hold potential for significant coding gains, few have been adopted in practice. Naccari and Pereira [24] propose a perceptual video coding architecture incorporating a JND model based on spatio-temporal masking used for rate allocation and rate–quality optimization. They suggest the use of decoder-based JND modeling to perceptually allocate the available rate. The performance reported for their approach indicates an average bit rate reduction of up to 30% compared to H.264/AVC High profile at the same quality level.

Zhang et al. proposed a perception-based quantization method for high dynamic range content that exploits luminance masking [26] in the HVS in order to enhance the performance of HEVC. Extending the idea in Ref. [25] a profile scaling is proposed, based on a tone-mapping curve computed for each HDR frame. The quantization step is then perceptually tuned on a Transform Unit (TU) basis. The proposed method has been integrated into the reference codec considered for the HEVC range extensions and its performance compared with the case without perceptual quantization. Using HDR-VDP-2 image quality metric, a bit rate reduction of 9% was reported.

An example of the quantization curve is shown in Figure 13.7.

FIGURE 13.7

Intensity-dependent quantization for HDR extensions to HEVC.

13.2.5 **Parameter interactions and the creative continuum**

So what is required to enable us to create an optimum viewing experience? In essence, we need:

- An understanding of the influence of, and interactions within, the extended visual parameter space.
- Generalized psycho visual measures of quality, separating narrative from format/medium, and environment.
- Use of these measures to characterize the impact of distortion, dynamic range, color palette, spatial and temporal resolution.
- Representations, formats, acquisition, display, and delivery processes that preserve or enhance immersive properties.
- Parameter optimization and adaptation methods that take account of the influences of content and use-context-dependent manner.
- Compression methods that minimize bit rate while preserving immersive properties.
- An understanding of the impact of cinematographic methods, e.g. shot length, framing, camera placement, camera and subject motion.
- New acquisition, delivery and display technologies.

A question that naturally arises is—should we maintain the idea of a fixed parameter set at all, or should we be looking at new formats that move away from the notion of fixed frame rates and resolutions? Perhaps the most important message is that we should no longer consider acquisition, production, compression, display, and assessment as independent processes, but should instead view these as a *Creative Continuum*. This intimately links production, cinematography, and acquisition to delivery, display, consumption, and quality assessment. Rather than an end-to-end delivery system we should instead consider a set of continuous relationships within an extended parameter space as indicated in Figure 13.8, where:

- The *Experience* must maximize engagement with the displayed content. Measuring it is essential if we are to fully understand the influences of the delivery and display processes.
- The *Delivery* processes must ensure that the content is delivered in a manner that preserves the immersive properties of the format.
- The *Capture* processes must employ formats and parameters that enable immersive experiences.
- The *Creation* processes must be matched to the acquisition formats to ensure optimum exploitation of the format in terms of sets, lighting, etc.

Importantly, an ability to measure the quality of the experience, not simply in terms of video quality, but in terms of how engaging it is, is critical in optimizing other aspects of the delivery chain.

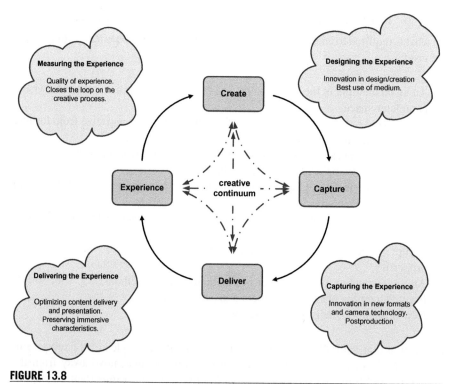

FIGURE 13.8

The creative continuum.

13.3 Challenges for compression

While the demand for new video services will, to some extent, be addressed through efficiency improvements in network and physical layer technology, the role of video compression remains of central importance in ensuring that content is delivered at an acceptable quality while remaining matched to the available bandwidth and variable nature of the channel. In the context of these growing demands for more content, at higher qualities and in more immersive formats, there is an urgent need for transformational solutions to the video compression problem, and these need to go well beyond the capabilities of existing standards.

All major video coding standards since H.261 have been based on incremental improvements to the hybrid motion-compensated block transform coding model. While this approach has produced impressive rate–distortion improvements, this author believes that more disruptive techniques can provide substantial additional gains. As we have seen, H.264/AVC is based on the picture-wise processing and waveform-based coding of video signals. HEVC is a generalization of this approach offering gains through improved intra-prediction, larger block sizes, more flexible ways of decomposing blocks for inter- and intra-coding, and better exploitation of long term correlations and picture dependencies.

Example 13.2 (The delivery challenge)

Let us for now put the above discussion in context by considering the effect that the immersive parameter set might have on bit rate. Let us assume that, for given content, optimum immersion is provided by the following fixed parameter values:

- Frame rate: 200 frames per second.
- Spatial resolution: UHDTV resolution at 7680×4320 pixels.
- Dynamic range: requiring 16 bits of dynamic range in R, G, and B.

Calculate the uncompressed bit rate for this format and compare its compression requirements with those for existing HDTV broadcasting and internet streaming systems.

Solution. Assuming no color sub-sampling, the overall uncompressed bit rate would be around 3×10^{11} bps. This is approximately 100 times greater than HDTV at 1080p50, 50,000 times greater than a typical current broadcast compressed bit rate and 100,000 times greater than high quality internet HD delivery.

Consider now Example 13.2. This clearly illustrates that, despite video compression advances providing a 50% reduction in bit rate every 10 years, such immersive parameter sets demand far greater bandwidths than are currently available. We thus not only need to discover the optimum video parameters, but also to understand the perceptual implications of compression in order to specify a compression ratio that, while exploiting psychovisual redundancy, exploits masking effects to minimize bit rate. It is clear that conventional compression techniques are unlikely to meet these requirements and hence new, perceptually driven methods will be necessary.

New approaches should be predicated on the assumption that, in most cases, the target of video compression is to provide good subjective quality rather than to minimize the error between the original and coded pictures. It is thus possible to conceive of a compression scheme where an analysis/synthesis framework replaces the conventional energy minimization approach. Such a scheme could offer substantially lower bit rates through region-based parameterization and reduced residual and motion vector coding. New and alternative frameworks of this type are beginning to emerge, where prediction and signal representations are based on a parametric or data-driven model of scene content. These often invoke a combination of waveform coding and texture replacement, where computer graphic models are employed to replace target textures at the decoder.

Such approaches can also benefit from the use of higher order motion models for texture warping and mosaicing or through the use of contextual scene knowledge. Preliminary work to date has demonstrated the potential for dramatic rate–quality improvements with such methods. It is clear however that a huge amount of research is still needed in order to yield stable and efficient solutions, before we can fully exploit the potential of these new approaches. For example, mean square error is no longer a valid objective function or measure of quality and emphasis must shift from rate–distortion to rate–quality optimization, demanding new embedded perceptually driven quality metrics (see Chapter 10).

The choice of texture analysis and synthesis models, alongside meaningful quality metrics and the exploitation of long-term picture dependencies, will be key if an effective and reliable system is to result. Furthermore, in order to rigorously evaluate these new methods, challenging test data (including high dynamic range, high resolution and high frame rate content) will be required. The hypothesis that underpins this approach is:

If consistent, representative spatio-temporal descriptions can be obtained for static and dynamic textures and a valid perceptual distortion measure can be defined, then these descriptions can be employed within a rate–quality optimized parametric framework to dramatically improve video compression performance.

13.4 Parametric video compression

Parametric video compression falls into the class of perception-based compression methods [16] and uses an analysis/synthesis framework rather than the conventional energy minimization approach. Parametric methods are employed to describe texture warping and/or synthesis as reported by Ndjiki-Nya et al. [8,9], Bosch et al. [10], Byrne et al. [11,12], Stojanovic et al. [22], Zhu et al. [23], and Zhang and Bull [13–15]. Although this approach has significant potential, a number of problems still need to be resolved.

We now focus on the work of Zhang and Bull [15], as an example of this class of codec. Their work combines dynamic and static texture synthesis using robust region segmentation, classification, and quality assessment. In a similar manner to that of other authors, the approach is hosted within a conventional block-based codec which is invoked wherever texture synthesis fails or is inappropriate. Textured regions in images or video frames have arbitrary shapes and these must first be segmented into homogeneous regions, each sharing similar texture properties. The authors employ a spatial texture-based image segmentation algorithm using an enhanced watershed transform [18] to obtain reliable textured regions.

To provide good texture coding performance, textures must be separated into static and dynamic classes. Classification rules are therefore employed to differentiate between them as well as to distinguish non-textured regions. Classification is based on an analysis of wavelet subband coefficients and motion information within each texture region using features from the Dual-tree Complex Wavelet Transform (DT-CWT) [19].

Texture synthesis is commonly employed in computer graphics, where synthetic content is generated using spatial and temporal methods [20]. Zhang and Bull apply different texture synthesis methods to static and dynamic textured regions according to their statistical and spectral characteristics. Static textures are coded using a perspective model, while dynamic textures are coded using a modified version of the synthesis method of Doretto et al. [21]. The primary benefits of this approach are that residual coding is generally not required and only side information (motion parameters and warping/synthesis maps) needs to be encoded.

One of the most challenging problems for synthesis-based coding is to create a reliable in-loop quality assessment measure with which to estimate subjective quality and detect any possible coding artifacts. Existing distortion-based metrics, such as peak signal-to-noise ratio (PSNR) and structural similarity (SSIM), are known to be inappropriate for this type of compression and other perceptual measures are too complex and perform poorly. A meaningful and simple objective video metric (AVM) was thus developed by Zhang and Bull which provides reliable quality estimates of synthesized textures, with low complexity due to extensive reuse of coding parameters (e.g. DT-CWT coefficients and motion vectors).

The overall architecture of the parametric compression scheme is shown in Figure 13.9. Figure 13.10 shows the results of the texture classification stage and

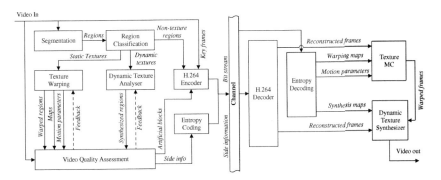

FIGURE 13.9

Parametric video compression architecture (Reproduced with permission from Zhang and Bull [15]).

FIGURE 13.10

Texture classification results for parametric video coding. Left: original frame. Right: classified regions for coding: Red—static, blue—dynamic, black—structural, or non-textured (Reproduced with permission from Ref. [15]). (For interpretation of the references to color in this figure legend, the reader is referred to the web version of this book.)

FIGURE 13.11

Coding results from parametric video coding. Left: H.264 frame. Right: proposed method. Center: coding mode employed (from Ref. [15]).

Figure 13.11 gives an example of a synthesized frame compared with H.264/AVC. The results presented in Ref. [15] indicate savings up to 60% over and above those offered by the host codec (H.264/AVC) alone for the same subjective quality.

13.5 Context-based video compression

Certain applications, such as sports broadcasting, are highly demanding due to their activity levels and the perceptual quality levels required. Such content is however often captured in a closed and well-defined environment such as a sports arena. The context-based coding method of Vigars et al. [17] exploits this to produce a hybrid CODEC framework which is able to exploit prior knowledge about the geometry of a scene. It applies a planar-perspective motion model to rigid planar and quasi-planar regions of the video. Unlike mosaic and sprite coding, this enables independent planar regions to be coded accurately without the need for registration, blending, or residual coding. It works well for both global and local perspective distortions. Non-planar or otherwise non-conforming regions are coded as normal by a host CODEC such as H.264/AVC or HEVC.

Figure 13.12 shows the matching process across two frames based on feature point matching. Firstly prior knowledge of the environment in which the video is captured is encapsulated into scene model. This can be done off- or on-line. Feature matching is then used to detect salient points in each frame, and to compose descriptors of neighborhoods for matching and tracking between images. This is then used to track known planar structures in the video. Vigars et al. use the SURF algorithm for this.

Planar structures in each frame of the video are thus located, facilitating perspective motion estimation between them. Foreground segmentation is then used to separate regions of the video which do not conform to the planar model. These may be foreground objects, reflective surfaces, dynamic textures, or other regions which do not conform. Such foreground regions are processed by the host codec. Vigars et al. report savings up to 50% for this method compared with conventional H.264/AVC encoding (see Figure 13.13).

FIGURE 13.12
SURF-based feature matching for context-based coding of planar regions. Left: reference frame. Right: current frame. Circles represent RANSAC inliers used for planar motion modeling.

FIGURE 13.13
Context-based video compression results. Left: H.264 coded. Right: context-based coding.

13.6 **Summary**

This chapter has summarized the likely future demands and challenges for video compression. It has been postulated that the increased bit rates demanded will, for many applications, outstrip those provided by advances in network technology and conventional compression methods. The hybrid codec, that has served us well for the past 30 years, will no doubt continue to dominate for the foreseeable future, but could be enhanced through the exploitation perception-based methods such as those described here.

We have seen that the Human Visual System (HVS) exhibits non-linear sensitivities to the distortions introduced by lossy image and video coding. There are several factors that contribute to this, including luminance masking, contrast masking, and spatial and temporal frequency masking. Coding methods in the future must attempt to exploit these phenomena to a greater extent. It is thus proposed that new approaches to compression, possibly based on increased contextual knowledge and the use of analysis/synthesis models, could provide the next step toward delivering content at the required quality for future immersive applications.

References

[1] Bureau of Labor Statistics, American Time Use Survey, 2010.

[2] <http://www.wireless-world-research.org/fileadmin/sites/default/files/publications/-Outlook/Outlook4.pdf>.

[3] Cisco Visual Networking Index: Forecast and Methodology, 2011–16 (update 2012–17). <http://www.cisco.com/en/US/netsol/ns827/networking_solutions_sub_solution.html>.

[4] Recommendation ITU-R BT.2020 (08/2012), Parameter Values for Ultra-High Definition Television Systems for Production and International Programme Exchange, ITU-R, 2012.

[5] S. Sakaida, N. Nakajima, A. Ichigaya, M. Kurozumi, K. Iguchi, Y. Nishida, E. Nakasu, S. Gohshi, The super HiVision codec, in: Proceedings of the IEEE International Conference on Image Processing, 2007, pp. I-21–I-24.

[6] R. Salmon, M. Armstrong, S. Jolly, Higher Frame Rates for More Immersive Video and Television, BBC White Paper WHP209, BBC, 2009.

[7] P. Hanhart, M. Rerabek, F. DeSimone, T. Ebrahimi, Subjective quality evaluation of the upcoming HEVC video compression standard, in: Applications of Digital Image Processing XXXV, vol. 8499, 2012.

[8] P. Ndjiki-Nya, C. Stuber, T. Wiegand, Texture synthesis method for generic video sequences, in: IEEE International Conference on Image Processing, vol. 3, 2007, pp. 397–400.

[9] P. Ndjiki-Nya, T. Hinz, C. Stuber, T. Wiegand, A content-based video coding approach for rigid and non-rigid textures, in: IEEE International Conference on Image Processing, 2006, pp. 3169–3172.

[10] M. Bosch, M. Zhu, E. Delp, Spatial texture models for video compression, in: IEEE International Conference on Image Processing, 2007, pp. 93–96.

[11] J. Byrne, S. Ierodiaconou, D.R. Bull, D. Redmill, P. Hill, Unsupervised image compression-by-synthesis within a JPEG framework, in: IEEE International Conference on Image Processing, 2008, pp. 2892–2895.

[12] S. Ierodiaconou, J. Byrne, D. Bull, D. Redmill, P. Hill, Unsupervised image compression using graphcut texture synthesis, in: IEEE International Conference on Image Processing, 2009, pp. 2289–2292.

[13] F. Zhang, D. Bull, N. Canagarajah, Region-based texture modelling for next generation video codecs, in: IEEE International Conference on Image Processing, 2010, pp. 2593–2596.

[14] F. Zhang, D. Bull, Enhanced video compression with region-based texture models, in: Picture Coding Symposium (PCS), 2010, pp. 54–57.

[15] F. Zhang, D. Bull, A parametric framework for video compression using region-based texture models, IEEE Journal of Selected Topics in Signal Processing 6 (7) (2011) 1378–1392.

[16] J. Lee, T. Ebrahimi, Perceptual video compression: a survey, IEEE Journal of Selected Topics in Signal Processing 6 (2012) 684–697.

[17] R. Vigars, A. Calway, D. Bull, Context-based video coding, in: Proceedings of the IEEE International Conference on Image Processing, 2013, pp. 1953–1957.

[18] R. O'Callaghan, D. Bull, Combined morphological-spectral unsupervised image segmentation, IEEE Transactions on Image Processing 14 (1) (2005) 49–62.

[19] N. Kingsbury, Complex wavelets for shift invariant analysis and filtering of signals, Journal of Applied and Computational Harmonic Analysis 10 (3) (2001) 234–253.

[20] V. Kwatra, A. Schodl, I. Essa, G. Turk, A. Bobic, Graphcut textures: Image and video synthesis using graph cuts, in: Proceedings of the SIGGRAPH, ACM, 2003, pp. 277–286.

[21] G. Doretto, A. Chiuso, Y. Wu, S. Soatto, Dynamic textures, International Journal of Computer Vision 51 (2) (2003) 91–109.

[22] A. Stojanovic, M. Wien, J.-R. Ohm, Dynamic texture synthesis for H.264/AVC inter coding, in: IEEE International Conference on Image Processing, 2009, pp. 1608–1611.

[23] C. Zhu, X. Sun, F. Wu, H. Li, Video coding with spatio-temporal texture synthesis and edge-based inpainting, in: Proceedings of the ICME (2008) 813–816.

[24] M. Naccari, F. Pereira, Advanced H.264/AVC-based perceptual video coding: architecture, tools, and assessment, IEEE Transactions on Circuits and Systems for Video Technology 21 (6) (2011) 766–782.

[25] M. Naccari, M. Mrak, D. Flynn, A. Gabriellini, Improving HEVC compression efficiency by intensity dependent spatial quantisation, JCTVC-J0076, 10th Meeting, Stockholm, SE, July 2012.

[26] Y. Zhang, M. Naccari, D. Agrafiotis, M. Mrak, D. Bull, High dynamic range video compression by intensity dependent spatial quantization, in: Picture Coding Symposium, December 2013.

Appendix A: Glossary of Terms

1-D   One-dimensional
2-D   Two-dimensional
3-D   Three-dimensional
4PMV+0   Four predicted motion vectors plus zero vector
3DTV   Three-dimensional television

AC   Alternating current. Used to denote all transform coefficients except the zero frequency coefficient
ACR   Absolute category rating
ACK   Acknowledge
A-D   Analog to digital
ADSL   Asymmetric digital subscriber line
AFD   Active format description
AMVP   Advanced motion vector prediction
ARQ   Automatic retransmission request
ASP   Advanced simple profile (of MPEG-4)
AV   Audiovisual
AVC   Advanced video codec (H.264)
AVM   Artifact-based video metric

B   Bi-predicted picture
BDM   Block distortion measure
BER   Bit error rate
BMA   Block matching algorithm
BMME   Block matching motion estimation
bpp   Bits per pixel
bps   Bits per second
BPSK   Binary phase shift keying

CABAC   Context-adaptive binary arithmetic coding
CAVLC   Context-adaptive variable length coding
CB   Coding block
CBP   Coded block pattern
CBR   Constant bit rate
CCD   Charge coupled device
CCIR   International radio consultative committee (now ITU)
CD   Compact disc
CDMA   Code division multiple access
CFF   Critical flicker frequency
CIE   Commission Internationale de L'Eclairage (the international color science commission)

Communicating Pictures. http://dx.doi.org/10.1016/B978-0-12-405906-1.00023-4
© 2014 Elsevier Ltd. All rights reserved.

CIF	Common intermediate format
CMY	Color primaries cyan, magenta, and yellow
codec	Encoder and decoder
CPB	Coded picture buffer
cpd	Cycles per degree
CRA	Clean random access
CRC	Cyclic redundancy check
CRT	Cathode ray tube
CSF	Contrast sensitivity function
CTB	Coding tree block
CTU	Coding tree unit
CU	Coding unit
CW-SSIM	Complex wavelet SSIM
DAB	Digital audio broadcasting
DC	Direct current. Refers to zero frequency transform coefficient.
DCT	Discrete cosine transform
DFD	Displaced frame difference
DFT	Discrete Fourier transform
DMOS	Difference mean opinion score
DoG	Difference of Gaussians
DPCM	Differential pulse code modulation
DS	Diamond search
DSCQS	Double stimulus continuous quality scale
DSIS	Double stimulus impairment scale
DSP	Digital signal processor
DST	Discrete sine transform
DT-CWT	Dual tree discrete wavelet transform
DVB	Digital video broadcasting
DVD	Digital versatile disc
DWHT	Discrete Walsh–Hadamard transform
DWT	Discrete wavelet transform
EBCOT	Embbeded Block Trunction Coding (EBCOT)
EBMA	External boundary matching algorithm
EBME	External block matching error
EBU	European Broadcasting Union
EECMS	Enhanced error concealment with mode selection
EG	Exp-Golomb probability distribution and entropy coder
EOB	End of block
EREC	Error-resilient entropy coding
EZW	Embedded zero-tree wavelet
FD	Frame difference
FEC	Forward error correction

FEF	Frontal eye field
FFA	Fusiform face area
FFT	Fast Fourier transform
FGS	Fine granularity scalability
FIR	Finite impulse response (filter)
FLC	Fixed length coding
FMO	Flexible macroblock ordering
fps	Frames per second
FR	Full reference
FRTV	Full reference TV VQEG database
FS	Full search
GIF	Graphics interchange format
GOB	Group of blocks
GOP	Group of pictures
HARQ	Hybrid ARQ
HDR	High dynamic range
HDTV	High definition television
HEVC	High efficiency video codec
HFR	High frame rate
HRD	Hypothetical reference decoder
HTTP	Hypertext transfer protocol
HVS	Human visual system
HEXBS	Hexagon-based search method
HM	HEVC test model
I	Intra-coded picture
IAR	Image aspect ratio
IDR	Instantaneous decoder refresh
IEC	International Electrotechical Commission
IEEE	Institute of Electrical and Electronic Engineers
i.i.d.	Independent and identically distributed
IIR	Infinite impulse response (filter)
IP	Internet protocol
ISDN	Integrated services digital network
ISO	International Standards Organization
IT	Inferior temporal cortex
ITU	International Telecommunications Union; -R, Radio; -T, Telecommunications
JM	H.264 Joint reference model
JND	Just-noticeable difference
JPEG	Joint Photographic Experts Group
JPSEC	JPEG2000 security extension
JPWL	JPEG2000 wireless extension

kbps	Kilobits per second
KLT	Karhunen–Loeve transform
LAN	Local area network
LCC	Linear correlation coefficient
LCD	Liquid crystal display
LCU	Largest coding unit
LDPC	Low density parity check (codes)
LGN	Lateral
LIVE	Laboratory for Image and Video Engineering
LSI	Linear shift invariant
LTE	Long term evolution (4G mobile radio technology)
LTI	Linear time invariant
LZW	Lempel–Ziv–Welch
MAC	Multiply and accumulate operation
MAC	Medium access control
MAD	Mean absolute difference
MAD	Most apparent distortion
MB	Macroblock
Mbps	Megabits per second
MC	Motion compensation
MCP	Motion-compensated prediction
MCS	Modulation and coding mode selection
MCTF	Motion-compensated temporal filtering
MDC	Multiple description coding
ME	Motion estimation
MEC	Motion estimation and compensation
MIMO	Multiple input multiple output
MMSE	Minimum mean squared error
MOS	Mean opinion score
MOVIE	Video quality metric from LIVE
MPEG	Motion Picture Experts Group
MRF	Multiple reference frame
MSB	Most significant bit
MSE	Mean squared error
MS-SSIM	Multiscale SSIM
MST	Medial superior temporal
MT	Medial temporal
MV	Motion vector
MVD	Differential motion vector
NACK	Negative acknowledge
NAL	Network abstraction layer
NALU	Network abstraction layer unit

NCCF	Normalized cross correlation coefficient
NR	No reference (metric)
NSS	*N*-step search
OBMC	Overlapped block motion compensation (or just OMC)
OR	Outlier ratio
P	Predicted picture
PAL	Phase alternating line
PB	Prediction block
PCA	Principal component analysis
PCM	Pulse code modulation
pdf	Probability density function
PER	Packet error rate
PhC	Phase correlation
PHY	Physical layer
PMR	Private mobile radio
PNG	Portable network graphics
PPS	Picture parameter set
PRF	Periodic reference frame
PSD	Power spectral density
PSNR	Peak signal to noise ratio
PU	Prediction unit
PVM	Perception inspired video metric
PVQ	Pyramid vector quantization
QAM	Quadrature amplitude modulation
QCIF	Quarter CIF resolution
QMF	Quadrature mirror filter
QOS	Quality of service
QP	Quantization parameter
QPSK	Quadrature phase shift keying
RAP	Random access point
RBER	Residual bit error rate
RDO	Rate–distortion optimization
RF	Radio frequency
RGB	Red, green, and blue color primaries
RPS	Reference picture selection
RPS	Reference picture set (HEVC)
RQO	Rate–quality optimization
RR	Reduced reference (metric)
RSE	Reed–Solomon erasure code
RSSI	Residual signal strength indication
RTCP	Real-time control protocol
RTP	Real-time transmission protocol

RTSP	Real-time streaming protocol
RVLC	Reversible variable length coding
SAD	Sum of absolute differences
SAMVIQ	Subjective Assessment Methodology for Video Quality
SAO	Sample adaptive offset
SATD	Sum of absolute transform differences
SDTV	Standard definition television
SE	Syntax element
SEC	Spatial error concealment
SEI	Supplemental enhancement information
SG	Study group (of ITU)
SG	Slice group
SI	Spatial information
SIFT	Scale invariant feature transform
SMPTE	Society of Motion Picture and Television Engineers
SNR	Signal to noise ratio
SPIHT	Set partitioning into hierarchical trees
SPS	Sequence parameter set
SROCC	Spearman rank order correlation coefficient
SSCQS	Single stimulus continuous quality scale
SSD	Sum of squared differences
SSIM	Structural similarity index
SSIS	Single stimulus impairment scale
ST-MAD	Spatio-temporal most apparent distortion
SURF	Speeded up robust features
SVC	Scalable video codec
SVD	Singular value decomposition
TB	Transform block
TCP	Transmission control protocol
TDL	Two-dimensional logarithmic search
TEC	Temporal error concealment
TI	Temporal information
TS	Transport stream
TSCES	Triple stimulus continuous evaluation scale
TSS	Three-step search
TU	Transform unit
TV	Television
UDP	User datagram protocol
UHDTV	Ultra high definition television
UHF	Ultra high frequency
UMTS	Universal mobile telecommunications system
URL	Universal resource locator

V1	Region of visual cortex (also V2–V5)
VBR	Variable bit rate
VCEG	Video coding experts group
VCL	Video coding layer
VDP	Visible difference predictor
VDSL	Very high bit rate digital subscriber line
VLC	Variable length coding
VLD	Variable length decoding
VQ	Vector quantization
VQEG	Video Quality Experts Group
VQM	Video quality metric
VSNR	Visual signal to noise ratio
VSTM	Visual short term memory
WBA	Warping-based algorithm (motion estimation)
WSS	Wide sense stationary
YC_bC_r	Color coordinate system comprising luminance, Y, and two chrominance channels, C_b and C_r
YUV	Color coordinate system comprising luminance, Y, and two chrominance channels, U and V

Appendix B: Tutorial Problems

Chapter 1: Introduction

Q1.1 What is the primary purpose of standardization in video compression? List two other advantages of standardization.

Q1.2 Using the example of a DVB-T2 terrestrial broadcast system transmitting HDTV video content to the home, explain why digital video compression is needed.

Q1.3 Consider the case of 4 K UHDTV, with the original video in 4:2:2 (a luma signal of 3840×2160 and two chroma signals of 1920×2160) format at 10 bits and a frame rate of 50 fps. Calculate the compression ratio if this video is to be transmitted over a DVB-T2 link with an average bandwidth of 15 Mb/s.

Chapter 2: The human visual system

Q2.1 Assuming that the field of view within the fovea is $2°$, compute the number of pixels that fall horizontally within the foveated visual field. Assume a 1 m wide screen with 1920 horizontal pixels viewed at a distance of $3H$.

Q2.2 The following table lists a number of important features of the human visual system (HVS). Complete the table by describing the influence each feature has on the design of a digital video compression system.

HVS characteristic	Implication for compression
HVS is more sensitive to high contrast image regions than low contrast regions	?
HVS is more sensitive to luminance than chrominance information	?
HVS is more sensitive to low spatial frequencies than high spatial frequencies	?
In order to achieve a smooth appearance of motion, the HVS must be presented with image frames above a certain minimum rate (and this rate depends on ambient light levels)	?
HVS responses vary from individual to individual	?

Q2.3 Calculate the normalized contrast sensitivity of the human visual system for a luminance-only stimulus at a spatial frequency of 10 cycles per degree.

Communicating Pictures. http://dx.doi.org/10.1016/B978-0-12-405906-1.00024-6

Chapter 3: Discrete-time analysis for images and video

Q3.1 Plot the sinusoidal signal $x(t) = \cos(20\pi t)$. Compute the frequency spectrum of this sinusoid and plot its magnitude spectrum.

Q3.2 Assume that the sinusoidal signal in Q3.1 is sampled with $T = 0.1$ s. Sketch the magnitude spectrum of the sampled signal and comment on any aliasing issues.

Q3.3 If an HDTV (full HD) screen with aspect ratio 16:9 has a width of 1.5 m and is viewed at distance of $4H$, what is the angle subtended by each pixel at the retina?

Q3.4 The impulse response for the Le Gall high-pass analysis filter is $h_1[n] = \{0.25 \quad -0.5 \quad 0.25\}$. Compute the output from this filter for the input sequence: $x[n] = \{1 \quad 2 \quad 3 \quad 0 \quad 0 \quad \ldots\}$.

Q3.5 Compute the z-plane pole zero plots and the frequency responses for the following filter pair: $H_0(z) = 1 + z^{-1}$ and $H_1(z) = 1 - z^{-1}$.

Q3.6 Perform median filtering on the following input sequence using a five-tap median filter and comment on the result:

$$x[n] = \{1, 3, 5, 13, 9, 11, 6, 15, 17, 19, 29\}$$

Q3.7 Compute the basis functions for the two-point DFT.

Q3.8 Consider the two 1-D digital filters:

$$\mathbf{h}_1 = [1 \quad 2 \quad 2 \quad 1]^T$$
$$\mathbf{h}_2 = [1 \quad -3 \quad -3 \quad 1]^T$$

Compute the equivalent 2-D digital filter where $\mathbf{h}_1$ performs horizontal filtering and $\mathbf{h}_2$ performs vertical filtering.

Q3.9 Compute biased and unbiased correlation estimates for the following sequence:

$$x[n] = \{1, 2, 5, -1, 3, 6, -4, -1\}$$

Q3.10 Plot the autocorrelation function, $r_v(k)$, for a white noise sequence $v[n]$ with variance σ_v^2. Form the autocorrelation matrix for this sequence for lags up to ± 3. What is the inverse of this autocorrelation matrix?

Q3.11 A feedback-based linear predictor with quantization uses a predictor $P(z) = (z^{-1} + z^{-2})/2$. Assume an input sequence as given below:

$$x[n] = \{1, 3, 4, 3, 5, 6 \ldots\}$$

and that quantization is performed as follows, with rounding of 0.5 values toward zero:

$$e_Q[n] = \text{rnd}\left(\frac{e[n]}{2}\right); \quad e_R[n] = 2e_Q[n]$$

compute the predictor output sequence $e_Q[n]$ and the reconstructed output signal $y[n]$ from the decoder. Comment on your results in terms of numerical precision.

Q3.12 Compute the entropies of the sequences $x[n]$ and $e_Q[n]$ from Q3.11. Comment on your result.

Chapter 4: Digital picture formats and representations

Q4.1 In video coding schemes it is usual to code the color components in the form Y, C_b, C_r rather than R, G, B. Explain why this approach is justified, its benefits in terms of compression and how it is exploited in image sampling. Explain how pictures can be efficiently stored or transmitted using a 4:2:0 format.

Q4.2 Compute YUV and YC_bC_r vectors for the following RGB vectors:

(a) $[R, G, B] = \begin{bmatrix} 128 & 128 & 128 \end{bmatrix}$

(b) $[R, G, B] = \begin{bmatrix} 255 & 255 & 255 \end{bmatrix}$

(c) $[R, G, B] = \begin{bmatrix} 100 & 0 & 0 \end{bmatrix}$

Q4.3 If a color movie of 100 min duration is represented using ITU-R.601 (720×576, 25 fps@8 bits, 4:2:0 format):

(a) What hard disk capacity would be required to store the whole movie?

(b) If the movie is encoded at a compression ratio CR = 50:1 and transmitted over a satellite link with 50% channel coding overhead, what is the total bit rate required for the video signal?

Q4.4 Given a video sequence with a spatial resolution of 1920×1080 at 50 fps using 10 bit color sampling, compute the (uncompressed) bit rates of 4:4:4, 4:2:0, and 4:2:2 systems.

Q4.5 If, for a given 1920×1080 I-frame in 4:2:0 format, the mean entropy of each coded 16×16 luminance block is 1.3 bits/sample and that for each corresponding chrominance block is 0.6 bits/sample, then estimate the total number of bits required to code this frame.

Q4.6 Calculate the MAD between the following original image block $\mathbf{X}$ and its encoded and decoded version $\tilde{\mathbf{X}}$:

$$\mathbf{X} = \begin{bmatrix} 2 & 4 & 4 & 6 \\ 3 & 6 & 6 & 6 \\ 5 & 7 & 7 & 8 \\ 3 & 7 & 7 & 8 \end{bmatrix}; \quad \tilde{\mathbf{X}} = \begin{bmatrix} 2 & 5 & 5 & 5 \\ 4 & 4 & 6 & 6 \\ 5 & 5 & 7 & 8 \\ 4 & 5 & 7 & 7 \end{bmatrix}$$

Q4.7 Assuming a 5 bit digital image block, $\mathbf{X}$, the reconstruction after image compression is given by $\mathbf{Y}$. Calculate the PSNR for the reconstructed signal and provide an interpretation of the results in terms of error visibility.

$$\mathbf{X} = \begin{bmatrix} 3 & 8 & 1 & 8 \\ 7 & 0 & 5 & 0 \\ 2 & 6 & 0 & 5 \\ 4 & 1 & 10 & 2 \end{bmatrix} ; \quad \mathbf{Y} = \begin{bmatrix} 4 & 10 & 1 & 9 \\ 7 & 0 & 6 & 1 \\ 3 & 6 & 0 & 4 \\ 5 & 1 & 12 & 3 \end{bmatrix}$$

Q4.8 Assuming an 8 bit digital image, **X**, the reconstructions due to two alternative coding schemes are given by $\mathbf{Y}_1$ and $\mathbf{Y}_2$ below. Calculate the Peak Signal to Noise Ratio (PSNR) for each of the reconstructions and give an interpretation of the results in terms of error visibility.

$$\mathbf{X} = \begin{bmatrix} 20 & 17 & 18 \\ 15 & 14 & 15 \\ 19 & 13 & 14 \end{bmatrix} ; \quad \mathbf{Y}_1 = \begin{bmatrix} 19 & 18 & 17 \\ 16 & 15 & 14 \\ 18 & 14 & 13 \end{bmatrix} ; \quad \mathbf{Y}_2 = \begin{bmatrix} 20 & 17 & 18 \\ 15 & 23 & 15 \\ 19 & 13 & 14 \end{bmatrix}$$

Q4.9 Describe a typical GOP structure used in MPEG-2 television broadcasting, explaining the different types of frame coding employed, the predictive relationships between all frames in the GOP, and the transmission order of the frames. If a transmission error affects the second P-frame in the GOP, how many pictures are likely to be affected due to error propagation?

Q4.10 An HDTV satellite operator allocates 10 Mb/s to each program in the DVB-S multiplex. State what color sub-sampling format will be used for this and calculate the required compression ratio for a 1080i25 system.

Q4.11 Compute the gamma corrected version of the following image block for the case where $\gamma = 0.45$. Assume an 8 bit wordlength.

$$\mathbf{X} = \begin{bmatrix} 20 & 17 & 18 \\ 15 & 14 & 15 \\ 19 & 13 & 14 \end{bmatrix}$$

Chapter 5: Transforms for image and video coding

Q5.1 Derive the 1-D two-point KLT for a stationary real-valued process, x, that has the following autocorrelation matrix:

$$\mathbf{R}_x = \begin{bmatrix} r_{xx}(0) & r_{xx}(1) \\ r_{xx}(1) & r_{xx}(0) \end{bmatrix}$$

Q5.2 Prove that the following vectors are an orthonormal pair:

$$\mathbf{a}_0 = \begin{bmatrix} \frac{\sqrt{2}}{2} & \frac{\sqrt{2}}{2} \end{bmatrix}^T ; \quad \mathbf{a}_1 = \begin{bmatrix} \frac{\sqrt{2}}{2} & -\frac{\sqrt{2}}{2} \end{bmatrix}^T$$

Q5.3 Prove that the four-point DWHT is a unitary transform.

Q5.4 Compute the basis functions for the eight-point 1-D DWHT. Compute the first four basis functions for the four-point 2-D DWHT.

Q5.5 Use the DWHT to transform the following 2-D image block, **S**. Assuming that all data and coefficients are represented as 6 bit numbers and that compression is achieved in the transform domain by selecting the four most dominant coefficients for transmission, compute the decoded data matrix and its PSNR.

$$\mathbf{S} = \begin{bmatrix} 5 & 6 & 8 & 10 \\ 6 & 6 & 5 & 7 \\ 4 & 5 & 3 & 6 \\ 8 & 7 & 5 & 5 \end{bmatrix}$$

Q5.6 Given the four-point DWHT basis functions, if the coefficients after transformation are: $c(0) = 1$; $c(1) = \frac{1}{4}$; $c(2) = -\frac{1}{2}$; $c(3) = 3$, plot the weighted basis functions and hence reconstruct the original signal waveform.

Q5.7 The 1-D discrete cosine transform is given by:

$$C(k) = \sqrt{\frac{2}{N}} \varepsilon_k \sum_{n=0}^{N-1} x[n] \cos\left(\frac{\pi(n+0.5)k}{N}\right) \quad 0 \leq n, k \leq N-1$$

where:

$$\varepsilon_k = \begin{cases} 1/\sqrt{2} & k = 0 \\ 1 & \text{otherwise} \end{cases}$$

Calculate the numerical values of the basis functions for a 1-D 4.4 DCT.

Q5.8 The DCT is an orthonormal transform. Explain the term orthonormal and describe what it means in practice for the transform. Write down the formula and the basis function matrix for the 1-D four-point DCT. Using this, show that, for the 1-D DCT:

$$\varepsilon_k = \begin{cases} 1/\sqrt{2} & k = 0 \\ 1 & \text{otherwise} \end{cases}$$

Q5.9 Compute the DCT transform coefficient vector for an input sequence, $\mathbf{x} = [1001]^T$.

Q5.10 Calculate the DCT of the following 2×2 image block:

$$\mathbf{X} = \begin{bmatrix} 21 & 19 \\ 15 & 20 \end{bmatrix}$$

Q5.11 Quantize the result from Q5.9 using the following quantization matrix:

$$\mathbf{Q} = \begin{bmatrix} 4 & 8 \\ 8 & 8 \end{bmatrix}$$

Q5.12 Perform inverse quantization and an inverse DCT on the output from Q5.10.

Q5.13 Compute the 2-D DCT of the following (4 × 4) image block:

$$
\mathbf{X} = \begin{bmatrix} 1 & 0 & 0 & 0 \\ 0 & 0 & 0 & 0 \\ 0 & 0 & 0 & 0 \\ 1 & 0 & 0 & 0 \end{bmatrix}
$$

Q5.14 Given the following block of DCT coefficients and the associated quantization matrix, compute the block of quantized coefficients. Perform zig-zag scanning to form a string of {run/value} symbols (where "run" is the number of zeros preceding a non-zero value) appropriate for entropy coding.

$$
\mathbf{C} = \begin{bmatrix}
128 & 50 & -20 & 22 & 12 & 27 & -5 & 7 \\
40 & -25 & 26 & 20 & -34 & -2 & 13 & -5 \\
-10 & 22 & 12 & 12 & 26 & 12 & 3 & 8 \\
12 & -2 & 16 & -7 & 9 & 3 & 17 & 17 \\
-32 & 6 & 21 & 9 & 18 & 5 & 4 & 7 \\
-10 & -7 & -14 & 3 & -2 & 13 & 18 & 18 \\
11 & -9 & -9 & 4 & 8 & 13 & 6 & 9 \\
-7 & 19 & 15 & 8 & 6 & -6 & 18 & 33
\end{bmatrix}
$$

$$
\mathbf{Q} = \begin{bmatrix}
8 & 16 & 19 & 22 & 26 & 27 & 29 & 34 \\
16 & 16 & 22 & 24 & 27 & 29 & 34 & 37 \\
19 & 22 & 26 & 27 & 29 & 34 & 34 & 38 \\
22 & 22 & 26 & 27 & 29 & 34 & 37 & 40 \\
22 & 26 & 27 & 29 & 32 & 35 & 40 & 48 \\
26 & 27 & 29 & 32 & 35 & 40 & 48 & 58 \\
26 & 27 & 29 & 34 & 38 & 46 & 56 & 69 \\
27 & 29 & 35 & 38 & 46 & 56 & 69 & 83
\end{bmatrix}
$$

Q5.15 Calculate the number of multiply and accumulate (MAC) operations required to compute a conventional (4 × 4) point 2-D DCT. Assume that the separability property of the 2-D DCT is exploited.

Q5.16 The complexity of the DCT can be reduced using "Fast" methods such as McGovern's Algorithm. Derive McGovern's algorithm for a four-point 1-D DCT. Compare its complexity (again assuming exploitation of separability) with that of the conventional approach for the case of a (4 × 4) point 2-D DCT.

Chapter 6: Filter banks and wavelet compression

Q6.1 For a filterbank downsampler, show that the frequency domain behavior of the output signal, $x_d[n]$, is related to that of the input $x[n]$ by:

$$
X_d(\Omega) = 0.5 \left[X \left(e^{j\Omega/2} \right) + X \left(e^{-j\Omega/2} \right) \right]
$$

Q6.2 The figure below shows a simple two-band analysis–synthesis filter bank and a representative input spectrum.

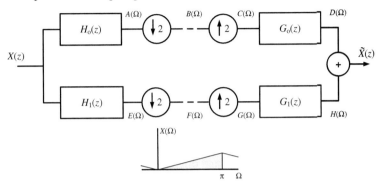

It can be shown that the output of this system is given by:

$$\tilde{X}(z) = \frac{1}{2}X(z)\left[G_0(z)H_0(z) + G_1(z)H_1(z)\right]$$
$$+ \frac{1}{2}X(-z)\left[G_0(z)H_0(-z) + G_1(z)H_1(-z)\right]$$

where the upsampler and downsampler relationships are given by:

$$C\left(e^{j\Omega}\right) = B\left(e^{j2\Omega}\right)$$
$$B\left(e^{j\Omega}\right) = 0.5\left[A\left(e^{j\Omega/2}\right) + A\left(-e^{j\Omega/2}\right)\right]$$

Using the upsampler and downsampler relationships given above, compute and sketch the spectra at points A to H. Assume that all the filters have ideal brickwall responses. Hence demonstrate graphically that the system is capable of perfect reconstruction.

Q6.3 Given the following subband filters:

$$H_0(z) = \frac{1}{\sqrt{2}}\left(1 + z^{-1}\right); \quad H_1(z) = \frac{1}{\sqrt{2}}\left(1 - z^{-1}\right)$$
$$G_0(z) = \frac{1}{\sqrt{2}}\left(1 + z^{-1}\right); \quad G_1(z) = \frac{1}{\sqrt{2}}\left(-1 + z^{-1}\right)$$

show that the corresponding two-band filter bank exhibits perfect reconstruction.

Q6.4 Demonstrate that the following filter relationships produce a filter bank that is alias-free and offers perfect reconstruction:

$$H_1(z) = zG_0(-z); \quad G_1(z) = z^{-1}H_0(-z)$$
$$P(z) + P(-z) = 2z^{-2}; \quad P(z) = H_0(z)G_0(z)$$

Assuming that $G_0(z) = H_0(z) = z^{-1}$, what is the output of this filter bank given an input sequence $\{1,1,0\}$?

Q6.5 Referring to the figure below, a signal, $x[n]$, is downsampled by 2 and then upsampled by 2. Show that, in the z-domain, the input–output relationship is given by:

$$\tilde{X}(z) = 0.5[X(z) + X(-z)]$$

Q6.6 Demonstrate that the following quadrature mirror filter relationships produce a filter bank that is alias-free and offers perfect reconstruction:

$$H_0(z) = z^{-2} + z^{-3}; \quad G_0(z) = H_1(-z);$$
$$H_1(z) = H_0(-z); \quad G_1(z) = -H_0(-z)$$

What are the limitations of the above filter bank? How, in practice, are more useful QMF filter banks designed?

Q6.7 A two-channel single stage 1-D QMF filter bank is constructed using the low-pass prototype filter $H_0(z) = (1 + z^{-1})$. Derive the other filters needed for this system and compute the output signal for an input: $x[n] = \{2, 3, 6, 4\}$.

Q6.8 A 1-D wavelet filter bank comprises two stages of decomposition and uses the same filters as defined in Q6.7 (but factored by $\frac{1}{\sqrt{2}}$ to ensure exact reconstruction). The input sequence is $x[n] = \{1, 1, 1, 1, 1, 1, 1, 1\}$. Using boundary extension, and assuming critical sampling between analysis and synthesis banks, compute the signal values at all internal nodes in this filter bank, and hence demonstrate that the output sequence is identical to the input sequence.

Q6.9 Repeat Q6.8, but this time assume that the bit-allocation strategy employed preserves the lower two subbands but completely discards the high-pass subband. Assuming a wordlength of 4 bits, what is the PSNR of the reconstructed signal after decoding?

Q6.10 Draw the diagram for a 2-D, three-stage wavelet filter bank. Show how this decomposition tiles the 2-D spatial frequency plane and compute the frequency range of each subband. Assuming that the input is of dimensions 256×256 pixels, how many samples are contained in each subband?

Q6.11 Prove that the following two diagrams are equivalent.

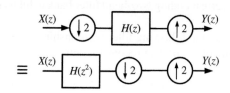

Chapter 7: Lossless compression methods

Q7.1 Consider the following codewords for the given set of symbols:

Symbol	C1	C2	C3
a_1	0	0	01
a_2	10	01	10
a_3	110	110	11
a_4	1110	1000	001
a_5	1111	1111	010

Identify which are prefix codes.

Q7.2 Derive the set of Huffman codewords for the symbol set with the following probabilities:

$$P(s_0) = 0.06; \ P(s_1) = 0.23; \ P(s_2) = 0.30; \ P(s_3) = 0.15;$$
$$P(s_4) = 0.08; \ P(s_5) = 0.06; \ P(s_6) = 0.06; \ P(s_7) = 0.06;$$

What is the transmitted binary sequence corresponding to the symbol pattern s_0, s_4, s_6? What symbol sequence corresponds to the code sequence: 000101110? Calculate the first order entropy of this symbol set, the average codeword length and the coding redundancy.

Q7.3 A quantized image is to be encoded using the symbols $I \in \{I_0 \cdots I_{11}\}$. From simulation studies it has been estimated that the relative frequencies for these symbols are as follows: I_0: 0.2; $I_1 \cdots I_3$: 0.1; $I_4 \cdots I_7$: 0.05; $I_8 \cdots I_{11}$: 0.075. Construct the Huffman tree for these symbols and list the resultant codewords in each case.

Q7.4 If the minimum variance set of Huffman codes for an alphabet, A, is as shown in the table below, determine the efficiency of the corresponding Huffman encoder.

Symbol	Probability	Huffman Code
a_1	0.5	0
a_2	0.25	10
a_3	0.125	110
a_4	0.0625	1110
a_5	0.0625	1111

Q7.5 Consider the following 4×4 matrix, **C**, of DCT coefficients, produced from a block-based image coder. Using zig-zag scanning and run length coding (assume a {*run, value*} model where *value* is the integer value of a non-zero coefficient and *run* is the number of zeros preceding it), determine the transmitted bitstream after entropy coding.

$$\mathbf{C} = \begin{bmatrix} 2 & 0 & 3 & 0 \\ 0 & 3 & 1 & 0 \\ 1 & 0 & 0 & 0 \\ 0 & 0 & 0 & 0 \end{bmatrix}$$

Use the following symbol to codeword mappings:

Symbol	Code	Symbol	Code
{0,1}	00	{1,3}	0111
{0,2}	10	{2,1}	0110
{0,3}	110	{2,2}	0101
{1,1}	1111	{2,3}	01001
{1,2}	1110	EOB	01000

Q7.6 Assuming that the DCT coefficient matrix and quantization matrix in Q5.14 form part of a JPEG baseline codec, derive the Huffman coded sequence that would be produced by the codec for the AC coefficients.

Q7.7 What is the Exp-Golomb code for the symbol index 132_{10}?

Q7.8 Show how the Exp-Golomb codeword 000010101 would be decoded and compute the value of the corresponding symbol index. What is the corresponding Golomb–Rice code for this index (assume $m = 4$)?

Q7.9 Given the symbols from an alphabet, $A = \{a_1, a_2, a_3, a_4, a_5\}$, and their associated probabilities of occurrence in the table below, determine the shortest arithmetic code which represents the sequence: $\{a_1, a_1, a_2\}$.

Symbol	Probability
a_1	0.5
a_2	0.25
a_3	0.125
a_4	0.0625
a_5	0.0625

Q7.10 Given the following symbols and their associated probabilities of occurrence, determine the binary arithmetic code which corresponds to the sequence: $\{a_1, a_2, a_3, a_4\}$. Demonstrate how an arithmetic decoder, matched to the encoder, would decode the bitstream: 010110111.

Symbol	Probability
a_1	0.5
a_2	0.25
a_3	0.125
a_4 (EOB)	0.125

Q7.11 Given the following symbols and their associated probabilities of occurrence, determine the arithmetic code which corresponds to the sequence: a_1, a_2, a_3 (where a_3 represents the EOB symbol).

Symbol	Probability
a_1	0.375
a_2	0.375
a_3	0.125
a_4	0.125

Show how the bitstream produced above would be decoded to produce the original input symbols.

Repeat this question using Huffman encoding rather than arithmetic coding. Compare your results in terms of coding efficiency.

Q7.12 Derive the arithmetic codeword for the sequence $s_1, s_1, s_2, s_2, s_5, s_4, s_6$, given a symbol set with the following probabilities:

$$P(s_0) = 0.065; \quad P(s_1) = 0.20; \quad P(s_2) = 0.10; \quad P(s_3) = 0.05;$$
$$P(s_4) = 0.30; \quad P(s_5) = 0.20; \quad P(s_6) = 0.10 = \text{EOB}$$

Q7.13 Consider an alphabet, $A = \{a_1, a_2, a_3, a_4\}$, where $P(a_1) = 0.6$; $P(a_2) = 0.2$; $P(a_3) = 0.1$; $P(a_4) = 0.1$. Compute the sequence, S, of three symbols which corresponds to the arithmetic code 0.58310.

Q7.14 Given an alphabet of two symbols $A = \{a, b\}$, where $P(a) = 0.25$; $P(b) = 0.75$, draw a diagram showing coding and probability intervals for the associated arithmetic coder. Derive a binary arithmetic code for the sequence $S = baa$.

Chapter 8: Coding moving pictures: motion prediction

Q8.1 Implement the full search BBME algorithm on the 6×6 search window, $\mathbf{S}$, using the current-frame template, $\mathbf{M}$, given below.

$$\mathbf{S} = \begin{bmatrix} 1 & 5 & 4 & 9 & 6 & 1 \\ 6 & 1 & 3 & 8 & 5 & 1 \\ 5 & 7 & 1 & 3 & 4 & 1 \\ 2 & 4 & 1 & 7 & 6 & 1 \\ 2 & 4 & 1 & 7 & 8 & 1 \\ 1 & 1 & 1 & 1 & 1 & 1 \end{bmatrix}; \quad \mathbf{M} = \begin{bmatrix} 7 & 7 \\ 7 & 7 \end{bmatrix}$$

Q8.2 Given the following reference window and current block, show how an N-step search algorithm would locate the best match (assume an SAD optimization criterion). Determine the motion vector for this block. Does this produce the same result as an exhaustive search?

$$
\begin{bmatrix}
1 & 4 & 3 & 2 & 1 & 2 & 3 & 2 & 2 \\
0 & 1 & 2 & 0 & 2 & 3 & 0 & 2 & 1 \\
0 & 2 & 0 & 0 & 1 & 2 & 0 & 1 & 0 \\
0 & 0 & 0 & 0 & 1 & 1 & 1 & 1 & 0 \\
1 & 2 & 1 & 1 & 4 & 4 & 0 & 0 & 1 \\
0 & 3 & 2 & 3 & 0 & 1 & 2 & 2 & 3 \\
0 & 3 & 3 & 1 & 0 & 1 & 1 & 2 & 2 \\
0 & 4 & 2 & 3 & 0 & 2 & 1 & 2 & 1 \\
0 & 1 & 2 & 1 & 2 & 3 & 2 & 2 & 0
\end{bmatrix}
\quad
\begin{bmatrix}
1 & 2 & 3 \\
1 & 2 & 0 \\
2 & 2 & 0
\end{bmatrix}
$$

Q8.3 Use bidirectional exhaustive search motion estimation, such as that employed in MPEG-2, to produce the best match for the following current block and two reference frames:

$$
\begin{bmatrix}
1 & 2 & 3 & 4 \\
1 & 2 & 2 & 3 \\
1 & 1 & 2 & 2 \\
2 & 2 & 2 & 1
\end{bmatrix}
\quad
\begin{bmatrix}
1 & 1 & 3 & 4 \\
1 & 1 & 0 & 3 \\
1 & 1 & 2 & 0 \\
2 & 1 & 0 & 1
\end{bmatrix}
\quad
\begin{bmatrix}
1 & 2 \\
1 & 2
\end{bmatrix}
$$

Q8.4 Explain how the Two-Dimensional Logarithmic (TDL) search method improves the search speed of block-based motion estimation. Illustrate this method for the case of a $\pm 6 \times \pm 6$ search window where the resultant motion vector is [2,5]. Quantify the savings for this particular example over a full search. What is the main advantage and disadvantage of this method?

Q8.5 Implement a hexagonal search block matching algorithm on the search window, **S**, using the current-frame template, **M**, as given below. Determine the motion vector for this search. Assume that any search points where the hexagon goes outside of the reference frame are invalid.

$$
S =
\begin{bmatrix}
0 & 1 & 2 & 3 & 22 & 19 & 18 & 23 \\
7 & 3 & 6 & 5 & 33 & 31 & 13 & 22 \\
4 & 6 & 3 & 7 & 23 & 23 & 15 & 26 \\
8 & 4 & 1 & 3 & 11 & 22 & 29 & 19 \\
2 & 8 & 9 & 7 & 8 & 14 & 16 & 18 \\
5 & 0 & 7 & 3 & 7 & 15 & 12 & 13 \\
7 & 4 & 6 & 6 & 9 & 8 & 8 & 12 \\
1 & 2 & 3 & 9 & 10 & 9 & 8 & 12
\end{bmatrix}
; \quad
M =
\begin{bmatrix}
1 & 4 \\
10 & 7
\end{bmatrix}
$$

Q8.6 Given the following current block P and its set of three adjacent neighbors A, B, C and D, with motion vectors as indicated, use motion vector prediction to initialize the search for the best motion vector for this block.

$\mathbf{d}_B=[1,1]$	$\mathbf{d}_C=[1,0]$	$\mathbf{d}_D=[1,0]$	
$\mathbf{d}_A=[1,2]$	P		

Q8.7 Motion vectors for six adjacent blocks in two partial rows of a frame are shown below (shaded). Using the prediction scheme $\hat{\mathbf{d}}_P = \mathrm{med}(\mathbf{d}_A, \mathbf{d}_C, \mathbf{d}_D)$, compute the predicted motion vectors for each of these blocks together with their coding residuals. State any assumptions that you make in computing predictions for blocks located at picture boundaries (see Chapter 12).

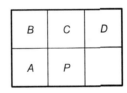

Chapter 9: The block-based hybrid video codec

Q9.1 The figure below shows a block diagram of a basic hybrid (block transform, motion compensated) video coding system.

Using the following assumptions and data at time n:

Image frame size:	12×12 pixels
Macroblock size:	4×4 pixels
Operating mode:	inter-frame
Motion estimation:	Linear translational model (4×4 blocks)

Transform matrix:
$$\mathbf{T} = \frac{1}{2} \begin{bmatrix} 1 & 1 & 1 & 1 \\ 1 & 1 & -1 & -1 \\ 1 & -1 & -1 & 1 \\ 1 & -1 & 1 & -1 \end{bmatrix}$$

Quantization matrix:
$$\mathbf{Q} = \begin{bmatrix} 1 & 1 & 4 & 4 \\ 1 & 2 & 4 & 4 \\ 4 & 4 & 4 & 4 \\ 4 & 4 & 4 & 4 \end{bmatrix}$$

Current input block:
$$\mathbf{A} = \begin{bmatrix} 2 & 2 & 0 & 0 \\ 2 & 2 & 0 & 0 \\ 0 & 0 & 0 & 0 \\ 0 & 0 & 0 & 0 \end{bmatrix} ; \text{ assume exact center of frame.}$$

Reference memory:
$$\mathbf{J} = \begin{bmatrix} 0 & 0 & 2 & 2 & 2 & 1 & 1 & 1 & 1 & 1 & 1 & 1 \\ 4 & 10 & 4 & 7 & 4 & 7 & 4 & 4 & 4 & 3 & 2 & 1 \\ 8 & 4 & 7 & 6 & 7 & 1 & 1 & 0 & 0 & 0 & 0 & 0 \\ 6 & 4 & 6 & 5 & 4 & 1 & 1 & 0 & 0 & 0 & 0 & 0 \\ 3 & 2 & 10 & 6 & 8 & 0 & 0 & 0 & 0 & 0 & 0 & 0 \\ 0 & 4 & 14 & 4 & 4 & 0 & 0 & 0 & 0 & 0 & 0 & 0 \\ 0 & 0 & 5 & 6 & 6 & 6 & 6 & 0 & 0 & 7 & 0 & 0 \\ 0 & 0 & 0 & 12 & 0 & 0 & 0 & 3 & 4 & 2 & 3 & 4 \\ 0 & 0 & 0 & 0 & 0 & 0 & 0 & 4 & 4 & 5 & 8 & 0 \\ 7 & 7 & 8 & 9 & 0 & 0 & 0 & 0 & 6 & 7 & 0 & 0 \\ 0 & 0 & 0 & 10 & 3 & 0 & 0 & 0 & 0 & 7 & 0 & 0 \\ 0 & 0 & 2 & 2 & 2 & 0 & 0 & 0 & 0 & 0 & 0 & 0 \end{bmatrix}$$

Compute:

1. The current motion vector, $\mathbf{H}$, at time n.
2. The motion-compensated output, $\mathbf{I}$, at time n.
3. The DFD for the current input frame, $\mathbf{B}$.
4. The transformed and quantized DFD output, $\mathbf{D}$.

Q9.2 If the encoded bitstream from Q9.1 is decoded by a compliant decoder, compute the decoder output at time n that corresponds to the input block, $\mathbf{A}$.

Q9.3 Compute the transformed and quantized output for the current input block given in Q9.1, but this time for the case of intra-frame coding.

Q9.4 Using horizontal, vertical, and DC modes only, produce the H.264/AVC intra-prediction for the highlighted 4 × 4 luminance block below:

1	2	3	4	2	3	5	4	7	8	9	5
2	4	4	6	7	8	4	6	4	4	3	2
4	3	5	6	3	3	3	4	4	4	3	3
4	4	5	6	3	3	3	4	4	4	3	3
1	1	4	6	**2**	**3**	**6**	**3**	×	×	×	×
4	3	3	5	**3**	**2**	**5**	**3**	×	×	×	×
2	2	3	5	**3**	**3**	**4**	**4**	×	×	×	×
1	3	5	3	**1**	**2**	**4**	**4**	×	×	×	×

Q9.5 Following the full search result from Q8.1, refine the motion vector for this search to half-pixel accuracy using the two-tap interpolation filter $(s_i + s_j)/2$ (where s_i and s_j are the horizontal or vertical whole-pixel locations adjacent to the selected half-pixel location).

Q9.6 Use the H.264/AVC subpixel interpolation filter to generate the half-pixel values for the shaded locations in the following search window:

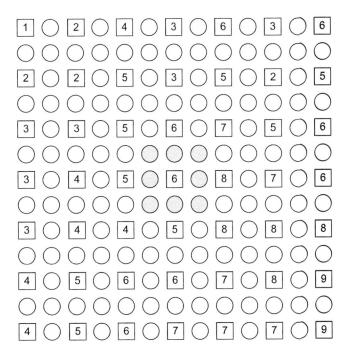

Q9.7 H.264 employs a 4 × 4 integer transform instead of the 8 × 8 DCT used in previous coding standards. Prove that the four-point integer approximation to the 1-D DCT transform matrix, $\mathbf{A}$, is given by:

$$\mathbf{A} = \begin{bmatrix} 1 & 1 & 1 & 1 \\ 2 & 1 & -1 & -2 \\ 1 & -1 & -1 & 1 \\ 1 & -2 & 2 & -1 \end{bmatrix} \otimes \mathbf{E}_f$$

where $\mathbf{E}_f$ is a 4×4 scaling matrix. State any assumptions made during your derivation.

Chapter 10: Measuring and managing picture quality

Q10.1 List the primary factors that should be controlled and recorded during subjective video assessment trials.

Q10.2 List the primary attributes that a good subjective database should possess.

Q10.3 Given the following three consecutive frames, compute the temporal activity (TI) for this sequence.

$$S_1 = \begin{bmatrix} 1 & 2 & 3 & 4 \\ 2 & 3 & 4 & 5 \\ 3 & 4 & 5 & 6 \\ 4 & 5 & 6 & 7 \end{bmatrix}; \quad S_2 = \begin{bmatrix} 1 & 2 & 3 & 5 \\ 2 & 3 & 4 & 6 \\ 3 & 4 & 5 & 7 \\ 4 & 5 & 6 & 8 \end{bmatrix}; \quad S_3 = \begin{bmatrix} 1 & 2 & 4 & 6 \\ 2 & 3 & 5 & 7 \\ 3 & 4 & 6 & 8 \\ 4 & 5 & 7 & 9 \end{bmatrix}$$

Q10.4 Consider the operation of a constant bit rate video transmission system at 25 fps with the following parameters:

$$R_0 = 500 \text{ kbps}; \quad B = 150 \text{ kb}; \quad F_i = 100 \text{ kb}; \quad \text{GOP} = 6 \text{ frames}$$

If the pictures transmitted have the following sizes, compute the occupancy of the decoder buffer over time and determine whether underflow or overflow occurs.

Frame no.	Picture size (kbits)
1	20
2–6	10
7	40
8–12	10

Chapter 11: Communicating pictures: delivery across networks

Q11.1 Using the Huffman codes in the following table, decode the following encoded bitstream: 0 0 1 0 1 0 0 1 1 1 1 1 1 1 0 1 0 0.

Symbol	Probability	Huffman Code
a_1	0.5	0
a_2	0.25	10
a_3	0.125	110
a_4	0.0625	1110
a_5	0.0625	1111

Assuming codeword 110 represents an EOB signal, what would be the effect of a single error in bit position 2 of the above sequence?

Q11.2 An image encoder uses VLC and Huffman coding of transform coefficients, based on a set of four symbols with the following mappings:

$$A \leftrightarrow 0; \quad B \leftrightarrow 10; \quad C \leftrightarrow 110; \quad D \leftrightarrow 111$$

Assuming that the sequence transmitted is $ABCDAC$, determine the received symbol sequences for the following scenarios:

1. An error occurring in the 3rd bit position.
2. An error occurring in the 1st bit position.
3. An error occurring in the 4th bit position.
4. An error occurring in the 8th bit position.

For each scenario, assuming that B represents the EOB symbol, comment on the impact that the error has on the final state of the decoder and on the reconstructed transform coefficients.

Q11.3 VLC codes for symbols $\{a, b, c\}$ are: $a = 0; b = 11; c = 101$, where: $P(a) = 0.5$; $P(b) = 0.25; P(c) = 0.25$. Comment on any specific property these codewords exhibit and on its benefits in a lossy transmission environment. Compare the efficiency of these codewords with that of a conventional set of Huffman codes for the same alphabet.

Q11.4 The sequence of VLC codewords produced in Q7.5 is sent over a channel which introduces a single error in the 15th bit transmitted. Comment on the impact that this error has on the reconstructed transform coefficients and any subsequent blocks of data.

Q11.5 Use EREC to code four blocks of data with lengths: $b_1 = 6, b_2 = 5; b_3 = 2; b_4 = 3$. Assuming that an error occurs in the middle of block 3 and that this causes the EOB code to be missed for this block, state which blocks in the frame are corrupted and to what extent.

Q11.6 A video coder uses VLC and Huffman coding based on a set of four symbols with probabilities 0.4, 0.3, 0.15, and 0.15. Calculate the average bit length of the resulting Huffman codewords. If this video coder is to employ reversible codes (RVLC) to improve error resilience, suggest appropriate codewords and calculate the bit rate overhead compared to conventional Huffman coding.

Q11.7 Assume that S_1 and S_2 below represent corresponding regions in two temporally adjacent video frames. Due to transmission errors, the central 4×4 block (pixels marked as "$\times$") in the received current frame S_2' is lost. Assuming that the codec operates on the basis of 2×2 macroblocks:

(a) Perform temporal error concealment using frame copying to provide an estimate of the lost block.

(b) Perform motion-compensated temporal error concealment, based on the BME measure. Assume that the candidate motion vectors obtained from the four adjacent blocks are either [0,1], [1,0] or [1,1].

$$
S_1 = \begin{bmatrix}
1 & 1 & 1 & 2 & 2 & 2 & 3 & 3 \\
2 & 2 & 2 & 3 & 3 & 3 & 4 & 4 \\
3 & 3 & 3 & 4 & 4 & 4 & 4 & 4 \\
3 & 3 & 3 & 4 & 4 & 4 & 5 & 5 \\
4 & 4 & 4 & 4 & 4 & 4 & 4 & 4 \\
5 & 5 & 5 & 6 & 6 & 7 & 6 & 6 \\
6 & 6 & 6 & 7 & 6 & 6 & 6 & 6 \\
7 & 7 & 7 & 7 & 8 & 8 & 8 & 8
\end{bmatrix}
\; ; \;
S_2' = \begin{bmatrix}
2 & 2 & 3 & 3 & 3 & 4 & 4 & 3 \\
3 & 3 & 4 & 4 & 4 & 4 & 5 & 5 \\
3 & 3 & 4 & 4 & 4 & 4 & 4 & 4 \\
4 & 4 & 5 & \times & \times & 4 & 4 & 6 \\
5 & 5 & 6 & \times & \times & 6 & 6 & 6 \\
7 & 6 & 7 & 6 & 6 & 6 & 6 & 7 \\
6 & 7 & 7 & 6 & 6 & 8 & 8 & 8 \\
9 & 9 & 9 & 7 & 8 & 7 & 8 & 8
\end{bmatrix}
$$

Q11.8 Assuming again that the central 2×2 block (pixels marked as "$\times$") in the received current frame S_2' (above) is lost, calculate the missing elements using spatial error concealment.

Q11.9 A block-based image coder generates a slice comprising five DCT blocks of data as follows:

Block Number	Block Data
1	01011011101111
2	10111001011101111
3	1100101111
4	101101101111
5	0001111

where the symbols and entropy codes used to generate this data are as follows (assume E is the End of Block Symbol):

Symbol	Huffman Code
A	0
B	10
C	110
D	1110
E	1111

Code this data using Error-Resilient Entropy Coding (EREC). Choose an appropriate EREC frame size and offset sequence for this slice and show all the encoding stages of the algorithm.

Q11.10 Perform EREC decoding on the bitstream generated from Q11.9.

If the slice data given in the above question is corrupted during transmission such that the values of the last 2 bits in each of block 2 and block 3 are complemented, how many blocks are corrupted after EREC decoding. Compute how many blocks would contain errors if the slice were transmitted as a conventional stream of entropy codewords without EREC.

Chapter 12: Video coding standards

Q12.1 Assuming a CIF format luminance-only H.261 sequence with synchronization codewords at the end of each GOB, calculate the (likely) percentage of corrupted blocks in a frame if bit errors occur in block 1 of macroblock 7 in GOB 2 and block 1 of macroblock 30 of GOB 10.

Q12.2 What parts of a video codec are normally subject to standardization in term of compliance testing? Why is this?

Q12.3 Discuss the concepts of profiles and levels in MPEG-2 indicating how these have enabled a family of compatible algorithms covering a range of applications and bit rates to be defined.

Q12.4 Compare the features of H.264/AVC and MPEG-2 and highlight those features that have enabled H.264 to approximately halve the bit rate of MPEG-2 for equivalent picture quality.

Q12.5 Compare the features of H.265/HEVC and H.264/AVC and highlight those features that have enabled HEVC to approximately halve the bit rate of H.264 for equivalent picture quality.

Q12.6 Why has H.265/HEVC increased its maximum block size to 64×64 pixels?

Chapter 13: Communicating pictures—the future

Q13.1 What are the primary challenges and demands for video compression in the future?

Q13.2 Consider the case of an 8 K resolution video signal in 4:2:2 format with 14 bits dynamic range and a frame rate of 300 fps. Calculate the total bit rate needed for this video.

Q13.3 How might video compression algorithms develop in the future in order to cope with the demands of formats such as that described in Q13.2?

Index

A

Absolute Category Rating (ACR), 327
Active Format Description (AFD), 110
Additive Gaussian white noise (AGWN), 337
Advanced Simple Profile (of MPEG-4), 14, 425
Algorithm
  diamond search, 283
  HEXBS complexity, 284
  NSS search, 281
  TDL search, 280
  video decoder
    inter-mode decoder operation, 296
    intra-mode decoder operation, 296
  video encoding loop
    inter-mode encoder operation, 295
    intra-mode encoder operation, 295
Arithmetic coding, 241
  advantages, 246
  binary arithmetic, 247
  context-adaptive, 250
  decoding algorithm, 244
  encoding algorithm, 241
  tag generation with scaling, 249
  vs. Huffman coding, 251
Aspect ratio
  angles of retina, 106
  field of view ratio, 109
  of film and TV, 106–107
  16:9 format, 108
  larger and closer screens, 108
  QCIF and UHDTV formats, 106–107
  widescreen format, 107–108
Automatic Repeat reQuest (ARQ), 362, 375
  advantage, 376
  delay constrained retransmission, 376
  problems, 376

B

Backward prediction, 260, 292
Bidirectional prediction, 261, 292
Bi-predicted (B) pictures, 104, 292
Binarization, 251
Bit error rate (BER), 365
Bit rates
  and compression ratio, 124
  management of, 351
  requirement, 3

Block-based hybrid video codec
  deblocking filter operations
    algorithm, 314
    effect of, 314–315
    one-dimensional edge sample, 314
    performance of, 313
  DFD signal, properties of, 293
  intra-frame prediction
    for large blocks, 300
    for small luminance blocks, 298
  motion estimation (*see* Motion estimation (ME))
  operation of
    video decoder, 295–296
    video encoding loop, 294–295
  picture types, 292
  prediction modes, 292
  tutorial problem, 491
  variable sized transforms
    DC coefficient transforms, 313
    integer transforms, 310
Block distortion measure (BDM), 270
Block matching motion estimation (BMME), 264
  full search algorithm, 272
  matching criteria
    block distortion measure, 270
    SAD *vs.* SSD, 271
  motion failure, 276–277
  properties of
    block motion field, 272
    effect of block size, 272
    motion residual error surface, 274
    motion vector probabilities, 275
    search grid size effect, 274
  restricted and unrestricted vectors, 277
  translational
    motion vector orientation, 269–270
    search window size, 269
Boundary Matching Error (BME), 401

C

Color processing, 56
  CIE 1931 chromaticity chart, 39
  opponent theories, 37–38
Color spaces
  for analog TV, 119
  CIE definition, 117
  for digital formats, 119–120
  sub-sampled

chroma sub-sampling, 121
common formats, 121–122
and compression, 123
luma processing, 121
transformations, 117, 122
Communicating pictures
compression, challenges for, 462
context-based video compression, 466
emerging formats, 452
extended video parameter space
creative continuum, 461–462
dynamic range, 459
generally, 451–452
influences, 453
parameter interactions, 461
spatial resolution, 453
temporal resolution, 455
immersive experiences, 452
overview, 467
parametric video compression, 464
rate-quality optimization, 451
synthesis-based coding, 465
texture synthesis, 464
visual experiences, 452
Complex cells, 30
Complexity *vs.* cost and performance, 10
Compression system
definition, 2
geometric interpretation of, 2–3
requirement of
bandwidth available, 4
picture formats and bit rate, 3
trade-off, 10
video (*see* Image and video compression)
Context Adaptive Binary Arithmetic Coding
(CABAC), 241
binarization, 251
binary arithmetic coding, 251
context modeling, 251
H.264/AVC and HEVC standards, 250–251
Context adaptive variable length coding (CAVLC)
in H.264/AVC, 236
video coding layer, 433
Context-based video compression, 466
Context modeling, 251
Continuous wavelet transform (CWT), 175
Contrast sensitivity function (CSF) model, 40
chart, 41
definition, 41
for luminance and chrominance, 42
against spatial frequency, 43
uses of, 43
Creative continuum, 461–462

Critical flicker frequency (CFF), 51
Current frames, 260
Cyclic Redundancy Check (CRC), 375

D

Deblocking filter
algorithm, 314
effect of, 314–315
one-dimensional edge sample, 314
performance of, 313
Decoded picture buffer (DPB), 438
Degradation Category Rating method (DCR), 327
Delay *vs.* performance, 10
Diamond search (DS), 262, 282–283
Difference Mean Opinion Scores (DMOS),
328, 335
vs. PSNR predicted, 335–336
Difference of Gaussian functions (DoG) model, 27
Digital cellular network, 363
Digital picture formats
coding tree units, 104
color descriptions and HVS
Bayer filter mosaic, 123
chromaticity diagrams, 118
color spaces (*see* Color spaces)
trichromacy theory, 116
formats and aspect ratios (*see* Video format)
frame types
bi-predicted (B) pictures, 104
intra-coded (I) pictures, 104
predicted (P) pictures, 104
gamma correction
between camera capture and displays
content, 114
examples of, 115
gamma characteristic of device, 114–115
Rec.709 format, 116
groups of pictures (GOPs)
intra- *vs.* inter-frame coding, 105–106
typical structure of, 105
macroblocks
definition, 103
typical structure of, 103
measuring and comparing picture quality (*see*
Picture quality)
moving pictures, 102
picture scanning (*see* Video scanning)
pixels/samples/pels
color image, 101
defined, 100
monochrome image, 100
at varying resolutions, 102

rate distortions
    characteristics of, 129–130
    optimization, 131
Dirac, 447
Discrete cosine transform (DCT), 13, 82, 148
  basis functions, 151, 154
  coefficients, quantization of, 155, 159
  complexity, 164
    reduction, 164
  derivation of, 149–151
  extension to 2-D, 153–154
  implementation
    demonstration, 167–168
    integer transforms, 167
    interlaced sequences, field *vs.* frame encoding
      for, 166
    McGovern algorithm, 164, 166
    transform block size, 163
  RDF of, 161–163
  transforms comparison, 160
  vs. DFT, 160
Discrete Fourier transform (DFT), 82
  and compression, 84
  linear transforms and, 82–83
  2-D, 84
Discrete-time analysis
  filtering and transforms (*see* Image filtering and
    transformations)
  images statistics
    correlation in natural images, 71
    histograms and distributions, 70
    mean values, 70
  information theory, 93
    entropy, 94
    self-information, 93
    symbols and statistics, 96
  linear predication
    feedback-based, 90
    feedforward analysis filter, 87
    Wiener filter, 92
    Wiener Hopf equations, 92–93
  picture sampling
    issues in, 64
    multidimensional sampling lattices, 69
    theorem, 65
  quantization
    basic theory, 84
    HVS weighting, 87
    non-uniform quantizers, 86
    vector quantization, 87
  tutorial problems, 480
Discrete Time Fourier Series (DTFS), 83, 144
Discrete Walsh-Hadamard transform (DWHT), 141

basis functions, 142
  1-D, basis function for, 142
  2-D
    basis function for, 143
    compression, 144
Discrete wavelet transform (DWT), 176
Displaced frame difference (DFD) signal, 257,
    260, 295
  properties of, 293
Double Stimulus Continuous Quality Scale
    (DSCQS), 129, 326–327, 330
Double Stimulus Impairment Scale (DSIS), 327
Dynamic range
  generally, 459
  high dynamic range (HDR), 459
  human visual system (HVS), 459
  perceptual compression methods and
    performance, 460

E

Elecard Streameye analyzer, 310
Embedded Block Coding with Optimal Truncation
    (EBCOT), 207
Embedded zerotree coding (EZW), 207
  structure for, 206
Empiricists, 19
End of Block (EOB) symbol, 230
Erasure errors, 365
Error resilience, 178
Error resilient entropy coding (EREC)
  operation principle, 389–390
    decoding process, 390–391
    encoding, 391
  performance
    benefits of, 393
    decoder, 392
    reconstructed images after, 392–393
    vs. conventional sequential encoding,
      392–393
European Broadcasting Union (EBU), 111
External Boundary Matching Error (EBME), 401
Extended video parameter space
  creative continuum, 461–462
  dynamic range
    generally, 459
    high dynamic range (HDR), 459
    human visual system (HVS), 459
    perceptual compression methods and
      performance, 460
  generally, 451–452
  influences, 453
  parameter interactions, 461

spatial resolution
    compression performance, 454
    effective spatial resolution, 453
    generally, 453
    HEVC performance, 455
    resolution *vs.* quality, 454
    Ultra High Definition Television (UHDTV)
        standard, 453–454
Eye movements, 53–54

F

Fast Fourier Transform (FFT), 164
FEC channel coding, 376
    cross packet FEC, 377–378
    erasure codes, 377
    rateless codes, 378
    unequal error protection and data partitioning,
        378
    use of, 376
Filter bank and wavelet compression
    coefficient quantization and bit allocation
        hierarchical coding, 207
        zonal coding and, 204
    multirate filtering, 182
        downsampling, 183, 185, 187–188
        perfect reconstruction, 189
        system transfer function, 186
        two-channel decomposition, spectral effects
            of, 190–191
        upsampling, 183–185
    multiscale process
        STFT and Gabor transform, 173
        subband filtering and wavelet nomenclature,
            173
        wavelet, 174
    perfect reconstruction
        1-D filter bank structure, 178
        filter and decomposition requirement, 178
        two-channel filter bank, development
            of, 179
    tutorial problem, 484
    useful filters and
        boundary extension, 201–202
        characteristics of, 191
        edge artifacts, 202
        finite length sequences, 202
        multistage decomposition, 197–199
        performance, 203–204
        quadrature mirror filters, 192
        separability and extension to 2-D, 200
        wavelet filters, 194

Filter-rectify-filter (FRF) process, 44
Finite impulse response (FIR), 77
Fixed-length codewords (FLC), 368
Flexible macroblock ordering (FMO), 403
Forward Error Correction (FEC) coding, 375
Forward prediction, 260, 292
Fovea, 25
Frame difference (FD), 260
    vs. motion compensated prediction, 260
Frame types
    B-frames, 380
    I-frames, 380
    intra-refresh, 380
    periodic reference frames, 381–382
    P-frames, 380
    reference picture selection, 381
Frontal eye field (FEF), 31
Fusiform face area (FFA), 31

G

Gamma correction
    between camera capture and displays content,
        114
    examples of, 115
    gamma characteristic of device, 114–115
    Rec.709 format, 116
GIF (Graphics Interchange Format), 216
Golomb coding
    advantage, 239
    exponential, 240
    Golomb-Rice codes, 239
    unary codes, 239
Goodput, 4
Grassman's Laws, 116
Groups of blocks (GOBs), 415, 419
Groups of pictures (GOPs), 292–293, 380
    digital picture
        Intra- *vs.* inter-frame coding, 105–106
        typical structure of, 105

H

H.120
    conditional replenishment strategy, 415
    history, 414
    primary features of, 415
H.261
    frame format, 415
    groups of blocks (GOBs), 415
    history, 415
    macroblock, 415
    primary features of, 416

H.263
  advanced prediction mode, 421
  extensions, 422
  groups of blocks (GOBs), 419
  history, 419
  independent segment decoding mode, 422
  macroblock, 419
  motion vector coding, 421
  overlapped motion compensation (OMC), 421
  PB-frames mode, 422
  picture format, 420
  primary features of, 419
  reference picture selection mode, 422
  slice structured mode, 422
  supplemental enhancement information
      mode, 423
  syntax-based arithmetic coding, 421
  temporal, SNR, and spatial scalability
      mode, 422
  unrestricted motion vectors, 421
Haar transform, 140
H.264/AVC, 379
  arithmetic coding, 250
  bitstream syntax, 427
  blocks, 429
  block sizes supported by, 310
  CABAC in, 250–251
  in CAVLC, 236
  Exp-Golomb coding, 240
  Foreman sequence coded using, 311
  history, 426
  improved coding efficiency, 426
  instantaneous decoder refresh (IDR), 428
  intra-prediction, 430
  layer structure, 427
  macroblock, 429
  multiview extensions, 435
  network abstraction Layer (NAL), 427–428
  performance, 435
  picture types, 429
  primary features of, 426–427
  profiles and levels, 434
  scalable extensions, 435
  simple syntax specification, 426
  slices and slice groups, 429
  syntax, 428
  video coding layer (VCL), 428, 430
    CABAC, 433
    CAVLC, 433
    coder control, 433
    deblocking operations, 433
    Exp-Golomb coding, 433
    inter-coding, 431

  intra-coding, 430
  variable length coding, 433
H.265/HEVC, 241, 310
  arithmetic coding, 250
  CABAC in, 250–251
  clean random access (CRA) pictures, 439
  coding tree blocks (CTBs), 439
  coding tree units (CTUs), 439
  decoded picture buffer (DPB), 438
  Exp-Golomb coding, 240
  extensions, 446
  high level syntax, 437
  intra-prediction, 298
  levels, 446
  main 10 profile, 446
  main profile, 446
  main still picture profile, 446
  NAL units, 437
  network abstraction, 437
  overview, 436
  performance gains for, 446
  pictures and partitions, 438
  prediction units (PUs), 439
  primary features of, 436
  quadtree CTU structure, 439
  random access points, 439
  reference picture set (RPS), 438
  slice structures, 437
  transform units (TUs), 439
  video coding layer (VCL)
    advanced motion vector prediction mode
        (AMVP), 441
    CABAC, variable length coding with, 444
    coefficient scanning, 443
    contexts and significance, 444
    deblocking filter (DBF), 445
    generally, 440
    in-loop filters, 445
    inter-coding, 441
    intra-coding, 440
    quantization, 442
    sample adaptive offset (SAO), 445
    subpixel motion compensation, 441
    transforms, 441
  video encoder architecture, 437
  video parameter set (VPS), 438
Hexagon-based Search Method (HEXBS),
    262, 284
Hotelling or Eigenvector transform. *see*
    Karhunen-Loeve transform (KLT)
Huffman coding
  adaptive methods
    elements, 226

entropy coding, 226
  limitations of, 225
  tree evolution, as example, 226–228
basic algorithm, 219
decoding process, 223
minimum variance, 221
modified, 224
properties
  Kraft inequality, 224
  length of, 225
  optimality, 225
tree-structured approach, 219
vs. arithmetic coding, 251
Human eye
  cross-section of, 21
  optical processing
    cornea, 23
    iris, 23
    lens, 23
    pupil, 23
  retinal photoreceptors and distribution
    cone cells, 24
    fovea, 25
    macula, 25
    optic disk and nerve, 26
    rod cells, 24
  retinal tissue layers, 21
    choroid, 22
    ciliary body, 22
    retina, 22
    sclera, 21
  visible spectrum, 20
  visual processing in retina, 26
Human visual system (HVS), 18
  color processing
    CIE 1931 chromaticity chart, 39
    opponent theories of color, 37–38
  depth cues
    and 3-D entertainment, 48
    examples, 48
    hollow mask illusion, 48–49
    list of, 47
  eye (*see* Human eye)
  eye movements, 53–54
  flicker and peripheral vision, 50
  principles and theories, 19
  saliency and attention, 52
  size and scale, 46
  spatial processing, 39–40
    CSF model (*see* Contrast sensitivity function (CSF) model)
    edge localization model, 44
    JND/contrast/Weber's law, 40–41

phase information, 45–46
    shape and object recognition, 44
    textures perception, 44–45
  spatio-temporal CSF, 50–51
  temporal CSF, 50
  tutorial problems, 479
  visual cortex
    adaptation processes, 29
    biased competition, 29
    dorsal and ventral streams, 30
    extrastriate areas, 31
    opponent coding, 29
    prestriate cortex (V2), 30
    primary visual cortex (V1), 30
    structure, 28
  visual fields and acuity
    brightness, 34
    field of view, 31
    light level adaptation, 35
    luma, 34
    luminance, 32–34
    radiant intensity and radiance, 33
  visual masking (*see* Visual masking)
Hybrid ARQ (HARQ), 379
Hypercomplex cells, 30
Hypothetical Reference Decoder (HRD), 352
  buffering and, 352–353

I

Image and video compression
  application drivers and markets
    business, manufacturing and automation, 8
    consumer video, 8
    healthcare, 9
    security and surveillance, 8
  bandwidths availability, 4
  basics of
    encoding video, 12
    measuring visual quality, 12
    still image encoding system, 11
  bit rate requirements, 3
  digital solution benefits, 9
  generic drivers, 6
  human visual system (*see* Human Visual System (HVS))
  influential factors, 56
  picture formats (*see also* Video format), 3
  requirements of, 10
  standardization of, 12
    basic facts, 12
    chronology of, 13–14
  transforms for (*see* Transform)

Image filtering and transformation
    convolution, 76
    discrete Fourier transform (DFT), 82–83
        and compression, 84
        2-D, 84
    discrete time linear systems
        linearity, 76
        shift invariance, 76
    frequency response, 79
    linear filters
        separability, 78
        2-D digital filters, 78
    non-linear filters, 81
        morphological filters, 82
        rank-order and median filters, 81
    practical filters examples
        H.264/AVC half-pixel interpolation filter, 80
        Le Gal wavelet analysis filters, 80
Image statistics
    autocorrelation/autocovariance, 71–72
        spatial, 73
        for Stampe_SV4 image, 73
        temporal, 75
    histograms and distributions
        image example, 70
        spatial and subband distributions, 70
    mean values, 70
Inferior temporal cortex (IT), 31
Infinite impulse response (IIR), 77
Information theory, 93
    entropy
        defined, 94
        and first order entropy, 95
    self-information, 93
        independent events, 94
    symbols and statistics, 96
Instantaneous decoder refresh (IDR), 428
Integer transforms, 310
    architecture for, 313
    reduced complexity, 313
International Standards Organization (ISO), 13, 207
International Telecommunications Union
        (ITU-T), 13
Internet, 362
Internet Protocol (IP), 362
Interpolation methods
    interpolation-free methods, 306
    H.264/AVC
        half-pixel estimation, 303
        quarter-pixel estimation, 303–304
Inter-prediction with subsequences, 381
Intra-coded (I) pictures, 104, 292
Intra-frame prediction

    for large blocks, 300
    for small luminance blocks, 298
        4 × 4, 299–300
        in H.264/AVC, 299

J

JPEG, 11, 167
    historical bases, 230
    lossless mode
        defined, 235
        prediction modes, 238
    matrix encoding, 233
    performance of, 235–237
    QM Coder used in, 250
    run/size value distribution, 231
    symbol encoding in
        AC coefficient, 230, 232–233
        baseline image encoder
            architecture, 230
        coefficient size categories, 232, 234
        DC coefficient, 230, 232
        run/size values, 231
JPEG2000
    aim of, 207
    benefits and status, 210
    bit planes and scalable coding
        architecture, 207
    coding performance, 208–209
    description, 207
    EBCOT (Embedded Block Coding with
        Optimized Truncation), 250
    ISO, 207
    region of interest coding, 208–209
    technical overview of, 208
Just noticeable differences (JND), 40

K

Karhunen-Loeve transform (KLT), 147–148

L

Laboratory for Image and Video Engineering
        (LIVE) database, 331
Lateral Geniculate Nucleus (LGN), 28
Lempel-Ziv-Welch (LZW), 216
Linear Correlation Coefficient (LCC), 332
Linear predication
    feedback-based, 90
    feedforward analysis filter, 87
        predictor dynamic range, 88
        with quantization, 89

Wiener filter, 92
Wiener Hopf equations, 92
Linear time-invariant (LTI) system, 76
Lossless compression methods
arithmetic coding, 241
advantages, 246
binary arithmetic, 247
context-adaptive, 250
decoding algorithm, 244
encoding algorithm, 241
tag generation with scaling, 249
vs. Huffman coding, 251
Golomb coding
advantage, 239
exponential, 240
and Golomb-Rice codes, 239
unary codes, 239
Huffman coding
adaptive methods, 225–228
basic algorithm, 219
decoding process, 223
minimum variance, 221
modified, 224
properties of, 224
tree-structured approach, 219
vs. arithmetic coding, 251
motivation for
applications, 215
approaches, 216
dictionary-based methods, 216
uses of, 214
symbol encoding
entropy/efficiency/redundancy, 217
generic model for, 217
prefix codes and unique decodability, 218
symbol formation and encoding
CAVLC in H.264/AVC, 236
in JPEG (*see* JPEG)
with sparse matrices, 229
tutorial problem, 487
Lossy compression
quantization, 155
vs. lossless compression, 10

M

Mach band effect, 32–33
Matlab® software, 167
McGovern algorithm, 164
Mean absolute difference
(MAD), 128

Mean Opinion Score (MOS), 128, 328
Mean square error (MSE), 92, 125
Medial superior temporal (MST), 31
Medium Access Control (MAC), 363
Modulation and coding schemes (MCS),
386–388
Monochrome image, 100–102
Morse code, 215
for English alphabet, structure of, 215
for MORSE CODE, 215
Most Apparent Distortion (MAD) model,
318, 340
Motion compensated prediction (MCP), 260
frame difference, 260
Motion compensated temporal replacement
(TEC_MCTR), 400–402, 404
Motion compensation (MC), 260
Motion estimation (ME), 260
approaches to, 262
block matching (*see* Block matching motion
estimation (BMME))
motion models and
affine, bilinear and perspective model, 263
node-based warping, 264–265
pixel-recursive methods, 266
problem formulation, 263
translation model, 265
using phase correlation (PhC), 267
multiple reference frame
performance and complexity, 308–309
properties, 307
reduced complexity, 308
set up of, 307
standardized in H.263, 307
reduced complexity
vs. block matching performance, 286
diamond search, 282
full search complexity, 278
hexagonal search, 284–285
initialization criteria, 285
N-step search, 281
pixel grids and search grids, 277–278
reducing search complexity, 279
termination criteria, 285
2-D logarithmic search (TDL), 280
sub-pixel
interpolation-free methods, 306
interpolation methods, 303
matching, 301
performance, 305
temporal correlation
apparent motion, 258

prediction, formation of, 260
projected motion, 258
between two adjacent video frames, 259
tutorial problem, 489
variable block size
influence of, 309
supported by H.264/AVC, 310
Motion tuned spatio-temporal quality assessment
(MOVIE), 318, 339
Motion vector
coding
entropy of, 287
prediction, 287
orientation, 269–270
predicted for initialization, 285
probabilities, 275–276
Moving Pictures Experts Group (MPEG), 13, 417
MPEG-2/DVB
encoder performance, 417
GOP structure, 418
history, 417
picture types, 417
primary features of, 417
profiles and levels, 418
MPEG-4
advanced simple profile (ASP), 425
audiovisual scene, 424
coding framework, 425
history, 423
picture types, 424
primary features of, 424
video object plane (VOP) coding, 425
Multiple description coding (MDC), 407
Multiple input-multiple-output (MIMO) wireless
technology, 408
Multiple reference frame motion estimation
(MRF-ME), 307
performance and complexity, 308–309
properties, 307
reduced complexity, 308
set up of, 307
standardized in H.263, 307
Multiscale process
STFT and Gabor transform, 173
wavelet
comparison of time and frequency
domain, 174
continuous wavelet transform, 175
discrete wavelet transform, 176
and filter bank properties, 177
theory, development of, 172–173
translation and dilation, 175

N

Near-threshold strategy, 337–338
Netflix, 113
Network abstraction layer (NAL), 379, 428
Network delivery
application layer solution
error tracking, 385
frame type, influence of (*see* Frame type)
network abstraction, 379
reversible VLC, 382
slice structuring, 384
synchronization codewords, 382
bitstream error effects
error resilient solutions, 374–375
subjective appearance of loss, 373
challenges and solution framework, 367
cross layer optimization
definition, 386
link adaptation, 386
effect of loss
header loss, 370
spatial error propagation, 370
synchronization failure, 368
temporal error propagation, 372
error characteristics
encoding, types of, 366
error types, 365
test data, 365
error concealment
benefits, 397
hybrid methods with mode selection, 403–404
missing information detection, 397
post-processing operation, 397
SEC, 398–399
temporal error concealment (*see* Temporal
error concealment)
error resilient entropy coding (*see* Error resilient
entropy coding (EREC))
performance, 392
error resilient, generic architecture of, 367
modern network characteristics
fixed IP network, 362
wireless network, 363
multiple description coding, 407
operating constraints, 365
pyramid vector quantization (*see* Pyramid vector
quantization (PVQ))
scalable video coding, 405–406
transmission types
downloads and streaming, 364
interactive communication, 364

multicast transmission, 364
    unicast transmission, 364
  transport layer solutions
    ARQ technique (*see* Automatic Repeat reQuest
      (ARQ))
    FEC channel coding (*see* FEC channel coding)
    hybrid ARQ (HARQ), 379
    packetization strategies, 379
  tutorial problem, 494
Normalized cross correlation (NCCF), 262, 271
N-Step Search (NSS), 262, 281
  algorithm, 281
  patterns, 281
  three step search, example, 282

O

Objective quality metrics, 319
  classification
    full-reference (FR) methods, 334
    no-reference (NR) methods, 334
    reduced-reference (RR) methods, 334
  future of, 341
  perception-based
    MAD and STMAD, 340
    MOVIE, 339
    PVM, 341
    reduced complexity metrics, 340
    results comparison, 341–342
    SSIM, 338
    VDP and VDP-2, 340
    VQM, 338
    VSNR, 340
  perceptual basis development, 337
  PSNR characterization, 335
  uses of
    algorithm development and benchmarking, 333
    rate-quality optimization, 333
    streaming control, 334
Overlapped motion compensation (OMC), 421

P

Packet error rate (PER), 373
Parametric video compression, 464
Peak signal to noise ratio (PSNR), 126, 318
  calculation of, 127
  for color images and video, 128
  quality comparisons for, 127
  same PSNR value, comparisons of, 127
  uses of, 126
Perception-based Video Metric (PVM),
      318, 341

Perfect reconstruction
  1-D filter bank structure, 178
    two-channel filter bank, development
      of, 179
  filter and decomposition requirement, 178
  multirate filtering, 189
P-frames, 257
Phase correlation (PhC), ME based
  advantages of, 267
  applications, 268
  example of, 269
  performance, 268
  principles, 267
Physical Layer (PHY), 363
Picture grammars, 20
Picture Parameter Set (PPS), 379
Picture quality
  compression ratio and bit rate, 124
  influences on
    artifact type, 320
    content type, 320
    evaluation of, 319
    human visual perception, 319
    viewing environment, 320
  objective distortion and quality metrics
    mean absolute difference, 128
    mean square error, 125
    peak signal to noise ratio (*see* Peak signal to
      noise ratio (PSNR))
  objective quality metrics, 319
    classification, 334
    future of, 341
    perception-based, 338
    perceptual basis development basis, 337
    PSNR characterization, 335
    uses of, 333
  rate control
    bit rate management, 350–351
    buffering and HRD, 352
    encoder and decoder, buffering at, 351
    playout schedule, 351–352
    in practice, 354
    and regions of interest, 357
  rate-distortion optimization
    coding modes and parameters, 348
    practical, 345
    to rate-quality optimization, 350
    theory, 343
  subjective assessment, 128–129
  subjective test database
    databases, 330
    metrics evaluation using public/
      private, 332
    MOS *vs.* objective metric, 331

subjective testing, 319
   consideration, 321
   evaluation process, 329
   methodology and recording of results, 324–325
   statistical analysis and significance testing, 328
   subject selection, 323
   testing environment, 324
   test sequences and conditions, 321
  tutorial problem, 494
Picture sampling
  issues in, 64
  multidimensional sampling lattices, 69
  theorem
   extension to 2-D, 67
   extension to 3-D, 67
   one dimension, 65
PNG (Portable Network Graphics), 216
Predicted (P) pictures, 104, 292
Prediction
  intra-frame (*see* Intra-frame prediction)
  motion (*see* Motion prediction)
Principal Component Analysis (PCA), 147
Principle of orthogonality, 92
Progressive downloads, 364
Pyramid vector quantization (PVQ)
  operation principle
   bit errors, resilient to, 394
   encoding algorithm, 394
   error propagation, prevention of, 394
   intra-mode, 394
   vector quantization, 394
  performance, 394–396

Q

Quadrature mirror filter (QMF), 172, 179, 192
  aliasing elimination, 192
  amplitude distortion, 192
  practical, 193
Quality of service (QoS), 362
Quantization
  of DCT coefficients, 155
  HVS weighting, 87
  non-uniform
   benefits, 86
   deadzone quantizer, 86
   Lloyd Max quantizer, 87
  theory of, 84
   uniform quantization, 85
  vector quantization, 87
Quantizer parameter (QP), 351

R

Random bit errors, 365
Rate control
  bit rate management, 350–351
  buffering and HRD, 352
  encoder and decoder, buffering at, 351
  playout schedule, 351–352
  in practice
   buffer model, 354
   coding unit bit allocation, 356
   complexity estimation, 354
   generic architecture, 354–355
   GOP bit allocation, 356
   QP initialization, 355
   QP values, 354
   rate-quantization model, 354
  and regions of interest, 357
Rate-distortion (RD) algorithm, 394
Rate-distortion optimization (RDO), 130–131, 163
  coding modes and parameters, 348
   Lagrangian multipliers approach, 348
  practical, 345
   Lagrangian methods, 346–347
   RDO complexity, 346
   statistics to parameterizable code, 346
  to rate-quality optimization, 350
  rate *vs.* distortion, 343
  theory, 343–344
   distortion measures, 344
   memoryless Gaussian source, 345
Rate-distortion performance (RDF), 276, 367, 374
  DCT comparisons, 161–163
Rate *vs.* quality, 10
Rationalists, 19
Real-Time Control Protocol (RTCP), 362
Real-Time Streaming Protocol (RTSP), 362
Real-time Transport Protocol (RTP), 362
RealVideo, 447
Redundant slices (RS), 385
Reed Solomon Erasure (RSE), 377
Reference frames, 260
Reference picture selection (RPS), 381
Reference picture set (RPS), 438
Residual bit error rate (RBER), 396
Residual signal strength indication (RSSI), 386
Retina, 22
  healthy, image of, 22
  photoreceptors distribution
   cone cells, 24
   fovea, 25
   macula, 25
   optic disk and nerve, 26
   rod cells, 24

visual acuity of, 32
visual processing in, 26
Reversible VLC (RVLC), 382
Run-length coding (RLC), 229

S

Scalability, 177
Scalable video coding (SVC), 405–406
Scene-grammar, 20
Selective recovery technique, 385
Sequence Parameter Set (SPS), 379
Set partitioning in hierarchical trees (SPIHT)
        approach, 207
Short-time Fourier transform (STFT), 173
Simple cells, 30
Single bit error
    on DCT/Huffman encoded image, effect of, 370
    loss of VLC synchronization, 368
Single stimulus methods, 327
Singular value decomposition (SVD), 408
Slice structure, 384
    FMO approach, 384
    redundant slices, 385
Soft Combining, 379
Spatial activity (SA), 404
Spatial error concealment (SEC), 398–399
Spatial error propagation
    due to VLC errors, 371
    single bit error, 370
Spatially multiplexed system, 408
Spatial processing, 39–40
    CSF model (see Contrast sensitivity function
        (CSF) model)
    edge localization model, 44
    JND/contrast/Weber's law, 40–41
    phase information, 45–46
    shape and object recognition, 44
    textures perception, 44–45
Spatial resolution
    compression performance, 454
    effective spatial resolution, 453
    generally, 453
    HEVC performance, 455
    resolution vs. quality, 454
    Ultra High Definition Television (UHDTV)
        standard, 453–454
Spatio-temporal CSF, 50–51
Spatio-temporal MAD (ST-MAD), 340
Spearman Rank-Order Correlation Coefficient, 332
Structural Similarity (SSIM), 338
Subband filterbanks, 177

Filterbanks and wavelet compression, 171
Subjective Assessment Methodology for Video
        Quality (SAMVIQ), 324, 327
Subjective test database
    databases, 330
        IRCCyN/IVC, 331
        LIVE, 331
        VQEG HDTV, 331
        VQEG multimedia Phase I, 331
        VQEQ FRTV, 330
    metrics evaluation using public/private
        linear correlation, 332
        outlier ratio, 333
        rank-order correlation, 332
    MOS vs. objective metric, 331
    uses of, 330
Subjective testing, 319
    consideration, 321
    evaluation process, 329
    methodology and recording of results
        algorithm, 330
        double stimulus methods, 325
        general principles of, 324, 326
        pair comparison methods, 328
        single stimulus methods, 327
        triple stimulus methods, 328
    statistical analysis and significance
        testing, 328
        confidence interval, 328
        mean score calculation, 328
        screening of observer, 329
    subject selection, 323
    testing environment, 324
    test sequences and conditions
        activity/information levels, 322–323
        test conditions, 322
        test material, 321
Sub-pixel motion estimation
    approaches, 302
    benefits of, 301
    concept of, 301
    half-pixel refinement, search grid with, 302
    interpolation-free methods, 306
    interpolation methods
        example, 304
        half-pixel and quarter-pixel in H.264, 303
        SAD values, 304
    with local refinement, 302
    performance
        full, half and quarter-pixel, comparison of,
        305–306
        on residual error, 306
Sum of absolute differences (SAD), 262

vs. sum of squared differences, 271
Sum squared differences (SSD), 262
   vs. sum of absolute difference, 271
Suprathreshold strategy, 337–338
Synchronization
   codewords, 382
      advantages, 382
      disadvantages, 382
   failure, 368
      single bit error, 368

T

Temporal activity (TA), 404
Temporal correlation
   apparent motion, 258
   prediction, formation of, 260
   projected motion, 258
   between two adjacent video frames, 259
Temporal CSF, 50
Temporal Differential Pulse Code Modulation
      (DPCM), 13
Temporal error concealment, 399
   TEC_MCTR, 400–402, 404
   temporal copying (TEC_TC), 400–401
Temporal error propagation
   due to prediction, 372
   example of, 372–373
   from single corrupted block, 372
Temporal resolution
   compression methods and performance, 457
   frame rates, 455–457
   generally, 455
   shutter angles, 455–457
   static *vs.* dynamic resolutions, 457–458
Texture synthesis, 464
Theories of vision, 19
3rd Generation Partnership Project (3GPP), 379
3-D imaging modalities, 9
Transforms
   basic
      discrete Fourier transform, 144–145
      Haar transform, 140
      Walsh-Hadamard, 141
   discrete cosine transform (DCT), 148
      basis functions, 151
      coefficient quantization of, 155, 159
      comparison of transforms, 160
      derivation of, 149
      implementation, 163
      rate-distortion performance of, 161–163
      2-D basis functions, 153–155
   JPEG, 167

optimum
   discarding coefficients, 146
   Karhunen-Loeve transform, 147–148
performance comparisons
   DCT *vs.* DFT revisited, 160
   KLT *vs.* DCT, 161–162
   on natural images, 160–161
principles of decorrelating
   image compression architecture, 135
   principle components and axis rotation, 135
tutorial problem, 482
unitary
   basis functions and linear combinations, 137
   orthogonality and normalization, 138
   two dimensional (2-D) signal, 139
Translational block matching
   motion vector orientation, 269–270
   search window size, 269
Transmission Control Protocol (TCP), 362
Triple Stimulus Continuous Evaluation Scale
      methodology (TSCES), 328
Triple stimulus methods, 328
Twitter, 112
2-D logarithmic search (TDL), 280

U

Ultra High Definition Television (UHDTV)
      standard, 4
   generally, 453
   compression ratio for, 6
   parameters, 453–454
Unitary transforms
   basis functions and linear combinations, 137
   orthogonality and normalization, 138
   two dimensional (2-D) signal, 139
UNIX compress, 216
User Datagram Protocol (UDP), 362

V

Variable length symbol encoding (VLC), 11
VC1, 447
Vector quantization (VQ), 87
Video coding layer (VCL)
   H.264/AVC, 428, 430
      CABAC, 433
      CAVLC, 433
      coder control, 433
      deblocking operations, 433
      Exp-Golomb coding, 433
      inter-coding, 431

intra-coding, 430
    variable length coding, 433
  H.265/HEVC
    advanced motion vector prediction mode
      (AMVP), 441
    CABAC, variable length coding with, 444
    coefficient scanning, 443
    contexts and significance, 444
    deblocking filter (DBF), 445
    generally, 440
    in-loop filters, 445
    inter-coding, 441
    intra-coding, 440
    quantization, 442
    sample adaptive offset (SAO), 445
    subpixel motion compensation, 441
    transforms, 441
Video coding standards
  chronology of, 412–413
  dirac, 447
  H.120
    conditional replenishment strategy, 415
    history, 414
    primary features of, 415
  H.261
    frame format, 415
    groups of blocks (GOBs), 415
    history, 415
    macroblock, 415
    primary features of, 416
  H.263
    advanced prediction mode, 421
    extensions, 422
    groups of blocks (GOBs), 419
    history, 419
    independent segment decoding mode, 422
    macroblock, 419
    motion vector coding, 421
    overlapped motion compensation (OMC), 421
    PB-frames mode, 422
    picture format, 420
    primary features of, 419
    reference picture selection mode, 422
    slice structured mode, 422
    supplemental enhancement information
      mode, 423
    syntax-based arithmetic coding, 421
    temporal, SNR, and spatial scalability
      mode, 422
    unrestricted motion vectors, 421
  H.264/AVC
    bitstream syntax, 427

blocks, 429
history, 426
improved coding efficiency, 426
instantaneous decoder refresh (IDR), 428
layer structure, 427
macroblock, 429
multiview extensions, 435
network abstraction, 427
network abstraction Layer (NAL), 428
performance, 435
picture types, 429
primary features of, 426–427
profiles and levels, 434
scalable extensions, 435
simple syntax specification, 426
slices and slice groups, 429
syntax, 428
video coding layer (VCL), 428, 430
H.265/HEVC
  clean random access (CRA) pictures, 439
  coding tree blocks (CTBs), 439
  coding tree units (CTUs), 439
  decoded picture buffer (DPB), 438
  extensions, 446
  high level syntax, 437
  levels, 446
  main 10 profile, 446
  main profile, 446
  main still picture profile, 446
  NAL units, 437
  network abstraction, 437
  overview, 436
  performance gains for, 446
  pictures and partitions, 438
  prediction units (PUs), 439
  primary features of, 436
  quadtree CTU structure, 439
  random access points, 439
  reference picture set (RPS), 438
  slice structures, 437
  transform units (TUs), 439
  video coding layer (VCL), 440
  video encoder architecture, 437
  video parameter set (VPS), 438
intellectual property and licensing, 414
MPEG-2/DVB
  encoder performance, 417
  GOP structure, 418
  history, 417
  picture types, 417
  primary features of, 417
  profiles and levels, 418

MPEG-4
  advanced simple profile (ASP), 425
  audiovisual scene, 424
  coding framework, 425
  history, 423
  picture types, 424
  primary features of, 424
  video object plane (VOP) coding, 425
  overview, 412, 447
  RealVideo, 447
  standard-compliant decoder, 413
  standard-compliant encoder, 413
  standardization
    convergence, 414
    divergence, 414
    focus of, 412
    process, 413
    requirements definition, 414
    verification, 414
  VC1, 447
  VP9, 447
Video encoder architecture, 437
Video parameter set (VPS), 438
VP9, 447
Video compression
  algorithms, 362
  block-based hybrid model (*see* Block-based
    hybrid model)
  prediction, classes of, 260
Video format
  aspect ratios
    angles of retina, 106
    field of view ratio, 109
    of film and TV, 106–107
    16:9 format, 108
    larger and closer screens, 108
    QCIF and UHDTV formats, 106–107
    widescreen format, 107–108
  displaying different formats
    2.39:1 format, 110
    4:3 format, 110
    pan and scan and AFD, 110
  picture/frame types
    B pictures, 104
    I pictures, 104
    P pictures, 104

Video Quality Experts Group (VQEG) FRTV
  Phase 1 programme, 330
Video Quality Metric (VQM), 338
Video scanning
  interlaced scanning
    effects of, 112
    vs. progressive scanning, 111
  standards conversion
    by motion compensation
      method, 114
    3:2 pull-down, 112–113
Visual acuity, 32
Visual cortex
  adaptation processes, 29
  biased competition, 29
  dorsal and ventral streams, 30
  extrastriate areas, 31
  opponent coding, 29
  prestriate cortex (V2), 30
  primary visual cortex (V1), 30
  structure, 28
Visual masking, 53
  edge masking, 55
  temporal masking, 55–56
  texture masking, 53–54
Visual short term memory (VSTM), 31
Visual Signal-to-Noise Ratio (VSNR),
  318, 340

W

Walsh-Hadamard transform, 141
Wavelet filters, 194
  Daubechies 9/7 filters, 196
  Filterbanks and wavelet compression, 171
  Le Gall 5/3 filters, 195–196
Weber's Law, 40
Wireless local area networks (WLANs),
  363, 386
Wireless network, 363

Z

Zig-zag scanning, 229
ZIP compression, 216